HUMAN FACTORS ISSUES IN COMBAT IDENTIFICATION

Human Factors in Defence

Series Editors:

Dr Don Harris, Cranfield University, UK
Professor Neville Stanton, Brunel University, UK
Professor Eduardo Salas, University of Central Florida, USA

Human factors is key to enabling today's armed forces to implement their vision to 'produce battle-winning people and equipment that are fit for the challenge of today, ready for the tasks of tomorrow and capable of building for the future' (source: UK MoD). Modern armed forces fulfil a wider variety of roles than ever before. In addition to defending sovereign territory and prosecuting armed conflicts, military personnel are engaged in homeland defence and in undertaking peacekeeping operations and delivering humanitarian aid right across the world. This requires top class personnel, trained to the highest standards in the use of first class equipment. The military has long recognised that good human factors is essential if these aims are to be achieved.

The defence sector is far and away the largest employer of human factors personnel across the globe and is the largest funder of basic and applied research. Much of this research is applicable to a wide audience, not just the military; this series aims to give readers access to some of this high quality work.

Ashgate's *Human Factors in Defence* series comprises of specially commissioned books from internationally recognised experts in the field. They provide in-depth, authoritative accounts of key human factors issues being addressed by the defence industry across the world.

Human Factors Issues in Combat Identification

EDITED BY

DEE H. ANDREWS
Air Force Research Laboratory, USA
ROBERT P. HERZ
Air Force Research Laboratory, USA
MARK B. WOLF
Air Force Research Laboratory and
Oak Ridge Institute for Science and Education, USA

ASHGATE

Published by
Ashgate Publishing Limited
Wey Court East
Union Road
Farnham
Surrey, GU9 7PT
England

Ashgate Publishing Company
Suite 420
101 Cherry Street
Burlington
VT 05401-4405
USA

www.ashgate.com

British Library Cataloguing in Publication Data
Human factors issues in combat identification. -- (Human factors in defence)
1. Friendly fire (Military science)--Safety measures.
2. Visual discrimination. 3. Cognitive psychology.
I. Series
355.4'22-dc22

ISBN: 978-0-7546-7767-3 (hbk)
ISBN: 978-0-7546-9515-8 (ebk)

Library of Congress Cataloging-in-Publication Data
Human factors issues in combat identification / by Dee H. Andrews, Robert P. Herz and Mark B. Wolf.
p. cm.
Includes bibliographical references and index.
ISBN 978-0-7546-7767-3 (hardback) -- ISBN 978-0-7546-9515-8 (ebook) 1. Friendly fire (Military science)--United States. 2. Combat--Psychological aspects. 3. Combat--Physiological aspects. 4. Psychology, Military. I. Herz, Robert P. II. Wolf, Mark B. III. Title.
U167.A53 2007
355.4'22--dc22

2009023765

Mixed Sources
Product group from well-managed forests and other controlled sources
www.fsc.org Cert no. SA-COC-1565
© 1996 Forest Stewardship Council

Printed and bound in Great Britain by
MPG Books Group, UK

Contents

List of Figures *ix*
List of Tables *xiii*
List of Contributors *xv*
Foreword *xxix*
Preface *xxxi*

1 Introduction 1
Katherine A. Wilson, Eduardo Salas and Dee H. Andrews

SECTION 1 COGNITIVE PROCESSES

2 Factors Affecting Speed and Accuracy of Response Selection in Operational Environments 31
Robert W. Proctor and Motonori Yamaguchi

3 Measuring Vigilance Abilities to Enhance Combat Identification Performance 47
Clark Shingledecker, David E. Weldon, Kyle Behymer, Benjamin Simpkins, Elizabeth Lerner, Joel Warm, Gerald Matthews, Victor Finomore, Tyler Shaw and Jennifer S. Murphy

4 Dimensions of Spatial Ability and their Influence on Performance with Unmanned Systems 67
Thomas Fincannon, A. William Evans III, Florian Jentsch and Joseph Keebler

SECTION 2 VISUAL DISCRIMINATION

5 The Effects of Conjunctive Search and Response Mappings on Automatic Performance in a Complex Visual Task 85
Cleotilde Gonzalez, Rick P. Thomas and Poornima Madhavan

6 What Visual Discrimination of Fractal Textures Can Tell us About Discrimination of Camouflaged Targets 99
Vincent A. Billock, Douglas W. Cunningham and Brian H. Tsou

7 A Cognitive Basis for Friend-Foe Misidentification of Vehicles in Combat 113
Joseph R. Keebler, Lee W. Sciarini, Florian Jentsch, Denise Nicholson and Thomas Fincannon

8 Preattentive Attributes in Visualization Design: Enhancing Combat Identification 129
Scott H. Summers

SECTION 3 SITUATION AWARENESS

9 Team Coordination and Shared Situation Awareness in Combat Identification 147
Cheryl A. Bolstad, Mica R. Endsley and Haydee M. Cuevas

10 The Use of Soft Sensors and I-Space for Improved Combat ID 161
David L. Hall and Stan Aungst

SECTION 4 TEAMS

11 Training Strategies to Mitigate Expectancy-Induced Response Bias in Combat Identification: A Research Agenda 173
Frank L. Greitzer and Dee H. Andrews

12 Comparing Individual and Team Judgment Accuracy for Target Identification under Heavy Cognitive Demand 191
Verlin B. Hinsz and Dana M. Wallace

13 A Team Training Paradigm for Better Combat Identification 205
Wayne Shebilske, Georgiy Levchuk, Jared Freeman and Kevin Gildea

14 Analysis of the Tasks Conducted by Forward Air Controllers and Pilots during Simulated Close Air Support Missions: Supporting the Development of the INCIDER Model 217
Beejal Mistry, Gareth Croft, David Dean, Julie Gadsden, Gareth Conway and Katherine Cornes

15 Team Cognition During a Simulated Close Air Support Exercise: Results from a New Behavioral Rating Instrument 227
Jerzy Jarmasz, Richard Zobarich, Lora Bruyn-Martin and Tab Lamoureux

SECTION 5 AUTOMATION

16 Evaluating Reliance on Combat Identification Systems: The Role of Reliability Feedback 249
Heather F. Neyedli, Lu Wang, Greg A. Jamieson and Justin G. Hollands

17 The Effects of Automation Bias on Operator Compliance and Reliance 265
Stephen Rice, Krisstal Clayton and Jason McCarley

18 An Examination of the Social, Cognitive, and Motivational Factors that Affect Automation Reliance 277
Mary T. Dzindolet, Linda G. Pierce and Hall P. Beck

19 On Fratricide and the Operational Reliability of Target Identification Decision Aids in Combat Identification 299
John K. Hawley, Anna L. Mares and Jessica L. Marcon

20 The Case for Active Fratricide Avoidance in Net-Centric C2 Systems 313
John Barnett

21 Mitigating Friendly Fire Casualties through Enhanced Battle Command Capabilities 327
Jean W. Pharaon

Index *355*

List of Figures

1.1	Components of simulation-based training	10
2.1	An illustration of a three-stage model of human information-processing and examples of operational factors that affect the processing stages	32
2.2	Examples of the stimulus-response mappings for (a) the SRC task and (b) the Simon task	32
2.3	The stimulus-response compatibility (SRC) effect as a function of response mode (yoke-turn vs. button-press) and mapping condition (pure mapping vs. mixed mapping)	34
2.4	Mean response times for the compatible and incompatible mappings as a function of hand-mapping assignment	35
2.5	The Simon effect as a function of practice amount)	36
2.6	The Simon effect for the horizontal and vertical tasks as a function of practiced spatial dimensions	37
2.7	The Simon effect for the joystick and keyboard Simon tasks without action effect as a function of practiced response mode	41
2.8	The Simon effect for the joystick and keyboard Simon tasks with action effect as a function of practiced response mode)	42
3.1	Vigilant Warrior model for developing a personnel-selection tool	51
3.2	Mean number of correct detections as a function of periods of watch for both simultaneous and successive conditions	53
3.3	Partial screen displays from the battlefield situation-monitoring task showing a non-signal event (left panel) and a simultaneous condition signal event (right panel)	58
3.4	Plot of cell means for SVT in validation study	60
3.5	Plot of cell means for laboratory-focused criterion task in validation study	61
4.1	Plot of the interaction between shared video manipulation and UAV spatial orientation on UAV	77
5.1	An example of the visual search component of Radar	89
5.2	Two-way interactions for the proportion of correct detections in Experiment 1	90
5.3	Two-way for detection time in Experiment 1	91
5.4	Three-way interactions for detection time in Experiment 1	91
5.5	Proportion of correct detections and detection time in Experiment 2	94
6.1	Fractal textures like those used in the experiments	101
6.2	Group averages for all four conditions	104
6.3	Individual data for all four observers for all four conditions	105

6.4	Discrimination contours for spatiotemporal fractal textures	108
7.1	Feature Integration Theory Model	116
8.1	Example of a simple search for targets based upon a difference in hue	132
8.2	Example of a search for a target based upon a difference in shape	132
8.3	An example of a search for a conjunction target	133
8.4	An example depicting variable symbology sizing to make a hostile air track more salient	135
8.5	An example contrasting two track symbology sets with regard to communicating target heading and platform	136
8.6	An example contrasting use and non-use of an icon designed to preattentively communicate track vertical speed trend information	137
8.7	An example illustrating the need to support history trails on not simply all tracks, but for individual tracks as well	138
8.8	An example depicting effectiveness of communicating friendly IFF/SIF information	139
8.9	An example illustrating results of layer brightness adjustment to optimize track presentation	140
9.1	Situation awareness forms the basis for decision-making and action	149
9.2	Format of GDTA	150
9.3	Example of mapping of GDTA for Logistics officer onto features of Synergy display	151
9.4	Full shared displays for Air Commander and Intelligence Officer	153
9.5	Sample screen from contingency planning module of SA in general aviation training program	157
10.1	Concept of human centered information fusion	165
10.2	Example of 3-D display for identifying key data among a large data set	167
10.3	NC2IF Extreme Events Laboratory	168
11.1	Framework for classifying teamwork breakdowns	177
11.2	Performance effects of stress	179
12.1	Predicted team proportion of correct responses given values for the proportion of correct individual responses	194
12.2	The layout of the grid for participants at the beginning of the scenario	195
13.1	Mean accuracy ratings for training objective 3 (TO3) for two sets of pre-tests and post-tests: Phase II training (II) and phase III training (III)	212
13.2	Communication dynamics for ISR and DTC	215
14.1	The INCIDER integrated analysis and experimentation programme	220
14.2	Laboratory layout	222
14.3	Summary task diagram for FAC-pilot tasks during a generic scenario with the synthetic environment	223

16.1 The effect of aid reliability on transformed FA rate 253
16.2 Response bias in each experimental condition 254
16.3 Mesh display: Separated and integrated and pie display versus separated and integrated 257
16.4 The main effect reliability level on miss rate 259
16.5 Main effect of display type on sensitivity, *d'* 259
16.6 Comparison of $\beta_{optimal}$ to β_{actual} for separated and integrated displays 260
17.1 A selective two-process model 267
17.2 Mandler's two-process model 267
17.3 Non-selective two-process model 268
17.4 Single-process model 268
17.5 *d'* as a function of bias and reliability 271
17.6 State Trace Analyses on a) Agreement Rates (%); and b) RTs (sec). Solid squares indicate miss-prone conditions, while clear circles indicate FA-prone conditions 273
18.1 Sample slide containing a soldier 279
18.2 Reliability rates focused on slide 281
18.3 Reliability rates focused on contrast detector's decision 281
18.4 Framework of Automation Use 283
18.5 Cognitive processes 284
18.6 Social processes 286
18.7 Motivational processes 290
21.1 The FBCB2 system hardware 330
21.2 Tadiran's RPDA-39 CDA version available to both U.S. and Israeli forces 333
21.3 General Dynamics' RPDA-57 CDA for the Land Warrior program 334
21.4 A CDA with external battery adapter 334
21.5 Raytheon's CDA version 335
21.6 SNC's Tacticomp 1.5 ruggedized PDA 336
21.7 SNC's Tacticomp 6 ruggedized PDA 336
21.8 Mean rating of each question for the FBCB2 system hardware and software overall acceptability 341
21.9 Mean rating of each question for the FBCB2 system training effectiveness 341
21.10 Mean rating of each question for the FBCB2 operational effectiveness 342

List of Tables

1.1	Understanding teamwork breakdowns and their causes	4–7
1.2	Formal training strategies	9
1.3	Informal training strategies	17
1.4	Training strategies and associated team outcomes	18
3.1	Factor analysis of personality traits: Factor pattern matrix ($N = 210$)	54
3.2	Correlations of intelligence and stress variables with performance	55
3.3	List of tests and items retained in the validation study	57
3.4	Regression coefficients (βs) for simultaneous and successive performance	62
4.1	Correlation matrix with means and standard deviations ($N = 41$)	75
4.2	Block rotation hierarchical regression analysis	76
4.3	Perspective taking hierarchical regression analysis	76
8.1	A partial listing of preattentive basic features	134
9.1	Effective and ineffective SA processes	156
11.1	Behavioral markers of teamwork breakdowns	175–176
12.1	Mean values and standard deviations of the measures for target identification tests 1 and 2	199
15.1	Dimensions of team cognition as adapted to the BARS	231
15.2	BARS items selection table	232
15.3	Sample BARS items for each team cognition category from the 'Transmit CAS brief and communicate remarks' task	233–234
15.4	CAS tasks to be rated during Northern Goshawk, grouped into generic CAS mission phases	235
15.5	Summary statistics of ratings for Phase 3 and 4 tasks in Exercise Northern Goshawk	237
15.6	Ratings for talk-on task of Incident 1	238
15.7	Ratings for talk-on and abort tasks for Incident 2	240
15.8	Ratings for talk-on task for Incident 3	240
16.1	The outcome matrix in the condition that the aid gave 'unknown' feedback	252
16.2	V and R^2 values for each display combination when the $\beta_{optimal}$ equation was fitted to β_{actual}	260
21.1	Historical American Fratricide rates in past United States conflicts	328
21.2	Attributed causes of U.S. deaths during the initial phase of OIF	328
21.3	FBCB2 system hardware and software overall acceptability questions	339
21.4	FBCB2 training effectiveness questions	339
21.5	FBCB2 operational effectiveness questions	340

List of Contributors

Dee H. Andrews is the Directorate Senior Scientist (ST) for the Human Effectiveness Directorate of the Air Force Research Laboratory (AFRL). As a Senior Scientist (ST), Dr. Andrews is the AFRL's principal scientific authority for training research. Dr. Andrews' responsibilities include sustaining technological superiority for training by planning and conducting theoretical and experimental studies. He is also responsible for mentoring and developing dedicated technical staff to assure quality in training research, and represents AFRL in training research matters to the external scientific and technical community.

Stan Aungst is a Professor of Practice for Security and Risk Analysis and Senior Research Associate for the Network-Centric Cognition and Information Fusion Center at Penn State University. Stan has been working in the computing area for the last 40 years and has held positions as an Assistant Professor of MIS and IST, DBA, database programmer, software engineer, and project manager.

John S. Barnett is a Research Psychologist with the U.S. Army Research Institute for the Behavioral and Social Sciences in Orlando, FL. His research interests include human-automation interaction, aviation, training, and human performance in extreme environments. He holds a Ph.D. in Applied Experimental and Human Factors Psychology from the University of Central Florida. He is a former U.S. Air Force navigator with extensive flying hours in B-52 and B-1B aircraft.

Hall P. Beck received his Ph.D. from the University of North Carolina-Greensboro in 1983. He is a Professor in the Department of Psychology at Appalachian State University in Boone, North Carolina. Much of his research has been within the areas of human-machine interaction and social influence.

Kyle J. Behymer received his M.S. in Human Factors Psychology from Wright State University in 2005 and is currently a Research Psychologist with JXT Applications, Inc. Mr. Behymer's research is focused on developing innovative Human-Computer User Interfaces that enhance decision-making and improve collaboration in both military and civilian environments. He has designed systems for a diverse set of domains, including emergency response, military command and control, and multi-crew cockpits. Mr. Behymer is also fluent in several programming languages and has led the software development for a wide variety of projects.

Herbert H. Bell is the Technical Advisor for the Air Force Research Laboratory's Warfighter Readiness Research Division. As the Division's Technical Advisor, he coordinates the development and execution of applied research and advanced development programs to improve Air Force training. His specific research interests focus on the integration of technology and cognitive science to improve the acquisition and maintenance of combat skills across a variety of mission areas. He received his Ph.D. in experimental psychology from Vanderbilt University and has held a variety of positions in academia, industry, and the military.

Vincent A. Billock is a biophysicist who takes a complexity theory approach to vision and theoretical neuroscience. He has a Ph.D. in Biophysics from the Ohio State University and a M.S. in Physics from Miami University. He specializes in applying complexity theory to human color and spatial vision and is fascinated by conditions that cause vision to misbehave. His book—'*Chaos Reigns When Vision Fails: Complexity and Catastrophe in the Perception of Color and Contour*'—will be published by Springer-Verlag in 2010. He is a Lead Scientist for General Dynamics, Inc. at Wright-Patterson Air Force Base.

Cheryl A. Bolstad is a Senior Research Associate with SA Technologies. She received a bachelor's degree in Computer Applications in Psychology and a Master's degree in Cognitive Psychology. She completed her doctoral degree in Psychology at North Carolina State University, specializing in cognition and aging. Dr. Bolstad has almost 20 years of experience in human factors and situation awareness (SA) research, conducting some of the earliest research on sources of individual differences in SA and on the design of shared displays to support team SA. Her recent research has focused on designing training programs for individual and team SA as well as developing methods for supporting team and shared SA in distributed environments.

Krisstal D. Clayton is currently a Ph.D. candidate at New Mexico State University. She received her Master of Arts degree in Psychology from New Mexico State University and will complete her Doctorate in Social Psychology in October 2009. She is an Instructor of Psychology at Western Kentucky University.

Gareth Conway is a Human Factors Psychologist with a particular interest in human function and performance in dangerous, complex, and safety-critical environments. Prior to joining Defence Science and Technology Laboratory (Dstl) in 2006, he obtained a Ph.D. at the University of Leeds under the supervision of Prof. Bob Hockey. Dr. Conway's dissertation considered the relationships between workload, effort, and mental fatigue in the context of complex task performance. He then completed a two year Post Doc under the supervision of Prof. Peter Hancock at the University of Central Florida. Dr. Conway also holds an MSc in Sport Psychology, and has recently spent 4 months embedded within an operational headquarters in Afghanistan with the British Army.

Katherine Cornes is a Cognitive Psychologist for the Defence Science and Technology Laboratory (Dstl) at Portsdown West, UK. She completed a Ph.D. in Cognitive Psychology and an MSc in Research Methods both at Southampton University under the supervision of Prof. Nick Donnelly.

Gareth Croft is a Human Systems Consultant with QinetiQ (Human Performance Solutions), specializing in Applied Cognition. Previously, Gareth was with the UK Ministry of Defence (MoD) Defence Science and Technology Laboratory (Dstl) as a Research Psychologist (2004–2008). Gareth received his Masters degree in Psychological Research Methods from the School of Psychology at Exeter University (UK) in 2004 and his Honours degree in Psychology from Cardiff University in 1999.

Haydee M. Cuevas is a Research Associate II at SA Technologies. She has a Ph.D. in Applied Experimental and Human Factors Psychology and a Bachelor of Arts degree in Psychology, both from the University of Central Florida. She has worked on projects funded by the National Science Foundation, Air Force Office of Scientific Research, Army Research Laboratory, Office of Naval Research, and Office of the Secretary of Defense. Her recent research has primarily focused on investigating how technology can best be used to support the attitudinal, behavioral, and cognitive processes of distributed teams performing in complex operational environments.

Douglas W. Cunningham is an experimental psychologist with a Ph.D. from Temple University. He specializes in the integration of computer graphics and visual psychophysics. He completed his habilitation at the University of Tübingen, where he is a Professor of Cognitive and Information Sciences.

David Dean is the Defence Science and Technology Laboratory (Dstl) Combat Identification Theme Advisor. He has worked for the Ministry of Defence (MoD) since 1986 as a procurement officer, working on warship communications projects, as a systems engineer, undertaking research in warchip whole lifecycle engineering and as a principal analyst and study leader in the field of Combat ID. He graduated with a B.Eng in Electrical and Electronic Engineering from Brunel University in 1991, and obtained an MSc in Systems Engineering for Defence from Cranfield University in 2006.

Mary T. Dzindolet is a Professor in the Department of Psychology at Cameron University in Lawton, Oklahoma. She earned her Ph.D. in Experimental Psychology from the University of Texas at Arlington in 1992, and a Masters of Applied Statistics from Louisiana State University in Baton Rouge, LA in 1987. Her research interests include group dynamics, group creativity, and automation reliance.

Mica R. Endsley is President of SA Technologies in Marietta, Georgia, a cognitive engineering firm specializing in the development of operator interfaces for advanced systems, including the next generation of systems for aviation, air traffic control, medicine, power transmission and distribution, and military operations. Dr. Endsley is a recognized world leader in the design, development, and evaluation of systems to support situation awareness (SA) and decision making. She has led numerous projects on a variety of issues related to SA, including investigations of human error, analyses of SA requirements, development of the SAGAT technique for measuring SA, development of training programs for enhancing SA among individuals and teams, investigations of the effect of free flight, studies of the effects of automation, and development of approaches for integrating humans and automated systems. Dr. Endsley received a Ph.D. in Industrial and Systems Engineering from the University of Southern California.

A. William ("Bill") Evans has been working with the Team Performance Laboratory since 1999 and has been a graduate student in the Applied Experimental/

Human Factors Ph.D. program at University of Central Florida since the fall of 2000. Expecting to graduate in 2009, Bill worked extensively with training and evaluating complex conceptual and perceptual knowledge, specifically mental models of complex automated systems. Since 2004, his research focus has shifted toward robotics and the semi-autonomous control of uninhabited military vehicles. Bill has become experienced in both the creation and utilization of scale military operations in urban terrain (MOUT) facilities and uses this experience to gain knowledge on the most effective human-robot teams. His recent research interests include target identification using unmanned vehicles and the effects of motion (or perceived motion) on target detection and identification.

Thomas Fincannon received a B.A. in psychology from the University of South Florida, Tampa, where he worked at the Center for Robot Assisted Search and Rescue. He is pursuing a Ph.D. in Applied Experimental Human Factors Psychology from the University of Central Florida, where he studies operator use of unmanned systems at the Team Performance Laboratory. His research interests include human-robot interaction focusing on remote operation, team effectiveness, and spatial abilities. He is a member of the Human Factors and Ergonomics Society, ACM SIGCHI, and American Psychological Association, divisions 19 and 21. His recent awards include 2006–2007 Institute for Simulation and Training Student Researcher of the Year and 2008 Human Factors and Ergonomics Society Student Member with Honors.

Victor S. Finomore Jr. is a Research Fellow with the Oak Ridge Institute for Science and Education (ORISE) and works at the Air Force Research Laboratory at Wright-Patterson Air Force Base in Dayton, OH. He received his doctorate in experimental psychology with a specialty in Human Factors from the University of Cincinnati in 2008.

Jared Freeman is Senior Vice President of Research at Aptima, Inc. Dr. Freeman is a cognitive scientist by training with a Doctorate in Human Learning and Cognition. He investigates problem solving and decision making in real-world settings, and designs training and job aids that address these challenges. Within the past several years, Dr. Freeman has served as Principal Investigator (P.I.) in research and development projects to analyze and develop training systems to support urban operations, information system security, imagery analysis, intelligence equipment maintenance, and air crew performance. Dr. Freeman has published more than 100 chapters, articles, and proceedings papers on the topics of critical thinking, task analysis, computational modeling of teams, performance measurement, training, and human factors. Dr. Freeman received a Ph.D. from Columbia University. He is a member of the Human Factors and Ergonomics Society, and a contributing editor to its journal: Human Factors.

Julie Gadsden is a Fellow of the UK Ministry of Defence's Defence Science and Technology Laboratory specializing in the analysis of complex military systems. She has a degree in Pure Mathematics from Nottingham University (1972) and is a Fellow of the UK's Operations Research Society.

Kevin Gildea received a Ph.D. and M.S. in Human Factors and Industrial/Organizational Psychology from Wright State University under an Air Force-sponsored multi-university research initiative fellowship. He received a B.A. in Psychology from Missouri Southern State University. He is an associate member of the Human Factors and Ergonomics Society and a member of the American Psychological Association. Kevin Gildea specialized in developing training systems for command and control environments and military senior leadership. Additional efforts included developing and testing methodologies for complex skill acquisition via the use of artificial partner-agents and adaptive-training protocols. Dr. Gildea's research encompassed team training and the effect of individual differences on performance.

Cleotilde Gonzalez is an Associate Research Professor and Director of the Dynamic Decision Making Laboratory (www.cmu.edu/ddmlab) in the department of Social and Decisions Sciences Department at Carnegie Mellon University in Pittsburgh, PA. She obtained a Ph.D. in Management Information Systems from Texas Tech University in 1996. Her research investigates how people make decisions in dynamic environments. Her research goal is to provide theories, methods, and computational tools that improve the way individuals make decisions in dynamic conditions across specializations of knowledge. She follows experimental methods using decision making games and cognitive computational models to explain and predict human decision processes.

Frank L. Greitzer is a Chief Scientist at the Pacific Northwest National Laboratory (PNNL), National Security Directorate. He holds a Ph.D. in Mathematical Psychology and a B.S. in Mathematics. Dr. Greitzer's R&D focus area of Cognitive Informatics addresses human factors and social/behavioral science challenges through modeling and advanced engineering/computing. With over thirty years of applied research and development experience in cognitive psychology, human information processing, user-centered design, and training system design, his research addresses behavioral modeling challenges to enhance decision making in diverse domains such as intelligence analysis, cyber security, counterintelligence, and electric power grid operations. He also conducts evaluation research to assess the effectiveness of decision aids, analysis methods, displays, and training approaches. In addition to his work at PNNL, Dr. Greitzer serves as an adjunct faculty member at Washington State University, where he teaches interaction design and human factors psychology. Representative publications and project descriptions may be found at http://www.pnl.gov/cogInformatics.

David L. Hall is a Professor of Information Sciences and Technology at the Pennsylvania State University, where he leads the Center for Network Centric Cognition and Information Fusion. He received his Ph.D. in Astrophysics from the Penn State University in 1976. He is an IEEE Fellow, and was awarded the Department of Defense Joe Mignona Award for his lifetime contributions in multisensor data fusion.

John K. Hawley is Chief of the U.S. Army Research Laboratory's Human Research and Engineering Field Element at Ft. Bliss, Texas. Since receiving his doctorate, Dr. Hawley has worked as an applied psychologist for more than 25 years in a variety of government and private-sector organizations. He recently served as project lead for an Army effort to examine human performance contributors to fratricides involving the Patriot air and missile defense system during the Second Gulf War and recommend potential solutions. Dr. Hawley is now working with the air defense community to implement and evaluate selected recommendations involving human-systems integration practices, test and evaluation methods, personnel assignment practices, and operator and crew training. The primary thread running through Dr. Hawley's professional experience is helping people and organizations manage the human side of transitions to new systems, processes, and technologies.

Verlin B. Hinsz received his Ph.D. in social-organizational psychology from the University of Illinois, Urbana-Champaign. He has been at North Dakota State University since earning his doctorate, where he is now Professor of Psychology. Professor Hinsz's research lies at the intersection of social and organizational psychology specializing on information processing and task performance in groups and teams.

Justin G. Hollands is Head of the Human Systems Integration Section at Defence Research and Development Canada-Toronto. He is also Adjunct Associate Professor at the University of Toronto, Department of Mechanical and Industrial Engineering. He received his Ph.D. in Psychology from the University of Toronto in 1993.

Greg A. Jamieson is Associate Professor of Mechanical & Industrial Engineering at the University of Toronto. He received his Ph.D. from the same institution in 2003.

Florian Jentsch is an associate scientist/scholar in the Department of Psychology at the University of Central Florida. He is also the director of the Team Performance Laboratory. Dr. Jentsch received his Ph.D. in human factors psychology from the University of Central Florida in 1997. He also holds master's degrees in aeronautical engineering from the Technical University of Berlin and in aeronautical science from Embry-Riddle Aeronautical University. His dissertation on training for junior commercial flight crew members won the American Psychological Association's 1998 George E. Briggs Award for the best dissertation in applied/experimental psychology, and he was also awarded the American Psychological Association's 2002 Earl Alluisi award for Early Career Achievement in applied/experimental psychology. Dr. Jentsch's research interests are in team training, human-robot-interaction, aviation human factors, cross-cultural research, research methodology, and simulation where he has co-authored over 150 publications, presentations, and technical reports. He has received grants and contracts from NAWC-TSD, ARL, RDECOM-STTC, ARO, and PEO-STRI, the Federal Aviation Administration, and the Transportation Security Administration. He also has consulted on system and software development projects for the FAA, the U.S. Navy, U.S. Army, NIH, NSF, and NASA.

Jerzy Jarmasz is a research scientist at Defence R&D Canada, the science and technology agency of the Canadian Department of National Defence. His work focuses on simulation-based training in a number of areas, including distributed mission training, IED detection and awareness, and decision making in complex, dynamic environments. His areas of expertise include basic cognitive processes, team communications and cognition, and the use of dynamic systems theory for understanding cognition and command-and-control.

Joseph R. Keebler is a Graduate Research Associate at Team Performance Laboratory, at the Institute for Simulation and Training, University of Central Florida (UCF). Joe received his bachelor's degree in 2005 from the University of Central Florida, and started in the Applied/Experimental Human Factors Psychology Ph.D. program the following year. His research interests include target identification/classification, augmented cognition, visual perception, human robot interaction, and research methodology. Joe currently serves as the President of the Human Factors and Ergonomic Society's Student Chapter at UCF.

Tab Lamoureux, at the time that this work was conducted, was a Senior Consultant and Team Manager with Humansystems Incorporated in Guelph, Ontario. He has a MSc from Cranfield University and over 16 years of experience conducting Human Factors research and development. Mr. Lamoureux has extensive experience in air traffic management, aviation, and military command and control systems, as well as nuclear power generation, and is drawn toward work involving high performing teams in high reliability organizations. Mr. Lamoureux is now an independent consultant.

Elizabeth Lerner received her M.S. in Human Factors Psychology at Wright State University in 2004 and is currently a Ph.D. candidate. Her dissertation focuses on memory and cognitive processes underlying reasoning about planning and scheduling representations on a team level. She has experience in experimental design and statistical analyses and was a Research Psychologist at JXT Applications, Inc.

Georgiy M. Levchuk is a Principal Engineer in Modeling and Simulation Division at Aptima, Inc. His research interests include multi-objective optimization, probabilistic models, and their applications in the domains of planning and reasoning under uncertainty. Dr. Levchuk has published over 50 book chapters, refereed journal and peer-reviewed conference papers in the areas of organizational design, resource allocation, team behavior analysis and simulation, and adversarial network and behavior identification. Within the past several years, Dr. Levchuk has served as the Principal Investigator in several research and development projects in human-centered engineering domains, including organizational design and optimization, behavior classification and analysis, adversarial reasoning, and planning under uncertainty. Prior to joining Aptima, he held a Research Assistant Position at the Institute of Mathematics (Kiev, Ukraine), a Teaching Assistantship at Northeastern University (Boston, MA), and a Research Assistantship at University of Connecticut (Storrs, CT). Georgiy Levchuk received his B.S. and M.S. degrees in Mathematics with Highest Honors from the National Taras

Shevchenko University, Kiev, Ukraine in 1995, and Ph.D. degree in Electrical Engineering from the University of Storrs, Connecticut in 2003.

Poornima Madhavan is an Assistant Professor of Human Factors in the Department of Psychology at Old Dominion University, Norfolk, VA, where she also holds an appointment at the Virginia Modeling, Analysis and Simulation Center. She received her Ph.D. in Human Factors (Engineering Psychology) from the University of Illinois at Urbana-Champaign in 2005, followed by a post-doctoral fellowship in the Dynamic Decision Making Lab at Carnegie Mellon University. Dr. Madhavan's Applied Decision Making Lab at Old Dominion University is devoted to the study of human learning and decision making processes in highly automated environments.

Jessica L. Marcon is a Ph.D. candidate in Experimental Psychology in the Department of Psychology at the University of Texas at El Paso. Ms. Marcon received her M.A. in Experimental Psychology from the University of Texas at El Paso in 2007. During the 2008 academic year, Ms. Marcon worked with the Army Research Laboratory's Ft. Bliss Field Element as a Field Placement intern. She is currently finishing her dissertation on a Dodson Dissertation Fellowship from the University and the Dodson Endowment.

Anna Lucia Mares received a Bachelor of Science in Electrical Engineering (BSEE) from the University of Texas at El Paso. She has worked for the United States Army Research Laboratory (USARL) for more than 15 years which has included developmental, analysis, and evaluation of numerous Joint and Army Programs. Initially her efforts focused on Ground, Soldier, and Munition Systems with more recent involvement centered on the human performance dimension of the Air and Missile Defense Systems In 2007, Ms. Mares received the prestigious Army Research Laboratory Award for Analysis for her analytical contributions in identifying significant issues involving the Patriot air and missile defense system usage during Operation Iraqi Freedom. Her efforts resulted in recommendations for material and training solutions which increased friendly protection involving the Patriot air and missile defense system.

Lora Bruyn Martin is a Senior Consultant and Team Manager with Humansystems Incorporated in Guelph, Ontario. She has over 7 years experience conducting Human Factors research relating to team decision making and planning, team collaboration, and command and control. Ms. Bruyn Martin has a MASc in Systems Design Engineering from the University of Waterloo.

Gerald Matthews is Professor of Psychology at the University of Cincinnati. He obtained his Ph.D. in Experimental Psychology from Cambridge University in 1984. His research focuses on human performance, cognitive models of personality, acute states of stress and fatigue, and emotional intelligence. He has published 10 books, and over 200 journal articles and book chapters on these topics. He is Secretary-Treasurer of the International Society for the Study of Individual Differences, and President-Elect of the Traffic and Transportation Psychology division of the International Association for Applied Psychology. He is also an associate editor for *Personality and Individual Differences*, and a consulting editor for *Journal of Experimental Psychology: Applied*.

Jason McCarley is currently an Assistant Professor at the University of Illinois, at Urbana-Champaign. He received his Ph.D. in Psychology from the University of Louisville.

Beejal Mistry is a psychologist working for the Defence Science and Technology Laboratory (Dstl) in Fareham, UK. After gaining her Ph.D. in Psychology at the University of Reading in 2000, she has worked as an analyst in the Private Sector before joining Dstl in 2003.

Jennifer Murphy is a research psychologist at the US Army Research Institute for the Behavioral and Social Sciences in Orlando, FL, where she works in the Technology-Based Research Unit. Currently, her research examines selection and training issues associated with target detection. Her other research interests include game-based training and training of cognitive skills. Dr. Murphy received her Ph.D. in Cognitive/Experimental Psychology in 2004 from the University of Georgia, where her research focused on the rapid visual categorization of natural scenes, inhibition of return, and sex-based differences in early visual processes.

Heather F. Neyedli is a masters student in the Cognitive Engineering Laboratory (CEL) in the Mechanical and Industrial Engineering Department at the University of Toronto. Ms. Neyedli received a Bachelor of Science (Hon.) degree in Kinesiology from Dalhousie University in Halifax, Nova Scotia in 2007.

Denise Nicholson is the Director of the Applied Cognition and Training in Immersive Virtual Environments Lab (ACTIVE) at University of Central Florida's Institute for Simulation and Training with additional affiliations in the Modeling and Simulation Graduate Program, Industrial Engineering and Management Department, and the College of Optics and Photonics/CREOL. She has a Ph.D. and M.S. in Optical Sciences from University of Arizona, a B.S. in Electrical Computer Engineering from Clarkson University and is a Certified Modeling and Simulation Professional. Dr. Nicholson's research focuses on human systems modeling, simulation and training includes virtual reality, augmented cognition, and adaptive human systems technologies for Department of Defense and Duel-Use applications. She joined UCF in 2005 with over 18 years of government service ranging from bench-level research at the Air Force Research Laboratory to leadership as the Deputy Director for Science and Technology at NAVAIR Orlando Training Systems Division.

Linda Pierce is Chief of the U.S. Army Research Institute-Aberdeen Proving Ground Research Unit. She leads a program of research to improve the training and performance of network-enabled teams and organizations. Dr. Pierce has almost 20 years of experience conducting research on collaboration and decision making in battle command teams. She earned a doctoral degree from Texas Tech University in 1987.

Jean Wadner Pharaon, at the time that this work was conducted, Jean was a Human-System Integration (HSI) Specialist for the Human Research and Engineering Directorate (HRED) of the U.S. Army Research Laboratory (ARL). His main effort consisted of implementing the Army's MANPRINT program in ground systems acquisition at the Army's Tank, Automotive and Armaments Command (TACOM) in Warren, MI. Currently, he is working for the Tank Automotive Research, Development and Engineering Center (TARDEC) as a Systems Engineer for the MaxxPro variant of the Mine Resistant Ambush Protected (MRAP) vehicles that are extensively being used in Iraq and Afghanistan to save Warfighters' lives. In addition, he is a Military Intelligence Officer in the Army Reserve. He received his Master's degree in Industrial and Operations Engineering from the University of Michigan, Ann Arbor and his Bachelor's degree in Mechanical Engineering from Florida International University, Miami, FL.

Robert W. Proctor is Distinguished Professor of Psychological Sciences at Purdue University, with a courtesy appointment in the School of Industrial Engineering. He received his Ph.D. in psychology from the University of Texas at Arlington in 1975. Dr. Proctor is co-author of the books, *Stimulus-Response Compatibility Principles: Data, Theory, and Application* and *Human Factors in Simple and Complex Systems*. He is a Fellow of the American Psychological Association, Association for Psychological Science, and Human Factors and Ergonomics Society.

Stephen Rice is an Assistant Professor of Psychology at New Mexico State University. He is the director of the Base Lab, which focuses on combat identification and UAV issues. He received his Ph.D. in Engineering Psychology from the University of Illinois at Urbana-Champaign in 2006.

Eduardo Salas is a University Trustee Chair and Pegasus Professor of Psychology at the University of Central Florida, where he also holds an appointment as program director for the Human-Systems Integration Research Department at the Institute for Simulation and Training. His expertise includes assisting organizations in how to foster teamwork, design and implement team training strategies, facilitate training effectiveness, manage decision making under stress, and develop performance measurement tools.

Lee W. Sciarini is a doctoral candidate the University of Central Florida, currently working on his dissertation for his Ph.D. in Modeling and Simulation. His research interests include training system development and effectiveness, human performance, human systems integration, team performance, unmanned systems, neuronergonomics, augmented cognition and how these areas can be leveraged to enhance future systems. Lee currently works at Applied Cognition and Training in Virtual Environments Lab (ACTIVE) at the University of Central Florida's Institute for Simulation and Training, and has also collaborated with Team Performance Laboratory.

Tyler Shaw is currently a Post Doctoral Research Fellow/Research Assistant Professor at George Mason University in Fairfax, Virginia. Tyler received his Ph.D. in experimental psychology/human factors from the University of Cincinnati in 2008. The focus of most of his graduate work was the study of the mechanisms underlying sustained attention. Tyler's current research interests involve the neurophysiological underpinnings of sustained attention and issues regarding reliability and trust in automation.

Wayne Shebilske is a professor in the Department of Psychology at Wright State University. He received from the University of Wisconsin a B.A. in 1969, M.S. in 1972, and Ph.D. in 1974. After nine years on the faculty of the University of Virginia Psychology Department (Assistant Professor, 1974–1979: Associate Professor 1979–1983), Dr. Shebilske served as Study Director for the Committee on Vision at the National Academy of Sciences (1983–1985) and a Full Professor at Texas A&M University (1985–1999). He has worked closely with government, military, and private agencies identifying critical issues for study in many areas including design and development of aerospace systems, visual display equipment, medical devices, standards for pilots and drivers, automated instruction for complex skills, virtual reality systems, and assistive technologies for people with disabilities. He has served on the Editorial Advisory Board for Human Factors and for Psychological Research.

Clark Shingledecker is an experimental psychologist whose research and teaching focus on the areas of applied cognitive psychology and human factors. He has 30 years of research and development experience in university, government, and industry settings. His recent work includes the development of personnel selection tools for vigilance skill (US Army), learning technologies for students with disabilities (NSF), and human factors in the design and evaluation of civil aviation and air traffic control systems (FAA). He currently serves as Research Professor of Technology-based Learning with Disability at Wright State University and is a consultant to government and industry in defense and aviation human factors.

Benjamin G. Simpkins is a Research Psychologist at JXT Applications Inc. who aided the data analysis for this research effort. He holds an M.S. in Experimental Psychology and was previously a laboratory and lecture instructor in the areas of Research Design and Statistical Analysis at Wright State University in Dayton, OH. Simpkins is also a software engineer and has worked extensively at JXT in the development of user interfaces and information displays.

Scott H. Summers is a Human Factors Engineer for Raytheon Solipsys, a position he has held since 2006. Prior to that, Scott was a Command and Control (C2) operations officer for the U.S. Air Force, serving in various operational capacities during his 22-year military career. Scott earned his Master's degree in Human Factors from Embry-Riddle Aeronautical University, and has served as the Human Machine Interface lead for several operational C2 systems, including the Battlespace Command and Control Center (BC3), currently employed by US forces in Iraq and Afghanistan.

Rick Thomas is an Assistant Professor in the Department of Psychology at the University of Oklahoma and directs the Decision Processes Laboratory (DPL). His primary research focus involves the application of mathematical memory models to study the role of memory processes in judgment and decision-making phenomena. In the laboratory, Dr. Thomas employs experimental methods and procedures from memory paradigms to help understand judgment processes. His current research investigates how a modified version of MINERVA-DM, HyGene, accounts for the effects of hypothesis generation on probability judgments, confidence judgments and hypothesis testing. Dr. Thomas also does work concerning the study of expertise, primarily in the areas of performance evaluation and the development of decision support tools.

Brian H. Tsou is a biophysicist who does research in human visual effectiveness under extreme conditions and its application to aviation sensor/display systems. He has a Ph.D. in Biophysics from the Ohio State University. He received the U.S. Air Force Harry G. Armstrong Award for Scientific Excellence in 1994 and the U.S. Air Force Paul M. Fitts Human Engineering Award in 1993 for his contributions to binocular head-mounted display design and evaluation. He is a Principal Scientist at the U.S. Air Force Research Laboratory at Wright-Patterson Air Force Base.

Dana M. Wallace received her undergraduate degree in Psychology from Jamestown College and her M.S. in psychology from North Dakota State University where she is completing her Ph.D. She will be joining the faculty of Jamestown College in 2009. Her research interests are in the exploration of informational and motivational influences on judgments in a variety of social contexts.

Lu Wang is a Ph.D. student in the Department of Systems and Information Engineering at the University of Virginia. Mrs. Wang received her Master's degree in Industrial Engineering from the University of Toronto in 2007.

Joel S. Warm is Senior Scientist in the 711th Human Performance Wing/CL, Warfighter Interface Division, Air Force Research Laboratory, Wright- Patterson Air Force Base, OH. He has recently retired as Professor of Psychology at the University of Cincinnati where he taught for 41 years and was a founder of his department's human factors program. Dr. Warm has done extensive research in the area of sustained attention or vigilance focusing on theoretical models and on problems in psychophysics, training and motivation, neuroergonomics, workload, and stress. He serves on the editorial boards of Human Factors and Theoretical Issues in Ergonomic Science, is a Fellow of AAAS, the American Psychological Association, the Association for Psychological Science, and the Human Factors and Ergonomics Society and is a member of the National Research Council Committee on Human Systems Integration. Dr. Warm received his Ph.D. in experimental psychology from the University of Alabama in 1966.

David E. Weldon holds a Ph.D. in Cognitive Social Psychology and is currently Chief Scientist and Senior Software Engineer at JXT Applications, Inc. and was the Project Manager/Principal Investigator for the research reported in this chapter. He has more than 35 years experience in advanced development research in social perception, social judgment, and computer simulation of human decision-making. He is also a software systems architect and engineer having designed and supervised Natural Language Processor implementations and distributed processing applications. Dr. Weldon previously held positions in advanced research and development at Infoglobe, HCI Technologies and NCR Corporation. Prior to that, he was Associate Professor of Computer Science at Winona State University and Assistant Professor of Social Psychology at Washington University, St. Louis.

Katherine A. Wilson is a Human Performance Investigator at the National Transportation Safety Board in the Office of Aviation Safety. Dr. Wilson received a Ph.D. in Applied Experimental and Human Factors Psychology in 2007 and a M.S. degree in Modeling & Simulation in 2002 from the University of Central Florida. Prior to joining the NTSB, she was a research assistant at the Institute for Simulation & Training. Her expertise includes teams, team training, and team performance in complex environments including aviation, the military and healthcare.

Motonori Yamaguchi is a Ph.D. candidate in cognitive psychology at Purdue University. He received bachelor's degrees in Computer Engineering from Tokyo University of Technology and in psychology from Indiana University South Bend, and a Master's degree in cognitive psychology from Purdue University. His research focuses on perceptual-motor relationships and mathematical modeling of human action selection.

Richard Zobarich, at the time that this work was conducted, was a Human Factors Consultant with Humansystems Incorporated in Guelph, Ontario. He has an MA in Human Factors and Applied Cognition from George Mason University along with over 8 years of experience in Human Factors research and development in military and commercial domains. Mr. Zobarich is now a Human Factors Consultant at CAE Professional Services (Canada) in Ottawa, Ontario.

Foreword

Warfare has always been a dangerous business and fratricide has often been an unfortunate part of that business. As warfare and technology evolve, the likelihood of fratricide appears to be increasing. Modern warfare is often conducted around-the-clock against fleeting targets of opportunity that must be engaged under conditions of limited visibility. Today's weapon systems allow for engaging these targets at longer ranges with a wide variety of advanced munitions. The precision of these munitions leaves little room for error as they rarely miss their targets, regardless of whether those targets are hostile or friendly. As a result, preventing the death or injury of friendly personnel from friendly fire has become a major challenge.

Fratricide is often the result of two factors that Clausewitz (1832–1984) called the fog of war and the friction of war. A wide range of elements including enemy and friendly tactics, rules of engagement, weather, terrain, fatigue, and technology underlie this fog and friction which produces a dynamic environment filled with uncertainty, fear, and confusion. The result is often limited awareness of the situation and difficulty in executing correct and timely military actions, which leads to situations in which friendly personnel may be targeted and engaged by other friendly forces.

Large investments have been made in technologies to improve our ability to correctly identify friendly forces. These technologies include marking vehicles with distinctive patterns, electronic systems that query air and ground vehicles for a friendly electronic code, and displays that seek to provide an accurate, common, up-to-date picture of how forces are dispersed within the battlespace.

While these technologies play an important role in reducing the probability of fratricide, warfare is ultimately a human endeavor and these technologies are embedded within complex systems that operated by and provide support to human warfighters. This volume provides a multidisciplinary view of how humans, acting as individuals and teams, are affected by the fog and friction of war and how an understanding of human capabilities and human-centric technologies can reduce the opportunities for fratricide produced by that fog and friction.

The chapters within this volume are organized in five sections that address many of the fundamental human elements of fratricide. The first chapter provides an introduction for the book by discussing fratricide and continuing work on mitigating potential teamwork breakdowns. The first two sections discuss humans as active information processors and the factors that affect their ability to operate under a variety of operational conditions. The third section focuses on situation awareness. The chapters in this section discuss the importance of shared situation awareness and describe the potential to develop new information based on interpersonal communication and virtual relationships. The fourth section looks at how teams perform combat identification as well as how they plan and conduct close air support. The final section examines the technical and social sides of automation in combat identification.

This volume provides an excellent overview of the human element and its role in fratricide. It points out how a wide variety of factors including fundamental human capabilities, training, social organization, and technology combine to provide opportunities for fratricide and how human-centric design based on those same factors can mitigate the likelihood of fratricide. This volume also provides a clear and concise starting point for both basic and applied research directed toward the systemic problems of combat identification and fratricide.

Herbert H. Bell

References

von Clausewitz, C. (1984). *On War*. (M. Howard & P. Paret, Trans., Eds.). Princeton, NJ: Princeton University Press. (Original work published 1832)

Preface

Friendly fire, the unintentional harming of a friendly force, a decidedly unfriendly act but a well-established reality of war, has been persistent throughout warfare. It is not realistic to think that it can ever be completely eliminated regardless of the best attempts to develop better technology and provide better training. If warfighters are to hold the enemy at risk, there is also always the threat that friendly forces may also accidentally be harmed. In addition, revolutions in electronic communications have made the nearly instant reporting of tragic neutral casualties, euphemistically called collateral damage, a major concern for military planners and commanders in ways not previously seen. The militaries of the west spend enormous resources to try and mitigate the risk of friendly fire and collateral damage. One leading cause is the inability of warfighters to identify friends, foes, and neutrals in combat. This book examines the issue of how human factors considerations in identifying, and eliminating, where possible, the risk of error in combat identification (CID) can best be performed in an increasingly complex battlespace.

In a January 10th, 1996 memorandum, then Undersecretary of Defense for Acquisition and Technology, Paul Kaminski, defined CID as follows, 'In current usage, Combat Identification is defined as the means to positively identify friendly, hostile, and neutral platforms to reduce fratricide due to misidentification, and to maximize the effective use of weapons systems.'

There are a variety of other definitions of the construct of combat identification that could be cited, but Dr. Kaminiski's definition encompasses the key elements of all the definitions. That is, all warfighters in combat seek to identify friend, foe, and neutral and then take appropriate action. The rules of engagement may change from battle to battle, but a primary rule is to avoid putting friendlies and neutrals at risk when possible.

During the Gulf War (Operation Desert Storm) in 1991, the friendly fire casualty rate exceeded 20 percent. This was considerably higher than the rates reported in previous wars, and the Department of Defense asked the Defense Science Board (DSB) to look into possible causes and remedies. The DSB (1996) reported the following:

> Combat ID does not result from a single device or process but results from the combination of many sources. Knowledge about the location and activities of friendly and enemy forces (situational awareness) comes from plans, reports, surveillance (often enhanced by distinctive uniforms and insignia) and necessarily includes identification. New technology for surveillance, processing, navigation, and networking is greatly increasing our ability to create and distribute accurate,

> timely situational information smoothing out the difference between situational awareness and combat ID. (p. v)

Clearly, the warfighter is at the center of both the causes and solutions of friendly fire. All the technology categories described by the DSB (1996) are designed to provide the human decision maker with the best possible information about where the enemy is and will be, and perhaps more importantly where the friendly forces and neutrals are.

Describing possible reasons for the relative increase in friendly fire incidents in Operation Desert Storm the DSB (1996) noted the changes in the battlefield of the modern, high-tech era. While there have been some key technological advancements in CID since Operation Desert Storm, the same challenges are also being seen in the recent wars in Iraq and Afghanistan. The DSB noted:

Several Trends Motivate Increased Attention

- Highly Mobile, Joint Service and Coalition Warfare (Air and Ground).
- Long Range, High Lethality Weapons.
- Overwhelming Blue Advantage Situations Very Likely; Reduced Tolerance; Improved Attributability.
- War Zone with Civil Activity; Common Red/Blue Assets (p. 1).
- The trends cited by the DSB highlight several themes, discussed below, relevant to the use of a human factors approach to enhance CID and reduce friendly fire incidents.
- The speed of the modern battlefield has greatly increased the chance that friendly units will encounter each other very quickly. Often, the time to decide via CID that a unit is friend or foe can be literally just seconds. In addition, weapon system advancements have made it much more likely that a warfighter will hit and kill or destroy their intended target. These weapons can be fired from incredibly long ranges that make CID, by any means, very difficult.
- Nations and loved ones now have less tolerance for fratricide incidents. In like manner, they have less tolerance for collateral damage that leads to deaths and injuries of neutrals. The press is now replete with detailed descriptions of all friendly fire incidents and resulting hearings and lawsuits. This was not the case in wars before the 1990s, but is a byproduct of our digital age. While it is sometimes difficult to know for certain that a fired weapon has hit a friendly or neutral, the DSB has correctly concluded that battle damage assessment is easier today than before the 1990s due to the capability to review refined data from weapons shots.
- It is important to consider the nature of asymmetric warfare and its effect on CID and friendly fire. Bockstette (2008) describes asymmetric warfare as:

> '... conflicts between parties that show essential quantitative and/or qualitative dissimilarities in the battle space dimensions: an imbalance in forces, a different determination/motivation, a different legitimization, a different application of methods and a difference in the quality or character of methods themselves. Asymmetrical conflicts are usually waged in a changing, asynchronous and unpredictable manner' (p. 7–8).

- In these types of conflicts the enemy is well known for taking advantage of their ability to 'blend' into the civilian population. That is their major asymmetric advantage. This brings the civilian population into direct risk as more traditional militaries must use all possible CID advantages to avoid collateral damage. In addition, when traditional military forces go into civilian areas to seek out the enemy, often in urban settings, the risk of friendly fire on their own forces are increased.
- Finally, as western weapons systems have either migrated to the rest of the world, or greatly influenced weapon system design in non-western countries, the look and function for blue and red forces have become more and more alike. In the past, it was not difficult to distinguish between friendly and enemy forces based on weapon systems' physical characteristics, it is now much more difficult to do so. This similarity has made the CID function increasingly problematic.

With that introduction, the reader may well agree with the editors that there are substantial challenges to overcoming CID issues and friendly fire incidents. The military has invested considerable resources toward improving CID and reducing friendly fire. In addition, researchers have contributed greatly to overcoming the challenges by examining fundamental sensation and perception issues. This book explores what the human factors community has to say about the CID construct. The book developed from a 2008 workshop in Gold Canyon, Arizona, that was sponsored by the Air Force Research Laboratory. The speakers presented research and experience that both directly and indirectly addressed what is known about the human factors of CID. The presenters came from government, industry, and academia, and represented a wide array of backgrounds and perspectives. Many of those speakers, as well as other experts in the field, agreed to write chapters for this book. As the reader will discover, there is a large variety of human factors areas that can and do impact the general topic of CID and friendly fire avoidance. A single book could not possibly cover them all; however, the authors of these select chapters touch on some very important aspects of human factors.

Improvement in automation has enhanced combat identification; however, research must also focus on the human. Human error is at the core of friendly fire incidents, in terms of the design of a combat identification system or of a target identification decision and the use of a system. In the past, most efforts have focused on technological solutions, but now more than ever we have realized that it is necessary to focus both on automation and the human element. The editors believe that equal attention and investment should be paid to research in developing human-centered solutions.

The book's twenty-one chapters are divided into five sections: Cognitive Processes, Visual Discrimination, Situation Awareness, Teams, and Automation. While many of the chapters relate to several or all of the five topics, they have been placed in the sections that best represents each chapter's key themes.

This collection of research focuses on fundamental research on human sensation and perception to applied attempts at developing practical solutions to CID challenges. In recent years the engineering community has developed, with the help of human factors specialists, impressive technological aids to improve CID and reduce friendly fire, some of which are described in this book. There is little doubt that these efforts have resulted in saved friendly and neutral lives on recent battlefields, but as the modern battlespace becomes more complex, it has been demonstrated that technological solutions alone will not mitigate tragedies. Focus on the human element with its attendant factors must be the central component if future scientists are to be successful at averting horrific consequences.

The human factors community has played a major role to improve battlefield awareness in the past and must continue this momentum to minimize opportunities for fratricide in the future. Researchers must target the gaps and deficiencies discovered through past combat identification incidents and build intervention strategies into training, interfaces, tactics, techniques, and procedures that further mitigate human error occurrences. This book presents an excellent foundation for the human-factors community to build upon and, along with military research funders around the world, we trust that they will continue to accelerate their efforts to over come the challenges of CID.

Dee H. Andrews
March 27, 2009

References

Defense Science Board. (1996). *Report of the Defense Science Board Task Force on Combat Identification.* Office of the Under Secretary of Defense for Acquisition and Technology. Retrieved February 23, 2009, from http://cryptome.org/iwd.htm.

Bockstette, Carsten. (2008). *Jihadist Terrorist use of Strategic Communication Management Techniques* (Occasional Paper Series No. 20). Germany: George C. Marshall European Center for Security Studies.

To all of the brave warfighters
who put themselves in harm's way to defend our liberty.

To my sons, Matthew and Clayton, who inspire me in so many ways.
-Dee

To my loving and inspirational wife Teri and our four amazing children
Tyler, Bobby, Amanda and Taylor, 143 always!
-Robert

To my wife Cady and my daughter Mira, I love you both.
-Mark

Chapter 1
Introduction
Preventing Errors in the Heat of Battle: Formal and Informal Learning Strategies to Prevent Teamwork Breakdowns

Katherine A. Wilson
Eduardo Salas
University of Central Florida

Dee H. Andrews
Air Force Research Laboratory

Introduction

Fratricide, or friendly fire, has been deemed an inescapable cost of war. The release of cockpit footage following a friendly fire attack on British troops by US jet fighters in 2004 brought to light this sensitive topic. No procedures were violated in this attack; this was simply a case of human error. Cognitive and physical task overload, poor communication, and failure to recognize British identification panels contributed to this tragic event in which one soldier was killed and four were injured (BBC News, 2007).

When incidents like this occur, the question arises, 'Why did this happen?' Wilson and her colleagues (2007) set out to address this by developing an initial taxonomy for investigating friendly fire incidents. The taxonomy proposed argues that friendly fire incidents result when there is a failure of shared cognition among team members. Contributing to this failure is a breakdown of individual, team, task, organization, and technology-based factors. The authors focused on one aspect of these breakdowns—teamwork—and discussed how breakdowns in communication, coordination, and/or cooperation may contribute to a failure of shared cognition and ultimately a friendly fire incident. While a number of taxonomies are available (e.g., HFACS [Wiegmann and Shappell, 2003], Swiss Cheese model [Reason, 1990]), the taxonomy created by these researchers further parses out why a breakdown in teamwork occurred. For example, a failure in teamwork cited as the cause of an incident is not diagnostic enough to address the problem. Rather, an understanding of what part of the teamwork failed—communication, coordination, or cooperation—is necessary (Wilson et al., 2007).

Given that a failure of teamwork in communication, coordination, and/or cooperation is a significant contributor to friendly fire incidents, the purpose of this chapter is fourfold. First, we review the taxonomy proposed by Wilson and her colleagues (2007). Next, we expand the existing taxonomy to further address why fratricide occurs and specifically address incidents caused by failures in teamwork. Third, methods are proposed to mitigate the likelihood of incidents and the severity of its consequences through both formal and informal training strategies. Specifically, a number of learning strategies have proven valid and reliable at improving team effectiveness in these complex environments. Several formal and informal strategies that can be used to reduce the risk of teamwork breakdowns on the battlefield are discussed. Finally, we provide a number or research-based principles to support the use of the formal and informal learning strategies discussed.

Expanding the Taxonomy

Wilson and her colleagues (2007) discussed three areas of teamwork that are critical on the battlefield—communication, coordination, and cooperation—which when in err may lead to fratricide. The authors discussed in detail these teamwork dimensions and provided behavioral indicators to help one understand why the breakdown occurred. First, critical on the battlefield is the communication of information—position of enemy and friendly troops, situation updates, plan of action, etc. When information is not transmitted in a timely manner, is incorrect, is not complete, or is misunderstood, communication breakdowns happen and the chance of fratricide increases. While communication is a broad term, Wilson and her colleagues discussed three types of communication that are important to reduce the risk of fratricide—information exchange (i.e., what information is passed), phraseology, and closed-loop communication.

In addition to communicating on the battlefield, US and allied troops must coordinate their actions. Coordination, therefore, is the behavioral mechanisms used by team members to synchronize team performance requirements. When these behaviors are not orchestrated properly, breakdowns occur which can lead to errors. Further complicating this concern is that the interdependent nature of these teams means that an error by one team member will likely translate into an error by another team member if the first error is not caught and corrected. Breakdowns in team coordination have been cited in numerous friendly fire incidents (OTA, 1993). As coordination, in a broad sense, does little to help us recognize where it broke down, there are several critical team coordination mechanisms that do—shared knowledge of the team, task, and environment; mutual performance monitoring and back up behavior; and adaptability (Xiao and Moss, 2001; Salas, Sims, and Burke 2005). While there are arguably additional coordination mechanisms critical to teams, these are most critical to maintaining shared cognition on the battlefield.

Finally, communication and coordination will not be successful without shared attitudes and beliefs among team members. When team members have compatible perceptions of the task and/or the environment, this will lead to better shared cognition, more effective decisions, and enhanced team performance. Breakdowns in team cooperation are the result of team members who lack the desire and the motivation to coordinate (e.g., do not anticipate or predict each other's needs). Cooperation, while important, is also difficult to measure as attitudes are not readily observable. However, they do manifest themselves through observable behaviors and it is those behaviors that can be measured.

The above discussion of critical teamwork dimensions and associated behavioral markers proposed by Wilson and her colleagues (2007) is a first step in analyzing why breakdowns occurred (see Table 1.1). Using the taxonomy proposed by Wilson and colleagues, one would use the behavioral markers to assess how poor teamwork contributed. For a taxonomy to be comprehensive, it must fully dissect what contributed to an incident. While identifying whether or not a behavior(s) occurred is useful, it is not practical as it does little to help in understanding how to prevent this failure to act in the future. Therefore, further understanding of why the failure occurred (or rather why a behavior did not occur) is necessary. We expand the original taxonomy by providing the answers to the question 'why not?' (see Table 1.1). For example, to identify if communication, namely information exchange, impacted the event, one could ask 'did team members pass information within a timely manner before being asked?' If the answer is 'yes' (i.e., this behavior did not impact the incident), the investigator can move on to the next question. If the answer is 'no,' it is important that the investigator understand why the information did not pass (i.e., why not?). Potential reasons include, but are not limited to, inadequate briefing/planning, information not available, information not reliable, or possible equipment failures which did not allow the information to be sent. This may not be an exhaustive list since it is not plausible to identify every possible cause. However, the answers provided to the question 'why not?' cover a wide range of explanations that will assist the investigator on his/her path to understanding why an incident occurred. For the remainder of this chapter, the focus will be on training strategies that can be used to help address the 'why not?' in the future.

Table 1.1 Understanding teamwork breakdowns and their causes (adapted from Wilson et al., 2007)

Communication		Why not?
Information Exchange	Did team members seek information from all available resources? Did team members pass information within a timely manner before being asked? Did team members provide 'big picture' situation updates?	Inadequate briefing/planning Failure to adhere to briefed/planned mission Failure to adhere to policies/procedures Poor leadership/supervision Unaware of resources Resources not available Time constraints Information (whole or partial) not available Information available but not read/reviewed Information not reliable Equipment failures/not available Equipment misused Queue delays Cultural language barriers Other
Phraseology	Did team members use proper terminology and communication procedures? Did team members communicate concisely? Did team members pass complete information? Did team members communicate audibly and ungarbled?	Failure to adhere to policies/procedures Poor leadership/supervision Incompatible verbal guidance Equipment failures/not available Equipment misused Time constraints Fatigue Situational constraints (e.g., noise) Information (whole or partial) not available Information available but not read/reviewed Cultural language barriers Other

Table 1.1 ***Continued***

Closed-Loop Communication	Did team members acknowledge requests from others? Did team members acknowledge receipt of information? Did team members verify information sent is interpreted as intended?	Information not sent Information not received Information received but not read/reviewed Equipment failures/not available Equipment misused Situational constraints (e.g., noise) Queue delays Time constraints Fatigue Non-standard/ambiguous language used Cultural language barriers Other
Coordination		**Why not?**
Knowledge Requirements	Did team members have a common understanding of the mission, task, team, and resources available to them? Did team members share common expectations of the task and team member roles and responsibilities? Did team members share a clear and common purpose? Did team members implicitly coordinate in an effective manner?	Lack of training/experience Inadequate briefing/planning Information (whole or partial) not sent Information not received Information received but not read/reviewed Queue delays Poor leadership/supervision Non-standard/ambiguous language used Information received but in error Conflicting team and/or organizational goals Task too complex Other

Table 1.1 ***Continued***

Cooperation		Why not?
Mutual Performance Monitoring	Did team members observe the behaviors and actions of other team members? Did team members recognize mistakes made by others? Were team members aware of their own and others surroundings?	Unaware of team member roles/responsibilities Task exceeded ability Time constraints Fatigue Punitive response for past behavior Behavior not encouraged Task too complex Other
Back up Behavior	Did team members correct other team member errors? Did team members provide and request assistance when needed? Did team members recognize each other when one performs exceptionally well?	Task exceeded ability Error not observed Error observed but not in a timely manner Error observed but failed to take appropriate action Time constraints Fatigue Punitive response for past behavior Behavior not encouraged Task too complex Other
Adaptability	Did team members reallocate workload dynamically? Did team members compensate for others? Did team members adjust strategies to situation demands?	Lack of training/experience Punitive response for past behavior Behavior not encouraged Fatigue All team members overloaded Unaware of situation demands Organizational constraints Other

Table 1.1 ***Concluded***

Team Orientation	Did team members put group goals ahead of individual goals? Were team members collectively motivated and did they show an ability to coordinate? Did team members evaluate each other, while using inputs from other team members? Did team members exhibit 'give-and-take' behaviors?	Inadequate briefing/planning Failure to adhere to briefed/planned mission Organizational constraints Poor leadership/supervision Conflicting team and/or organizational goals Cultural barriers Other
Collective Efficacy	Did team members exhibit confidence in fellow team members? Did team members exhibit trust in others and themselves to accomplish their goals? Did team members follow team objectives without opting for independence? Did team members show more and quicker adjustment of strategies across the team when under stress based on their belief in their collective abilities?	Organizational constraints Poor leadership/supervision Lack of trust Unaware of goals (team and mission) Cultural barriers Task too complex Other
Mutual Trust	Did team members confront each other in an effective manner? Did team members depend on others to complete their own tasks without 'checking up' on them? Did team members exchange information freely across team members?	Organizational constraints Poor leadership/supervision Punitive response for past behavior Behavior not encouraged Cultural barriers Equipment failures/not available Other
Team Cohesion	Did team members remain united in pursuit of mission goals? Did team members exhibit strong bonds and desires to want to remain a part of the team? Did team members resolve conflict effectively? Did team members exhibit less stress when performing team tasks?	Lack of training/experience Poor leadership/supervision Punitive response for past behavior Behavior not encouraged Unaware of goals (team and mission) Conflicting team and/or organizational goals Cultural barriers Task too complex Other

How Can Learning Strategies Help?

The fog of war is not easy to overcome. Teams on the battlefield are faced with unimaginable challenges. US and allied troops rely on teams to overcome these challenges. However, effective teams do not just happen and teamwork breakdowns occur. Therefore, teams require strategies to develop the knowledge, skills, and attitudes to accomplish their tasks. By understanding why breakdowns occur, strategies to overcome these breakdowns can be proposed. A brief review of training in general gives the reader an understanding of the underlying structure of these learning strategies. Learning can be defined as the systematic acquisition of knowledge (i.e., what we need to know), skills (i.e., how we need to act), and attitudes (i.e., what we need to feel; KSAs). Together, these KSAs lead to enhanced performance in a given environment (e.g., the battlefield). Important to note is that learning is about both cognitive and behavioral change, and this change occurs when instructional strategies are designed based on the principles of learning and the science of training. In addition, an environment is needed in which trainees can: (1) learn the necessary KSAs through information (e.g., lecture) and demonstration (e.g., videos), (2) practice applying what they have learned (e.g., simulations), and (3) receive timely, constructive feedback that helps them improve performance in the future. This in turn is accomplished through sound instructional strategies focused on the specific needs of the organization (Salas and Cannon-Bowers, 1997; Salas, Bowers, and Edens, 2001).

The strategies used by organizations to improve learning can be formal or informal. Formal learning often involves a structured approach in which classroom training provides the necessary knowledge and skills followed by the application of these knowledge and skills in structured manner (e.g., simulations). Informal learning, on the other hand, does not typically follow a structure. Rather, informal learning is typically provided on the job with an experienced employee guiding a less experienced or newly hired employee. Both types of learning have pros and cons, and it is up to the organization to decide which type of strategy will help them to best accomplish their needs. In the following sections, several formal and informal learning strategies that may be useful at improving teamwork on the battlefield are discussed.

What Formal Learning Strategies are Available?

There are a number of formal learning strategies discussed in the literature. Formal learning is defined as being structured through information, demonstration, practice, and feedback. While some are familiar within the military domain (e.g., scenario-based training, team coordination training), others are less familiar but nonetheless relevant (e.g., guided error training, demonstration-based training). In this section, eight formal learning strategies that may be useful at improving teamwork on the battlefield to reduce friendly fire (see Table 1.2) are discussed.

Not one strategy in isolation is the answer to address all breakdowns but rather a combination of strategies is likely necessary to have the greatest impact.

Table 1.2 Formal training strategies

Strategy	Definition	References
Scenario-Based Training	Provides a meaningful framework from which learning events can be embedded into training scenarios.	Oser et al., 1999
Stress Exposure Training	Provides strategies to manage stress and maintain performance.	Driskell and Johnston, 1998
Guided Error Training	Guides trainees to errors allowing them to see the consequences of such errors and develop strategies for mitigating the consequences.	Lorenzet et al., 2005
Team Training	Provides trainees with the requisite knowledge, skills and attitudes to communicate and coordinate effectively as a team.	Salas and Cannon-Bowers, 2000a
Cross Training	Develops a shared mental model of team members' roles and responsibilities ranging from a basic understanding to ability to accomplish each other's tasks.	Volpe et al., 1996
Game-based Training	Provides the opportunity to practice the necessary knowledge and skills in a simulated environment.	Whitehall and McDonald, 1993; Ricci et al., 1996
Demonstration-Based Training	Provides the opportunity to observe (1) a person performing a task, (2) components of a task (e.g., via video), or (3) characteristics of a task environment that are critical for learning targeted competencies.	Rosen et al., 2008
Cross-Cultural Training	Provides the knowledge, skills, and attitudes to effectively communicate and coordinate with persons of another culture.	Littrell et al., 2006; Littrell and Salas, 2005

Scenario-Based Training

Principle 1. Scenario-based training provides trainees with the opportunity to learn and refine targeted knowledge and skills needed on the battlefield through guided, meaningful practice.

Principle 2. Scenario-based training, in conjunction with additional training strategies, can maximize the learning potential of trainees.

Scenario-based training, also referred to as event-based training, is one instructional strategy that can be used to train troops on the battlefield to reduce the risk of teamwork breakdowns (Fowlkes, Dwyer, Oser, and Salas, 1998). What sets scenario-based training apart from other instructional strategies is that with scenario-based training there is no formal curriculum. Rather, the scenarios, defined a priori from critical incidents, act as the curriculum. What this means is that scenario-based training uses carefully crafted training scenarios in which 'trigger' or learning events are embedded. These trigger events serve to elicit targeted knowledge and skills through structured and guided practice (Dormann and Frese, 1994; Karl, O'Leary-Kelly, and Martocchio, 1993; Ivancic and Hesketh, 1995). Furthermore, greater control is gained by defining the scenarios prior to the training, allowing for more effective and accurate performance measurement. It is through this guided practice that trainees are provided with a meaningful framework from which to learn and refine targeted knowledge and skills (Fowlkes et al., 1998; Salas and Cannon-Bowers, 2000b).

To effectively utilize scenario-based training, a six step lifecycle should be followed (see Figure 1.1; Oser, Gualtieri, Cannon-Bowers, and Salas, 1999). First, and critical to scenario-based training (as well as any training program), is determining what tasks and competencies the training will focus on (circle 1). Based on this, training objectives are developed (circle 2) which drive the development of the training scenarios (i.e., events, exercises, and curriculum; circle 3). Once scenarios are embedded within training to elicit desired knowledge and skills, it is important that performance measures are incorporated (circle 4) so that the knowledge and skills can be evaluated. The evaluation of this data should then be translated into constructive and timely feedback that can be given to trainees (circle 5). Finally, this information is incorporated into future training programs so that each iteration of training will build upon each other (circle 6).

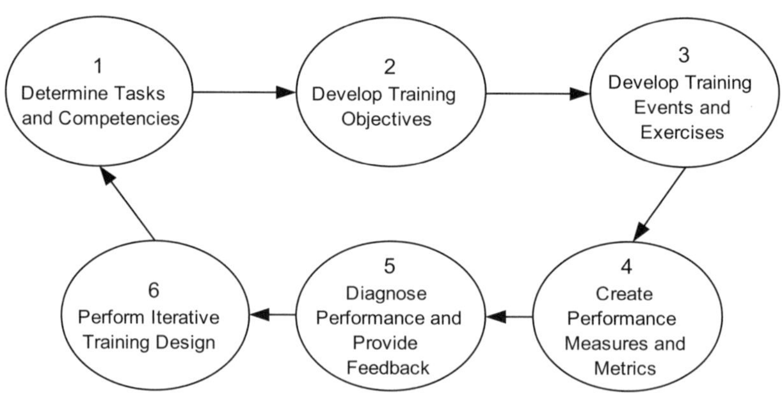

Figure 1.1 Components of simulation-based training (adapted from Oser et al., 1999)

One of the many challenges faced on the battlefield is the novelty of the events that occur. Scenario-based training allows trainees to experience these events in a realistic yet simulated (i.e., 'safe') environment. The flexible architecture of scenario-based training helps to build expertise and response repertoires that are crucial when making decisions under extreme stress (Richman, Staszewski, and Simon, 1995; Klein, 1997). Furthermore, scenario-based training is versatile in that it can supplement other instructional strategies in order to maximize learning potential. For example, scenario-based training could be used in conjunction with cross training to train team members to anticipate inputs from others and how to respond or step up when those inputs are absent. This combination of training benefits team members by recognizing the importance of team coordination as well as creating an awareness of one's surroundings (e.g., team members, environment) to catch an error before it escalates.

Stress Exposure Training

Principle 3. Stress exposure training offers trainees a realistic preview of stress on the battlefield as well as strategies for minimizing the consequences (i.e., reducing the risk of fratricide) and maintaining performance.

It should be no surprise warfighters face insurmountable levels of stress on the battlefield (e.g., noise, fatigue/sleep deprivation, time pressure, workload, danger; Orasanu and Backer, 1996). One instructional strategy, stress exposure training (SET), is available to provide troops with strategies to manage this stress and maintain performance (Driskell and Johnston, 1998). This is important because high levels of stress can lead to missed cues in the environment or missed procedures, both of which can lead to an error such as fratricide. There are three phases of stress exposure training—information provision, skill acquisition, and application and practice. The first phase of training, information provision, involves providing trainees with the basics about stress, types of stress, and its effects on performance. While the first phase if primarily information or knowledge-based, phase 2 (skill acquisition) seeks to help trainees acquire the specific behavioral and cognitive skills necessary to manage and adapt to the stressful environment. In the last phase, application and practice, trainees are able to apply what they have learned in a simulated environment, as well as receive feedback on positive and negative areas of performance. Strategies should also be discussed for those areas that are in need of improvement.

Although stress exposure training has typically been studied as an individual-level instructional strategy, implications at the team level are evident. As previously mentioned, the effects of stress on an individual can be of great consequence—temperature extremes leading to impaired judgment, excessive noise leading to missed communications, time pressure or workload leading to tunnel vision or poor decision-making (Orasanu and Backer, 1996). Each of these effects has consequences for the team, and as such, training to reduce the effects of these stressors is important to avoid errors on the battlefield.

Guided Error Training

Principle 4. By guiding trainees to and providing feedback regarding errors, they will be better prepared to handle errors which can lead to fratricide on the battlefield.

Human fallibility dictates that errors occur. But not all errors lead to detrimental consequences such as fratricide. In order to help troops catch errors before they escalate to an incident, guided error training can be provided. The purpose of guided error training is to use errors that commonly occur on the battlefield as a learning experience for trainees in a simulated environment (Frese and Altman, 1989; Lorenzet, Salas, and Tannenbaum, 2005). In other words, training is designed to intentionally guide trainees to errors so as to encourage learning.

Most training strategies take one of three approaches to errors—(1) avoid errors (training is designed to prevent errors), (2) allow for errors (training is designed to allow for errors but they are not triggered; e.g., Gully, Payne, Koles, and Whiteman, 2002), or (3) induce errors (training is designed to evoke errors; e.g., Dormann and Frese, 1994). Furthermore, when these errors occur, trainees are left to work though errors alone (e.g., Frese et al., 1991). Guided error training, on the other hand, actually guides trainees to errors where they can see the consequences and learn the skills necessary for mitigating them while being supported by instructors (Lorenzet et al., 2005). The results of research by Lorenzet and his colleagues (2005) indicate that trainees who are guided to errors and supported throughout the correction process performed better than those who received training in which errors were avoided. We would argue that allowing troops to practice making common errors that lead to fratricide, and to learn and practice strategies to correct these errors, would reduce the risk of fratricide on the battlefield.

Team Training

Principle 5. Team training can reduce the risk of fratricide incidents by training teams to communicate and coordinate effectively on the battlefield.

One of the most commonly used training strategies is team training which focuses on improving the coordination efforts between team members (Salas and Cannon-Bowers, 2000a). There are several types of team training strategies, such as crew resource management (CRM) training and team adaptation and coordination training. Crew resource management training is widely used throughout commercial and military aviation, and its purported success has led to its implementation in other domains (e.g., healthcare, nuclear power, offshore oil production). Crew resource management training was introduced over 25 years ago as a way to train cockpit crews to use all available resources (i.e., people, equipment, and information) through better communication and coordination (Wiener, Kanki, and Helmreich, 1993; Salas, Bowers, and Edens, 2001). Crew resource management training uses both lecture and practice-based instruction to improve teamwork skills. However, what skills to train are generally left to the discretion of the organization, based on the task teams perform. A review of the crew resource management training literature revealed a widespread number of skills categorized as 'CRM' (over 50 different skills found; Salas, Wilson, Burke, and Wightman, 2006). However, the most commonly cited skills trained were communication, situation awareness/assessment, decision-making, leadership/followership, preflight brief/planning, stress awareness/management and assertiveness. Orasanu and Backer (1996) discussed how crew resource management training can be a strategy for reducing the negative effects of stress including team members serving as redundancy when fatigued (e.g., monitoring and assisting, stimulating conversations to maintain alertness) and sharing responsibilities when overloaded. It is also suggested that improved teamwork may build cohesion and reduce battlefield stress (Stewart and Weaver, 1988, as cited in Orasanu and Backer, 1996). When designed and delivered systematically, crew resource management training has been shown to lead to positive reaction, learning, and safe behaviors (see Salas, Burke, Bowers, and Wilson, 2001; Salas et al., 2006). Through teamwork, warfighters can communicate, coordinate, and cooperate to reduce errors and the risk of fratricide.

Cross-Training

Principle 6. Cross-training creates a shared understanding of each other's roles and responsibilities thereby increasing the team's ability to communicate and coordinate (explicitly and implicitly) on the battlefield.

On the battlefield, team members must not only be concerned with their own tasks but also be concerned with overall team performance. This requires that team members have a shared mental model (especially interpositional knowledge) so that they communicate and coordinate effectively (Baker, Salas, Cannon-Bowers, and Spector, 1992). Interpositional knowledge is knowledge about team member roles, responsibilities, and requirements. This knowledge helps to generate expectancies of what team members should be doing to assist in monitoring performance and providing backup behaviors (for example) to minimize the risk of errors. Cross training is one instructional strategy that can help to develop interpositional knowledge by training each team member on the duties of their teammates (Volpe, Cannon-Bowers, Salas, and Spector, 1996).

Depending on the depth of knowledge and interdependency required by team members to complete their tasks, there are three types of cross training—positional clarification, positional modeling, and positional rotation (Blickensderfer, Cannon-Bowers, and Salas, 1998). Positional clarification (the lowest level of interpositional knowledge) develops a general awareness and knowledge of each team member's role and responsibilities. This type of training is recommended for teams with little interdependence (e.g., minimal communication, coordination, and leadership required). At the next level, positional modeling focuses on training team members to discuss and observe the roles and responsibilities of each team member, as well as a discussion on how these duties relate to and affect the duties of other members. It is suggested that teams requiring medium levels of interdependence (e.g., fairly constant communication, some coordination, feedback provided at team level) should receive positional modeling. At the greatest depth is positional rotation. This training technique involves providing team members with direct, hands on practice of each other's roles and responsibilities in order to build a working knowledge of each member's specific tasks and how they interact. Positional modeling is recommended for teams who are highly interdependent and must coordinate and communicate to accomplish their tasks. Depending on the tasks to be accomplished on the battlefield and the requirements of team members determines which level is appropriate. As most tasks likely require medium to high levels of interdependence, positional modeling and rotation are the more relevant levels of training to provide to warfighters. For example, Cannon-Bowers and colleagues (Volpe et al., 1996; Cannon-Bowers, Salas, Blickensderfer, and Bowers, 1998) found that aviation and command and control teams provided with cross training resulted in better team and task performance, better communication, and higher quality teamwork. In high stress situations, the understanding that warfighters will share following training will reduce the amount of explicit communication needed as well as assist in the monitoring and supporting of team member tasks to mitigate the likelihood and severity of errors.

Game-Based Training

Principle 7. Games offer a low cost alternative to train team members the necessary knowledge and skills necessary to reduce the risk of fratricide.

Computer-based games are being more readily used for educational and training purposes. The benefit of games over other large scale simulators is that games offer a low cost alternative to learning and retention of critical knowledge and skills (e.g., Whitehall and McDonald, 1993; Ricci, Salas, and Cannon-Bowers, 1996). Furthermore, games offer similar benefits as simulators including active participation by the learner, the learner becomes engaged, the task difficulty can be varied, and immediate feedback is provided, to name a few (e.g., Driskell and Dwyer, 1984; Ricci et al., 1996). A number of off-the-shelf games have proved successful—e.g., Microsoft *Flight Simulator* for basic flight training skills (e.g., Dennis and Harris, 1998); Tom Clancy's *Rainbow Six* for training teamwork skills (e.g., Bowers and Jentsch, 2001); and *Rogue Spear* for infantry training (Woodman, 2006). In addition, games have been shown to improve learner motivation (e.g., Ricci et al., 1996), declarative knowledge (e.g., Veale, 1999), and transfer of training (e.g., Gopher, Weil, and Bareket, 1994). It is important to note that games alone do not lead to learning. The science of learning and training must still be applied and the learner must be guided if games are to result in learning. Given this, games are a great way to train troops to work as a team (e.g., communicate, coordinate; Jentsch and Bowers, 1998; Proctor, Panko, and Donovan, 2004) as well as to retain these skills over time. This is critical, as trainees must retain and transfer what they have learned to the task environment if performance is to improve. Like that of scenario-based training, game-based training offers an engaging, safe environment in which to train teamwork skills in order to reduce the risk of fratricide on the battlefield.

Demonstration-Based Learning

Principle 8. Demonstration-based learning develops within trainees a mental model of how tasks on the battlefield should be performed.

Drawing from observational learning and behavioral modeling theories, demonstration-based learning involves the learner observing (1) another person (or team) as they perform a task, (2) components of a task (real time or pre-recorded [e.g., video, computer generated medium]), or (3) characteristics of a task environment that are critical for learning, with the purpose of building targeted knowledge, skills, and attitudes (Rosen, Salas, and Upshaw, 2008). Furthermore, demonstration-based learning can be passive or active. Passive learning involves the learner simply observing an activity where learning is reliant on the content provided in the demonstration. While passive observation is typically unguided (e.g., Austin and Laurence, 1992; Berry, 1991; Blandin and Proteau, 2000), learners can be provided with tips prior to watching the demonstration (i.e., to guide them) highlighting content to be particularly aware of without requiring any action from the learner (e.g., Jentsch, Bowers, and Salas, 2001). Active demonstration, on the other hand, requires that the learner engage in the demonstration with the purpose of increasing learning. There are four types of active demonstration—preparatory, concurrent, retrospective, and prospective (Rosen et al., 2008). Active-preparatory demonstrations provide the learner with activities that will help to orient and focus the learner during the actual demonstration (beyond what is provided by passive guided demonstrations). Examples of this preparatory information includes imagery to increase motivation (e.g., Cumming, Clark, Ste-Marie, McCullagh, and Hall, 2005) and instruction on goal setting and self-regulatory skills needed for observing (e.g., Ferrari, 1996). Active-concurrent demonstrations provide activities to engage the learner during the demonstration, such as note taking (e.g., Lozano, Hard, and Tversky, 2006). Some demonstrations also provide activities to engage the learner following the demonstration, such as open ended group discussions focused on targeted competencies (e.g., Johnson, Johnson, and Stanne, 1985; Prislin, Jordan, Worchel, Tschan-Semmer, and Shebilske, 1996) and activities focused on how to transfer what has been learned to the real world (e.g., learner generates own practice scenarios; Wexley and Latham, 2002). These post-demonstration activities are referred to as active-retrospective and active-prospective, respectively. Demonstration-based learning is useful at building knowledge about teamwork breakdowns. The ability to watch positive and negative examples of teamwork (e.g., via videos) is critical so that they can be recognized on the battlefield and mitigated. In addition, this knowledge will serve as a foundation for building the necessary teamwork skills.

Cross-Cultural Training

Principle 9. Teams must be trained on not only the technical skills, but also the cross-cultural interpersonal skills necessary to effectively communicate, coordinate, and cooperate with foreign forces.

Training for diversity has been around since the 1970s (Caudron, 1993). Despite this, US organizations lose over $2 billion annually resulting from failed international assignments, for example expatriates (Noe, 2002). One of the main culprits is a lack of cultural preparation (i.e., individuals lack the skills to interact with those of another culture). Given the above and the high consequence of failure (i.e., loss of life), US and allied troops need to be prepared for cross-cultural interactions as well. In Scales (2006) article, it was stated that more attention needs to be paid to the development of the individual soldier in a number of areas: (1) cultural awareness, (2) building alien armies/alliances, (3) perception shaping, (4) inculcate knowledge and teach wisdom, (5) tactical intelligence, (6) psychological and physiological tuning, (7) developing high performing soldiers and small units, (8) leadership and decision-making, and (9) intuitive battle command. To accomplish this, troops must be equipped with the knowledge and skills to effectively communicate and coordinate with troops from foreign nations.

Two reviews by Littrel and her colleagues (2005, 2006) indicated that organizations most successful at training employees for overseas operations used a combination of strategies—such as attribution training (i.e., training individuals to interpret behaviors similarly to foreign nations), cultural awareness training (i.e., understanding one's own culture to appreciate cultural differences), didactic training (i.e., providing trainees with factual information about foreign nation), and interaction training (i.e., on-the-job training). The benefit of using multiple strategies is that both informational and experiential (i.e., practice-based) learning opportunities are combined. In addition, some training is now focused on implementing successful verbal communication methods in intercultural operations (Archer, 1997). Just as important as verbal communication is nonverbal communication (i.e., body language and gestures). Researchers have begun to investigate nonverbal communication to identify which cues are universal across cultures and which are culturally contingent (Elfenbein and Ambady, 2002). Furthermore, it is also suggested that nonverbal cues are misunderstood more frequently than verbal statements (Oludaja, 1992). As such, warfighters need to be trained to recognize and distinguish verbal and nonverbal cues to successfully interact with troops and civilians of foreign nation. The consequence of a misinterpretation (e.g., as a hostile gesture rather than a friendly one) could be costly and result in fratricide.

What Informal Learning Strategies are Available?

Typically performed in the task environment, informal learning strategies can be just as beneficial as formal learning strategies when they follow the science of learning and training. In other words, they too must be designed and developed systematically to ensure effectiveness (Goldstein, 1993). Informal learning can take place in any context at any time, as people interact with others (Rossett and Sheldon, 2001). It can take the form of employees talking to one another, sharing opinions and stories, observing others performing a task, mentally playing out a task, or giving and receiving feedback. Furthermore, informal learning can have clearly defined objectives (e.g., getting up to speed on using a new piece of equipment) to less defined objectives (e.g., understanding a unit's culture; Rossett and Sheldon, 2001). In this next section several informal strategies for learning (some old, some new) that may be useful at improving teamwork on the battlefield (see Tables 1.3 and 1.4) are discussed.

Table 1.3 Informal training strategies

Strategy	Definition	References
On-the-Job Training/ On-the-Job Learning	Provides trainees with an opportunity to observe and practice actual required behaviors needed to do task in the actual task environment.	Goldstein, 1993; Rossett and Sheldon, 2001
Storytelling	Allows for the sharing of past experiences and future expectations to encourage communication and coordination.	Denning, 2004, 2006
Coaching	Builds a personal relationship between a more and a less experienced person to assist in the assessment, intervention, follow up and evaluation of performance.	Kampa-Kokesch and Anderson, 2001
Mental Practice	Mental or cognitive rehearsal of a task without physical movement.	Driskell et al., 1994

Table 1.4 Training strategies and associated team outcomes

	Information Exchange	Phraseology	Closed-Loop Comm.	Knowledge Requirements	Mutual Perf. Monitoring	Backup Behavior	Adaptability	Team Orientation	Collective Efficacy	Mutual Trust	Team Cohesion
Scenario-Based Training	+	+	+	+	+	+	+				
Stress Exposure Training					+	+	+			+	
Guided Error Training	+	+	+		+	+	+				
Team training			+		+	+	+	+	+	+	+
Cross Training				+	+	+	+	+	+		+
Game-Based Training				+	+	+	+				
Demonstration-Based Training	+	+	+	+	+	+	+	+			+
Cross-Cultural Training	+	+	+		+	+		+		+	+
On-the-Job Training/Learning	+	+	+	+	+	+	+				
Storytelling	+							+		+	+
Coaching	+	+	+	+	+	+		+		+	+
Mental Practice	+	+	+	+	+	+	+				

On-the-Job Training/On-the-Job Learning

Principle 10. On-the-job training/learning provides trainees with hands on practice of the knowledge and skills needed on the battlefield in the actual environment while under the supervision of an expert.

On-the-job training (OJT), also referred to as on-the-job learning (OJL), is one of the most widely used informal types of training in organizations and can be used in conjunction with a formal strategy (such as those mentioned above) or as the sole strategy (Goldstein, 1993). The premise behind on-the-job training/on-the-job learning is that an experienced employee or supervisor trains specific skills related to the job (e.g., what to do, how to do it) to an employee while in the actual task environment (i.e., physical and social environment; Sacks, 1994; Wehrenberg, 1987; De Jong and Versloot, 1999). There are several benefits to using this type of training. First, transfer of training to the job is very likely as the necessary skills are being learned in the same environment where they must be applied. In addition, on-the-job training/on-the-job learning allows for the practice of skills that trainees will be expected to perform while under the supervision of an expert. Given this, trainees can receive feedback on what they are doing properly as well as areas where improvement is needed on-the-job training/on-the-job learning also provides a safer environment in which to practice these skills in that an expert is on hand to catch an error, if one should occur, before it becomes detrimental.

There are two types of on-the-job training/on-the-job learning that could be used for training battlefield skills. First, apprenticeship training, begins with classroom-based instruction followed by supervision by an experienced employee on the job. At the beginning, the trainee will shadow the more experienced employee. As the knowledge and skills are learned, the trainee will be able to work on a task under supervision until finally being allowed to perform a task with minimal to no supervision (Goldstein, 1993; Hendricks, 2001; Lewis, 1998). This type of training is beneficial in that it begins by developing the knowledge and skills for the job followed by the opportunity to practice what has been learned. It is only when the experienced employee feels that the 'apprentice' has mastered the knowledge and skills to a point where mistakes and errors will be unlikely that the apprentice is allowed to perform alone.

Similar to apprenticeship training is mentoring, the second type of on-the-job training/on-the-job learning relevant for training battlefield skills. The process of mentoring involves creating a relationship between an experienced individual and a trainee (Wilson and Johnson, 2001). It is through this relationship of providing guidance and feedback that the trainee will learn to perform a task successfully (Scandura, Tejeda, Werther, and Lankau, 1996). Mentoring can take an even further informal role by an experienced employee sharing opinions and past experiences over an occasional lunch (Rossett and Sheldon, 2001). The research that has been conducted on the effectiveness of mentoring indicates that this strategy improves communication, job satisfaction, and success in an organization (Mobley, Jaret, Marsh, and Lim, 1994; Forret, Turban, and Dougherty, 1996). While this research has been conducted in organizational settings, we would argue that similar results would be possible on the battlefield and that these improvements would improve teamwork as well. Important to the success of mentorship is identifying a 'mentor' who values and respects what is being trained, follows policies and procedures of the organization, and encourages trainees to succeed. It has also been argued that mentoring improves not only trainee skills but also the skills of the mentor, especially those that may have diminished over time (Forret et al., 1996). This form of training is especially useful for the battlefield environment where there may not be sufficient time to conduct formal training (Rossett and Sheldon, 2001).

Storytelling

Principle 11. Storytelling allows trainees to relate past experiences to a current context thus allowing them to learn from what worked and what didn't work.

In order to help team members make sense of a complex situation, storytelling is one strategy that can be used (Bruner, 1990). Depending on the goals of the learning experience, there are a number of different stories that can be told (Denning, 2004, 2006). For example, stories can be told to *spark action* in team members by discussing a successful change or previously used strategy that will allow team members to imagine how that same or a similar strategy could work in their current situation. In addition, stories can be told to *foster collaboration* by telling a story that encourages team members to recall a similar experience and to share their own story. This will in turn establish communication between team members. Similarly, stories are useful for *sharing knowledge* regarding errors that have been made in the past, what strategies worked or didn't work in resolving the error, and why the strategy was a success or not. Finally, stories can *lead people into the future* by sharing expectations for the future that encourages team members to want to help build that future. While there are additional types of stories that can be told (e.g., communicating who you are), those discussed are most relevant to building teamwork on the battlefield.

One of the challenges faced with storytelling is that the stories must be guided by the learning objectives. It is not enough to just be a good storyteller (Denning, 2006). It is important that the trainer have flexibility in what and how the story is told, but at the same time it is important that the story engage team members. Their response in order to make change happen is what is most critical. In order to improve the benefit of storytelling, this strategy can also be used with formal methods of training (such as scenario-based training) to help generate learning scenarios to make stories more meaningful.

Coaching

Principle 12. Experienced team members serve as coaches to less experienced members to help them build the necessary knowledge and skills for warfighting.

Another strategy of informal learning involves coaching. Much has been learned from the sports and psychotherapy literatures about how coaches can be used to improve leadership and teamwork in organizations (Kampa-Kokesch and Anderson, 2001; Mulec and Roth, 2005). Executive coaching has been defined as 'a facilitative, one-to-one, mutually designed relationship between a professional coach and a key contributor who has a powerful position in the organization' (International Coaching Federation Conference 2000, as cited in Kampa-Kokesch and Anderson, 2001, p. 209). While coaching in organizations has been primarily used for improving upper level management inter- and intra-personal skills, some have begun to use coaching to improve team performance (e.g., Mulec and Roth, 2005). Coaches can be either internal or external to the organization, each offering benefits such as knowledge of organizational culture or a wider range of ideas, respectively (Hall, Otazo, and Hollenbeck, 1999). Most researchers in this area agree that there are five stages of executive coaching: relationship building, assessment, intervention, follow-up, and evaluation (Kampa-Kokesch and Anderson, 2001). Critical throughout these stages is that the coach provides reliable feedback and good action ideas (Hall et al., 1999). Furthermore, when selecting a coach, Brotman and his colleagues (1998) suggested a number of competencies—approachability, comfort around top management, compassion, creativity, customer focus, integrity and trust, intellectual horsepower, interpersonal savvy, listening, dealing with paradox, political savvy, and self-knowledge. This strategy is especially useful for building skills in higher ranking warfighters, such as tactics for leading and motivating their teams on the battlefield.

Mental Practice

Principle 13. Mental practice allows team members to play out 'if-then' scenarios in their minds prior to making a decision which could lead to errors and fratricide.

In addition to the aforementioned learning strategies, mental practice has also been shown to be effective at improving performance. Mental practice can be defined as 'the cognitive rehearsal of a task in the absence of overt physical movement' (Driskell, Copper, and Moran, 1994, p. 481). In other words, mental practice is a 'technique in which the skills or behaviors required to perform a task are mentally rehearsed before performance' (Driskell et al., 1994, p. 489). Mental practice (or simulation) can also be used to construct a story or causal account of how a current situation may have evolved or will evolve over time (Klein, 1997). This can be especially critical when making decisions, as it allows the decision maker to play out not only the events that s/he believed led to the current state and also how the decision made will play out. Based on a meta-analysis, Driskell and his colleagues (1994) offered a number of suggestions for when and how to use mental practice. First, mental practice is especially effective at improving performance on tasks which involve a cognitive element (e.g., decision-making). In addition, mental practice must be reinforced. The authors argue that there is a decrease in the impact of mental practice over time and that refresher training should be provided every week to two weeks. Third, mental practice has a different impact on experts versus novices. Specifically, experienced trainees benefit from mental practice regardless of task type, whereas novices benefited from it more on cognitive tasks. They suggest that mental practice may be more beneficial to novices if schematic knowledge is provided prior to mental practice of a physical task. Finally, increasing the time in which trainees use mental practice does not necessarily lead to enhanced performance. Rather, it is suggested that 20 minutes of total mental practice time may be optimal.

Conclusions

While friendly fire incidents have been deemed an inescapable cost of war, there are a number of learning strategies available in the literature that have been proven effective at enhancing communication, coordination, and cooperation. By further understanding why breakdowns occur, the appropriate training strategies can be selected to target specific deficiencies. Depending on the time and resources available, a range of formal and informal learning strategies that could be effective at reducing teamwork breakdowns that lead to fratricide (Tables 1.2 and 1.3) are offered. Of course, these strategies require the expertise of both learning and domain professionals to be designed, developed, and delivered effectively. Furthermore, no one learning strategy alone will be successful but rather several strategies in combination are needed (Table 1.4). The 'ingredients' to reduce teamwork breakdowns offered here will help learning and domain experts create the 'recipe' that fits their needs. We hope that what we have presented here will encourage more thinking in this area as well as encourage researchers to expand their thinking to some less customary learning strategies.

Acknowledgements

This research was supported by the U.S. Air Force Research Laboratory, Mesa, Arizona. All opinions expressed in this chapter are those of the authors and do not necessarily reflect the official opinion or position of the University of Central Florida, the US Air Force Research Laboratory or the Department of Defense.

References

Archer, D. (1997). 'Unspoken diversity: Cultural differences in gestures'. *Qualitative Sociology, 20*, 79–105.

Austin, S., and Laurence, M. (1992). An empirical study of the SyberVision golf videotape. *Perceptual and Motor Skills, 74*, 875–881.

Baker, C. V., Salas, E., Cannon-Bowers, J. A., and Spector, P. E. (1992, May). *The Effects of Inter-positional Uncertainty and Workload on Teamwork and Task Performance.* Paper to the Society for Industrial and Organizational Psychology, Montreal, Quebec.

Bainbridge and S. A. Ruiz (eds). (1989). *Developing Skills with Information Technology* (pp. 65–86). New York: Wiley.

BBC News. (2007). 'Friendly fire' probes differ widely. *BBC News.* Retrieved February 10, 2007, from http://news.bbc.co.uk/1/hi/uk/6348811.stm

Berry, D. C. (1991). The role of action in implicit learning. *Quarterly Journal of Experimental Psychology, 43*, 881–906.

Blandin, Y., and Proteau, L. (2000). On the cognitive basis of observational learning: Development of mechanisms for the detection and correction of errors. *Quarterly Journal of Experimental Psychology, 53A*, 846–867.

Blickensderfer, E., Cannon-Bowers, J. A., and Salas, E. (1998). Cross-training and team performance. In J. A. Cannon-Bowers and E. Salas (eds), *Making Decisions under Stress: Implications for Individual and Team Training* (pp. 299–311). Washington, DC: American Psychological Association.

Bowers, C. A., and Jentsch, F. (2001). Use of commercial, off-the-shelf, simulations for team research. In E. Salas (ed.), *Advances in Human Performance and Cognitive Engineering Research* (Vol. 1, pp. 293–317). New York: Elsevier Science.

Brotman, L. E., Liberi, W. P., and Wasylyshyn, K. M. (1998). Executive coaching: The need for standards and competence. *Consulting Psychology Journal: Practice and Research, 50*, 40–46.

Bruner, J. (1990). *Acts of Meaning.* Cambridge, MA: Harvard University Press.

Cannon-Bowers, J.A., Burns, J. J., Salas, E., and Pruitt, J. S. (1998). Advanced technology in scenario-based training'. In J. A. Cannon-Bowers and E. Salas (eds), *Making Decisions Under Stress: Implications for Individual and Team Training* (pp. 365–374). Washington, DC: American Psychological Association.

Cannon-Bowers, J. A., Salas, E., Blickensderfer, E., and Bowers, C. A. (1998). The impact of cross-training and workload on team functioning: A replication and extension of initial findings. *Human Factors, 40*, 92–101.

Caudron, S. (1993). Training can damage diversity efforts. *Personnel Journal, 72*, 51–62.

Cumming, J., Clark, S. E., Ste-Marie, D. M., McCullagh, P., and Hall, C. (2005). The functions of observational learning questionnaire (FOLQ). *Psychology of Sport and Exercise, 6*, 517–537.

De Jong, J. A., and Versloot, A. M. (1999). Job instruction: Its premises and its alternatives. *Human Resource Development International*, *2*, 391–404.

Denning, S. (2004). Telling tales. *Harvard Business Review*, *82*, 122–153.

Denning, S. (2006). Effective storytelling: Strategic business narrative techniques. *Strategy and Leadership*, *34*, 42–48.

Dennis, K. A., and Harris, D. (1998). Computer-based simulation as an adjunct to ab initio flight training. *International Journal of Aviation Psychology*, *8*, 261–276.

Dormann, T., and Frese, M. (1994). Error training: Replication and the function of exploratory behavior. *International Journal of Human Computer Interaction*, *6*, 365–372.

Driskell, J. E., Copper, C., and Moran, A. (1994). Does mental practice enhance performance? *Journal of Applied Psychology*, *79*, 481–492.

Driskell, J. E., and Dwyer, D. J. (1984). Microcomputer videogame based training. *Educational Technology*, *24*, 11–17.

Driskell, J. E., and Johnston, J. H. (1998). Stress exposure training. In J. A. Cannon-Bowers and E. Salas (eds), *Making Decisions Under Stress: Implications for Individual and Team Training* (pp. 191–217). Washington, DC: American Psychological Association.

Elfenbein, H. A., and Ambady, N. (2002). On the universality and cultural specificity of emotion recognition: A meta-analysis. *Psychological Bulletin*, *128*, 203–235.

Entin, E. E., and Serfaty, D. (1999). Adaptive team coordination. *Human Factors*, *41*, 312–325.

Ferrari, M. (1996). Observing the observer: Self-regulation in the observational learning of motor skills. *Developmental Review*, *16*, 203–240.

Forret, M. L., Turban, D. B., and Dougherty, T. W. (1996). Issues facing organizations when implementing formal mentoring programmes. *Leadership and Organization Development Journal*, *17*, 27–30.

Fowlkes, J., Dwyer, D. J., Oser, R. L., and Salas, E. (1998). Event-based approach to training (EBAT). *International Journal of Aviation Psychology*, *8*, 209–221.

Frese, M. and Altman, A. (1989). The treatment of errors in learning and training. In L. Bainbridge and S.A. Ruiz (eds), *Developing Skills with Information Technology* (pp. 65–86). Quintanilla. Chichester: John Wiley and Sons.

Frese, M., Brodbeck, F., Heinnbokel, T., Mooser, C., Schleiffenbaum, E., and Thiemann, P. (1991). Errors in training computer skills: On the positive function of errors. *Human Computer Interaction*, *6*, 77–93.

Goldstein, I. L. (1993). *Training in Organizations* (3rd edn). Pacific Grove, CA: Brooks/Cole Publishing.

Gopher, D., Weil, M., and Bareket, T. (1994). Transfer of skill from a computer game trainer to flight. *Human Factors*, *36*, 387–405.

Gully, S. M., Payne, S. C., Koles, K. L. K., and Whiteman, J. K. (2002). The impact of error training and individual differences on training outcomes: An attribute-treatment interaction perspective. *Journal of Applied Psychology, 87*, 143–155.

Hall, D. T., Otazo, K. L., and Hollenbeck, G. P. (1999). Behind closed doors: What really happens in executive coaching. *Organizational Dynamics, 27*, 39–53.

Hendricks, C. C. (2001). Teaching causal reasoning through cognitive apprenticeship: What are the results from situated learning? *Journal of Educational Research, 94*, 302–311.

Ivancic, K., and Hesketh, B. (1995). Making the best of errors during training. *Training Research Journal, 1,* 103–125.

Jentsch, F., and Bowers, C. A. (1998). Evidence for the validity of pc-based simulations in studying aircrew coordination. *International Journal of Aviation Psychology, 8*, 243–260.

Jentsch, F., Bowers, C. A., and Salas, E. (2001). What determines whether observers recognize targeted behaviors in modeling displays? *Human Factors, 43*, 496–507.

Johnson, R. T., Johnson, D. W., and Stanne, M. B. (1985). Effects of cooperative, competitive, and individualistic goal structures on computer-assisted instruction. *Journal of Educational Psychology, 77,* 668–677.

Kampa-Kokesch, S., and Anderson, M. Z. (2001). Executive coaching: A comprehensive review of the literature. *Consulting Psychology Journal: Practice and Research, 53*, 205–228.

Karl, K. A., O'Leary-Kelly, A. M., and Martocchio, J. J. (1993). The impact of feedback and self-efficacy on performance in training. *Journal of Organizational Behavior, 14*, 379–394.

Klein, G. A. (1997). *Developing Expertise in Decision Making*. Unpublished manuscript.

Lewis, M. (1998). What workers need to succeed. *American Machinist, 142*, 88–98.

Littrell, L. N., and Salas, E. (2005). A review of cross-cultural training: Best practices, guidelines, and research needs. *Human Resource Development Review, 4*, 305–334.

Littrell, L. N., Salas, E., Hess, K. P., Paley, M., and Riedel, S. (2006). Expatriate preparation: A critical analysis of 25 years of cross-cultural training research. *Human Resources Developmental Review, 5*, 355–388.

Lorenzet, S. J., Salas, E., and Tannenbaum, S. I. (2005). Benefiting from mistakes: The impact of guided errors on learning, performance, and self-efficacy. *Human Resource Development Quarterly, 16*, 301–322.

Lozano, S. C., Hard, B. M., and Tversky, B. (2006). Perspective taking promotes action understanding and learning. *Journal of Experimental Psychology: Human Perception and Performance, 32*, 1405–1421.

Mobley, G. M., Jaret, C., Marsh, K., and Lim, Y. Y. (1994). Mentoring, job satisfaction, gender, and the legal profession. *Sex Roles: A Journal of Research, 31,* 79–98.

Mulec, K., and Roth, J. (2005). Action, reflection, and learning-coaching in order to enhance the performance of drug development project management teams. *RandD Management, 35*, 483–491.

Noe, R. A. (2002). *Employee Training and Development* (2nd edn). New York: McGraw-Hill.

Office of Technology Assessment (OTA). (1993). *Who Goes There: Friend or Foe?* (OTA-ISC-537). Washington, DC: US Government Printing Office.

Oludaja, B. J. (1992, March). *Listening to Nonverbal Cues Across Cultures*. Paper presented at the 13th International Listening Association conference, A Listening Mosaic, Seattle, WA.

Orasanu, J., and Backer, P. (1996). Stress and military performance. In J. E. Driskell and E. Salas (eds), *Stress and Human Performance* (pp. 89–125). NJ: Lawrence Erlbaum.

Oser, R. L., Cannon-Bowers, J. A., Salas, E., and Dwyer, D. J. (1999). Enhancing human performance in technology-rich environments: Guidelines for scenario-based training. In E. Salas (ed.). *Human/Technology Interaction in Complex Systems* (vol. 9, pp. 175–202). Greenwich, CT: JAI Press.

Prislin, R., Jordan, J. A., Worchel, S., Tschan-Semmer, F., and Shebilske, W. L. (1996). Effects of group discussion on the acquisition of complex skills. *Human Factors, 38,* 404–416.

Proctor, M. D., Panko, M., and Donovan, S. (2004). Considerations for training team situation awareness and task performance through PC-gamer simulated multi-ship helicopter operations. *International Journal of Aviation Psychology, 4*, 191 -205.

Reason, J. T. (1990). *Human Error*. New York: Cambridge University Press.

Ricci, K. E., Salas, E., and Cannon-Bowers, J. A. (1996). Do computer-based games facilitate knowledge acquisition and retention? *Military Psychology, 8*, 295–307.

Richman, H. B., Staszewski, J. J., and Simon, H. A. (1995). Simulation of expert memory using EPAM IV. *Psychological Review, 102*, 305–330.

Rosen, M. A., Salas, E., and Upshaw, C. L. (2008, April). *Understanding Demonstration-based Training: A Definition, Framework, and Some Initial Guidelines*. Poster presented at the 23rd Annual Conference of the Society for Industrial and Organizational Psychology, San Francisco, CA.

Rossett, A., and Sheldon, K. (2001). *Beyond the Podium: Delivering Training and Performance to a Digital World.* San Francisco: Jossey-Bass/Pfeiffer.

Sacks, M. (1994). *On-the-job Learning in the Software Industry. Corporate Culture and the Acquisition of Knowledge.* Westport, CT: Quorum Books.

Salas, E., Bowers, C. A., and Edens, E. (eds). (2001). *Improving Teamwork in Organizations: Applications of Resource Management Training*. Mahwah, NJ: Lawrence Erlbaum Associates.

Salas, E., Burke, C. S., Bowers, C. A., and Wilson, K. A. (2001). Team training in the skies: Does crew resource management (CRM) training work? *Human Factors, 43*, 641–674.

Salas, E., and Cannon-Bowers, J. A. (1997). Methods, tools, and strategies for team training. In M.A. Quinones and A. Ehrenstein (eds), *Training for a Rapidly Changing Workplace: Applications of Psychological Research* (pp. 249–279). Washington, DC: American Psychological Association.

Salas, E., and Cannon-Bowers, J. A. (2000a). The anatomy of team training. In S. Tobias and J. D. Fletcher (eds), *Training and Retraining: A Handbook for Business, Industry, Government, and the Military* (pp. 312–335). New York: Macmillan Reference.

Salas, E., and Cannon-Bowers, J. A. (2000b). Designing training systems systematically. In E. A. Locke (ed.), *The Blackwell Handbook of Principles of Organizational Behavior* (pp. 43–59). Malden, MA: Blackwell Publisher.

Salas, E., Sims, D. E., and Burke, C. S. (2005). Is there 'big five' in teamwork? *Small Group Research, 36*, 555–599.

Salas, E., Wilson, K. A., Burke, C. S., and Wightman, D. (2006). Does CRM training work? An update, extension, and some critical needs. *Human Factors, 48*, 392–412.

Scales, R. H. (2006). Clausewitz and World War IV. *Armed Forces Journal, 143*, 22–29.

Scandura, T. A., Tejeda, M. J., Werther, W. B., and Lankau, M. J. (1996). Perspectives on mentoring. *Leadership and Organizational Development Journal, 17,* 50–56.

Serfaty, D., Entin, E. E., and Johnston, J. H. (1998). Team coordination training. In J. A. Cannon-Bowers and E. Salas (eds), *Making Decisions Under Stress: Implications for Individual and Team Training* (pp. 221–246). Washington, DC: American Psychological Association.

Veale, T.K. (1999). Targeting temporal processing deficits through Fast For Word: Language therapy with a new twist. *Language, Speech and Hearing Services in Schools, 30*, 353–362.

Volpe, C. E., Cannon-Bowers, J. A., Salas, E., and Spector, P. E. (1996). The impact of cross-training on team functioning: An empirical investigation. *Human Factors, 38*, 87–100.

Wiegmann, D. A., and Shappell, S. A. (2003). *A Human Error Approach To Aviation Accident Analysis: The Human Factors Analysis and Classification System*. Aldershot: Ashgate.

Wehrenberg, S. B. (1987). Supervisors as trainers: The long-term gains of OJT. *Personnel Journal, 66*, 48–51.

Wexley, K. N., and Latham, G. P. (2002). *Developing and Training Human Resources in Organizations* (2nd edn). Upper Saddle River, NJ: Prentice Hall.

Whitehall, B., and McDonald, B. (1993). Improving learning persistence of military personnel by enhancing motivation in a technical training program. *Simulation and Gaming, 24,* 294–313.

Wiener, E. L., Kanki, B. G., and Helmreich, R. L. (eds). (1993). *Cockpit Resource Management*. New York: Academic Press.

Wilson, P. F., and Johnson, W. B. (2001). Core virtues for the practice of mentoring. *Journal of Psychology and Theology, 29*, 121–130.

Wilson, K. A., Salas, E., Priest, H. A., and Andrews, D. H. (2007). Errors in the heat of battle: Taking a closer look at shared cognition breakdowns through teamwork. *Human Factors, 49*, 243–256.

Woodman, M. D. (2006). *Cognitive Training Transfer Using a Personal Computer-based Game: A Close Quarters Battle Case Study*. Unpublished doctoral dissertation, University of Central Florida, Orlando, FL.

Xiao, Y., and Moss, J. (2001, October). Practices of high reliability teams: Observations in trauma resuscitation. Paper presented at the Human Factors and Ergonomics Society 45th Annual Meeting, Minneapolis, MN.

SECTION 1
Cognitive Processes

Decades of human-factors research has given us a solid foundation of both theories and data that can be applied to the combat identification (CID) challenge. As with all basic research, results from basic research on fundamental human cognitive processes do not lead directly to solutions for better CID and the reduction of the risk of friendly fire, but provide solution developers with theories that can help to explain and predict which solution approaches have the best chance of accomplishing CID and friendly fire mitigation. There is a huge body of literature related to basic cognitive processes that has developed over decades. A key question is whether CID solution developers are making good use of that literature.

The three chapters in this section are excellent examples of research on cognitive processes that can be used by CID solution developers. Obviously, they only represent a fraction of the human factors research literature. However, they provide examples of the kinds of research that can be very helpful in understanding basic human factors principles vital in CID research and development.

In Chapter 2, Robert W. Proctor and Motonori Yamaguchi examine models of human information processing as they relate to the prediction of human performance. The authors tell us that a typical model breaks task performance into three broad processes: stimulus encoding, response selection, and motor execution. The response-selection process is the most critical in determining the speed and accuracy of performance in a complex operational environment. Their research investigated factors that affect speed and accuracy of response selection. Implications of the research for interface designs and training of operational skills are discussed. Their findings led them to conclude, 'human interfaces should be designed in a way that maximizes the compatibility of information presentation and actions that are taken in response to the information, if the operational environment is to be made more efficient' (p. 42).

A key factor in many friendly fire incidents is a lack of vigilance on the part of warfighters. CID developers should consider vigilance issues in all aspects of their designs. Clark Shingledecker, David Weldon, Kyle Behymer, Benjamin Simpkin, Elizabeth Lerner, Joel Warm, Gerald Matthews, Victor Finomore, Tyler Shaw, and Jennifer Murphy (Chapter 3) state, 'The overarching purpose of the research program described … is to identify a set of individual difference variables that predict vigilance performance, to assemble selection measures to assess these variables, and to validate the measures for use in practical, military personnel-selection settings' (p. 48).

The last few years have seen a substantial increase in the use of unmanned systems (air, land, and sea) in operational settings. CID issues are a particular problem in these systems because the operator is removed from the actual platform. Thomas Fincannon, A. William Evans III, Florian Jentsch, and Joseph Keebler (Chapter 4) take a detailed look at the human factors issues that should be addressed by CID developers, especially in the area of spatial abilities. They report on an interesting study that focuses on 'how spatial ability influenced the actual communication of visual information from one teammate to another' (p. 73) in an unmanned system task setting.

Chapter 2

Factors Affecting Speed and Accuracy of Response Selection in Operational Environments

Robert W. Proctor
Motonori Yamaguchi
Purdue University

Since the rapid growth of cognitive theories in the mid 1950s, human performance has been analyzed from an information-processing perspective (Proctor and Vu, 2006b), which decomposes task performance into distinct processing components (e.g., Card, Moran, and Newell, 1983). Information-processing models typically consist of three major subsystems—stimulus encoding, response selection, and response execution (see Figure 2.1)—of which the most critical component in determining the speed and accuracy of performance is that of response selection (e.g., Welford, 1960). Consequently, factors that affect response-selection processes have been of theoretical and practical concern. They include, for example, the number of alternative choices (Hick, 1952), the interval between tasks (Smith, 1967), cognitive load (Baddeley, 1998), task sequence (Monsell, 2003), and compatibility between stimuli and responses (Fitts and Seeger, 1953).

Among those factors, the compatibility between stimuli and responses has been recognized as one of the most important in designing operational systems to provide optimal task performance. In basic psychological research, the effect of compatibility is termed *stimulus-response compatibility* (SRC; Proctor and Vu, 2006a), whereas in applied human factors research it is often called *display-control compatibility* (Andre and Wickens, 1990). The SRC effect occurs when responses are mapped to stimuli for which there are some corresponding attributes, or dimensional overlap. These stimulus and response attributes may be their spatial locations (e.g., a signal displayed on the left side of the monitor and a response device placed on the left of the operator), or symbology of displays used to indicate controlling actions (e.g., arrow pointing to the left as a signal to turn a vehicle to the left). When stimuli are mapped to the corresponding responses, the stimulus-response (S-R) mapping is called *compatible;* whereas when they are mapped to noncorresponding responses the mapping is called *incompatible* (see Figure 2.2).

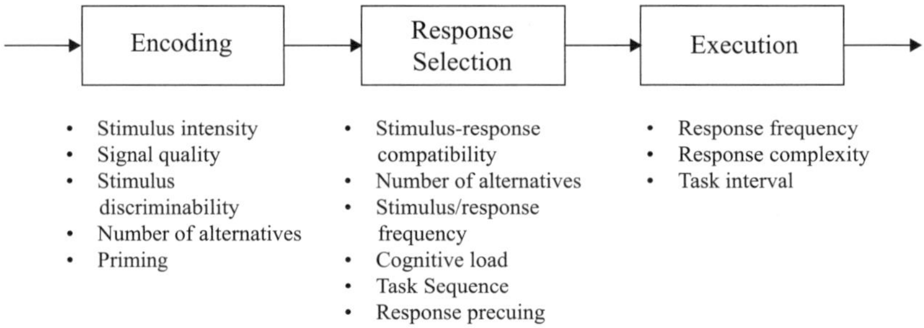

Figure 2.1 An illustration of a three-stage model of human information-processing and examples of operational factors that affect the processing stages

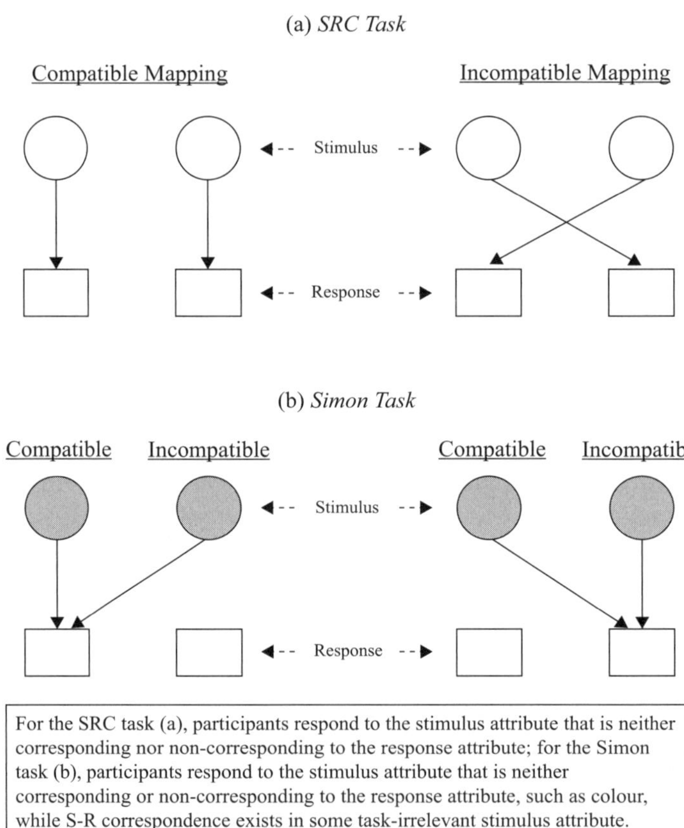

Figure 2.2 Examples of the stimulus-response mappings for (a) the SRC task and (b) the Simon task

The SRC effect is often attributed to population stereotypes (Fitts and Seeger, 1953) or long-term associations (Barber, and O'Leary, 1997) acquired through experience. A traditional description of the SRC effect is that stimuli are more easily translated into responses when they correspond than when they do not. The advantage for compatible stimulus and response sets remains intact after several days of practice with the task (Dutta and Proctor, 1992; Fitts and Seeger, 1953), and the SRC effect is robust in a variety of task settings (Proctor and Vu, 2006a). Nevertheless, in recent studies we identified several task conditions in which the advantage of the compatible mapping is reduced or eliminated. In the following sections, we describe several lines of research conducted for a Multidisciplinary University Research Initiative (MURI) project on training (see Acknowledgments) that include such conditions.

SRC Effect in Mixed-Mapping Task

In the first line of research, we examined the SRC effect in a flight simulator (Yamaguchi and Proctor, 2006, 2007b). In one study, participants monitored the primary flight display and banked the aircraft by turning the yoke to the left or right according to the location of a signal presented on the display. They were to bank the aircraft in the direction corresponding to the signal location in one condition and in the direction opposite the signal location in another condition. That is, participants always responded compatibly or incompatibly to the signal locations depending on the mapping condition. Response times (RTs) and percentage errors (PEs) for the two conditions showed a relatively large SRC effect. Of more interest, in another condition, participants were required to choose on each trial whether to bank the aircraft compatibly or incompatibly to the signal location on the basis of its color. When the signal was green, participants were to bank the aircraft toward the signal location, but when the signal was red, they were to bank away from the signal location. Thus, the compatible and incompatible mappings were intermixed in a single task condition.

Responses were generally slower and less accurate when the two mappings were mixed than when they were not, indicating a *mixing cost*. More important, the cost of mixing was larger for trials with compatible mapping than those with incompatible mapping, which reduced the SRC effect. Moreover, when the yoke-turn responses were replaced with presses of buttons on the left and right handles of the flight yoke, the SRC effect with mixed mappings was completely eliminated (see Figure 2.3). In a follow-up study (Yamaguchi and Proctor, 2007a), we observed that the elimination of the SRC effect in a mixed-mapping task was stable with the button-press responses but not with the yoke-turn responses, with yoke turns, the SRC effect appeared when responses were made quickly but less accurately, whereas it was eliminated when responses were made slowly but more accurately; thus, this observation seems to reflect speed-accuracy tradeoffs. A possibility is that modes of responding differently affect participants' strategy, but it is also possible that the observation is due to the fact that these response modes allow different types of response preparation. In fact, a previous study suggested that preparatory states can affect the reduction/elimination of the SRC effect in a mixed-mapping task (De Jong, 1995).

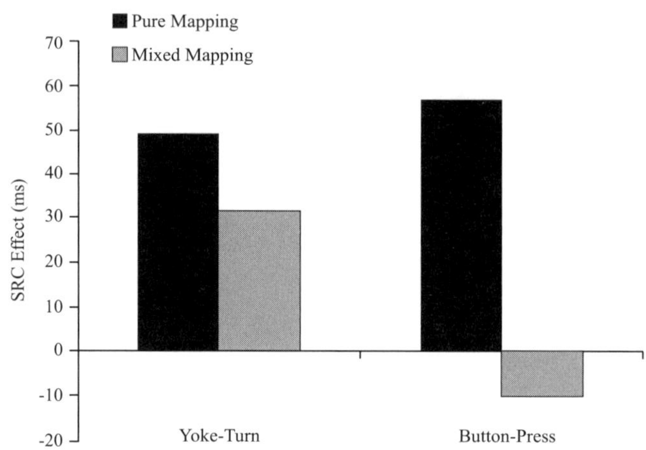

Figure 2.3 The stimulus-response compatibility (SRC) effect as a function of response mode (yoke-turn vs. button-press) and mapping condition (pure mapping vs. mixed mapping) in Yamaguchi and Proctor (2006, Experiments 3 and 4)

We also conducted a series of experiments using a mixed-mapping task with four alternative responses (Proctor and Vu, 2009), rather than only two alternatives as in the studies discussed above. Stimuli were presented on the left or right of the screen in white or red, with color indicating whether the mapping of stimulus locations to responses was compatible or incompatible. Two keys were operated by the middle and index fingers of the left hand for one mapping, and another two keys were operated by the index and middle fingers of the right hand for the other mapping. Thus, while participants were required to select the appropriate mapping on each trial, each response was assigned to only a single mapping. In this case, the SRC effect was neither eliminated nor reduced by mixing the two mappings (see Figure 2.4), most likely because the mapping of stimulus location to each response was consistent rather than variable. Though it is still left to future research to determine exactly which environmental manipulations can influence the pattern of SRC effects in mixed-mapping tasks, both the two- and four-alternative choice tasks suggest that the nature of responses modulates the SRC effect in mixed-mapping tasks.

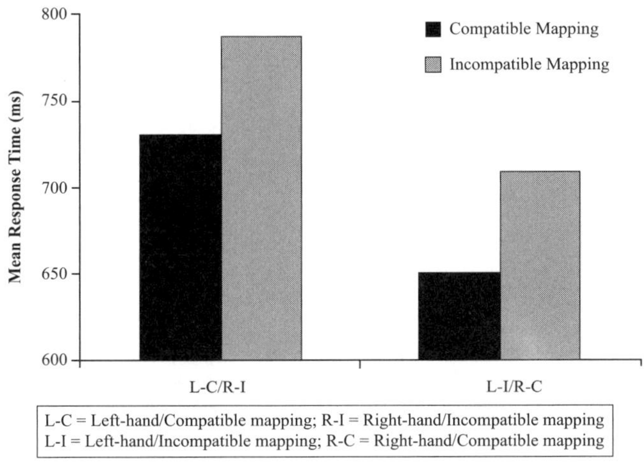

Figure 2.4 Mean response times for the compatible and incompatible mappings as a function of hand-mapping assignment in Proctor and Vu (2009, Experiment 3)

Transfer of Learning: 1. Sensory Modality

Robustness of the SRC effect is also implied by the fact that responses are typically faster and more accurate when a 'task-irrelevant' stimulus attribute overlaps with a response attribute. The SRC effect on the basis of a task-irrelevant stimulus dimension is called the *Simon effect*, and the task that produces this effect is called the Simon task (Lu and Proctor, 1995; Simon, 1990). In a typical Simon task, participants are asked to make a left or right key press in response to the color of the stimulus that appears on the left or right of the screen (see Figure 2.2b). Even though participants are instructed to ignore the spatial dimension of the stimuli, responses are faster and more accurate when they correspond to the spatial stimulus attributes than when they do not, yielding the Simon effect. Although the Simon effect is also known to persist through several days of practice for the task (e.g., Proctor and Lu, 1999; Simon, Craft, and Webster, 1973), another line of research we have conducted for the MURI project shows that the Simon effect can be reduced, eliminated, or even reversed in certain task contexts.

In the first study of this type, Proctor and Lu (1999) had participants complete 930 trials of a two-choice reaction task with the spatially incompatible mapping over three consecutive days and then perform the Simon task on the fourth day. For that task, the typical Simon effect was reversed such that responses to noncorresponding stimuli were faster than responses to corresponding stimuli. Subsequently, Tagliabue, Zorzi, Umiltà, and Bassignani (2000) reported elimination of the Simon effect after participants had performed only 72 trials of the incompatible-mapping practice task.

Instead of using visual stimuli for the two tasks, Vu, Proctor, and Urcuioli (2003) used auditory stimuli for both and found that transfer of the incompatible mapping to the Simon task did not occur. We followed up this study, assessing whether the lack of transfer effect was due to the nature of the auditory stimuli or insufficient practice (Proctor, Yamaguchi, and Vu, 2007). Groups of participants performed 84, 300, or 600 trials of the incompatible-mapping task and then transferred to the Simon task, with both tasks using auditory stimuli. The Simon effect was significantly reduced after 300 and 600 practice trials, but not after 84 practice trials, compared to a control group that performed only the Simon task (see Figure 2.5). These results indicate that auditory stimuli produce a stronger tendency to make spatially corresponding responses than visual stimuli and more extended practice is required for transfer of learning with auditory stimuli.

In an unpublished study, we also had participants perform the same amounts of practice with visual stimuli and transfer to the Simon task with auditory stimuli. In this case, there was no transfer effect even after 600 or 1,200 practice trials. In contrast, in Vu et al.'s (2003) study (see also Tagliabue, Zorzi, and Umiltà, 2002), there was a significant transfer effect when participants practiced with auditory stimuli for 84 trials and then transferred to the visual Simon task, indicating an asymmetric pattern of transfer between visual and auditory stimuli. The studies suggest collectively that the amount of practice strengthens the transfer of learning, but there is specificity of transfer that is difficult to overcome even with extended practice.

These results led us to investigate factors responsible for the specificity of transfer, which is also of particular interest in the study of training principles in the

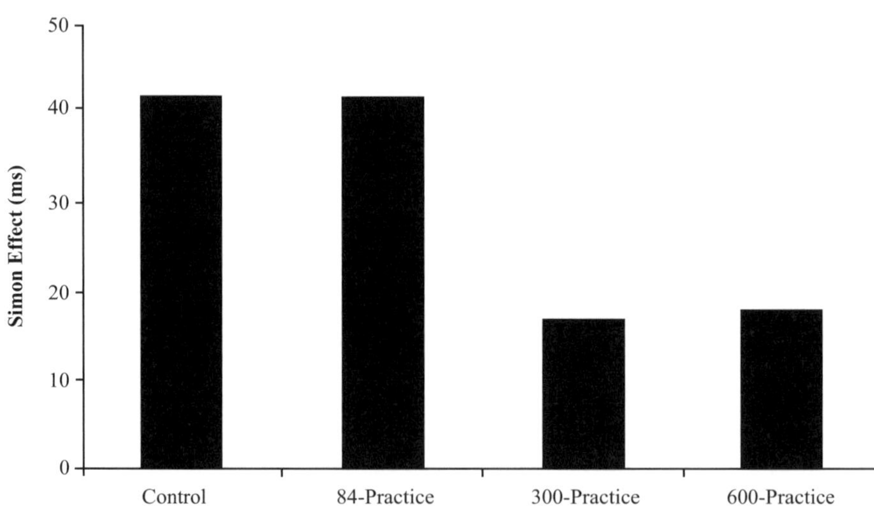

Figure 2.5 The Simon effect as a function of practice amount in Proctor, Yamaguchi, and Vu (2007, Experiment 1)

MURI project (Healy, Wohldmann, Parker, and Bourne, 2005; Healy, Wohldmann, Sutton, and Bourne, 2006). We reported the first experiment in our transfer study with auditory stimuli described above (Proctor et al., 2007), which showed that the transfer effect was specific to the practiced spatial dimension (see also Vu, 2007). When participants performed a task with stimuli and responses arranged horizontally, there was a significant transfer effect when the stimuli and responses in the subsequent task were also arranged horizontally but little transfer effect when they were arranged vertically. The same was true when participants practiced with vertically arranged stimuli and responses; transfer was significant when the subsequent task involved vertically arranged stimuli and responses, but not when it involved horizontally arranged stimuli (see Figure 2.6). In other words, there were transfer effects within the same spatial dimension but not between different spatial dimensions. There was also evidence of between-dimension transfer for visual stimuli, though, when extended practice was provided (Vu, 2007). These outcomes suggest that the transfer effect with short practice depends on S-R associations that are specific to the practiced spatial dimension, whereas extended practice not only strengthens the association but also can lead to acquisition of an abstract rule (i.e., 'respond opposite') for S-R transformation.

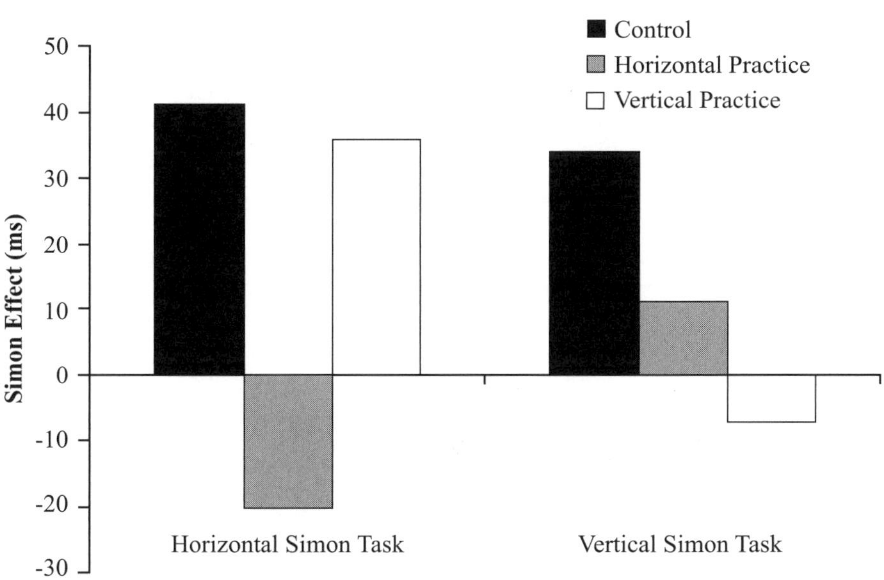

Figure 2.6 The Simon effect for the horizontal and vertical tasks as a function of practiced spatial dimensions in Proctor, Yamaguchi, and Vu (2007, Experiments 3 and 4)

Transfer of Learning: 2. Stimulus Codes

It has been acknowledged that the SRC and Simon effects depend on how stimulus information and response information are encoded rather than their physical attributes (e.g., Proctor and Cho, 2006). The notion of stimulus and response codes is implicated because SRC and Simon effects can occur on the basis of S-R correspondence that emerges on their salient feature dimensions (Proctor and Reeve, 1985), spatial locations of stimulus and response keys rather than responding hands (Wallace, 1971), and as a consequence of responding (e.g., task goal; Hommel, 1993). Spatial information can be conveyed by symbols (e.g., arrows pointing to a direction) or words (e.g., 'LEFT' or 'RIGHT'), as well as by physical stimulus location (e.g., a circle on the left or right of the display). These different manners of conveying spatial information also yield the Simon effect (Lu and Proctor, 1995). Some models attribute the SRC effect to spatial codes of a general nature that are independent of specific stimuli or responses (e.g., Zhang, Zhang, and Kornblum, 1999). However, it is also possible that the SRC effect depends on spatial codes that are specific to the types of stimuli and responses. This issue can be examined using the transfer paradigm by varying stimuli used in the practice and transfer sessions. Thus, in the next series of experiments, we examined the specificity of transfer in terms of the types of stimulus code that affects response-selection processes (Proctor, Yamaguchi, Zhang, and Vu, 2009).

First, we tested whether the transfer effect could be observed for three types of stimuli, physical locations (circles on the left or right of the display), arrow directions (pointing to the left or right), and location words ('LEFT' or 'RIGHT'). A significant transfer effect occurred in the Simon task after 84 practice trials of the incompatible-mapping task for physical location and arrow directions but not for location words. Subsequently, we varied the stimuli used in the practice and transfer sessions and found that nearly perfect transfer occurred between physical locations and arrow directions, indicating that they shared the same stimulus codes. However, there was no significant transfer effect between location words and physical locations or arrow directions. Therefore, the results are indicative of two types of stimulus codes; *visual-spatial* for physical locations and arrow directions, and *semantic-spatial* for location words. The lack of transfer effect from physical locations and arrow directions to location words is consistent with the distinction between visual- and semantic-spatial codes. Nevertheless, the lack of transfer from location words to physical locations or arrow directions could have been due to the fact that the amount of practice was simply insufficient to yield a significant transfer effect, because practice with location words did not show transfer within the same stimulus codes.

As in the case of auditory stimuli, therefore, we tested whether extended practice could lead to transfer of the incompatible mapping for location words (Proctor et al., 2008). The amounts of practice were 84, 300, 600 trials with location words, which were compared to the control group who did not perform the practice session. The results were consistent with those for auditory stimuli;

significant transfer appeared after 300 and 600 practice trials, whereas there was no transfer after 84 trials. Further, we examined whether extended practice with location words could produce a transfer effect to physical locations or arrow directions. The results suggested that there was significant transfer from location words to arrow directions but not to physical locations. In fact, there was some evidence in that study that arrow stimuli have both visual- and semantic-spatial characteristics, probably due to the fact that they have both perceptual and semantic (symbolic) spatial attributes. Similarly, when participants were given extended practice with arrow directions and then transferred to the Simon task with location words, there was a clear tendency of transfer effect. In contrast, even when they had extended practice with physical locations, little transfer to location words was evident.

In summary, we observed that there was code-specific transfer of the incompatible mapping to the Simon task when stimuli varied between the practice and transfer sessions. In particular, the results of our studies are consistent with the distinction between visual-spatial and semantic-spatial codes. Transfer can occur with a relatively small amount of practice for visual-spatial codes, whereas extended practice is required for semantic-spatial codes. Interestingly, arrow stimuli seem to involve both types of spatial codes. Hence, transfer of learning depends on the type of stimulus codes rather than on a general spatial code.

Transfer of Learning: 3. Response Codes

So far, we have only considered the influences of stimulus domain on transfer of learning. As discussed in the section on mixed-mapping tasks, however, response domain is also an important factor that affects response selection. Consequently, we examined the influences of this domain on transfer of the incompatible mapping (Yamaguchi and Proctor, 2009).

Though simply called 'response,' it is a complex composite of several different factors. For instance, the response device used in performing a task provides the immediate perceptual attributes of responding. Also, a specific action (response mode) required for a task provides the immediate motor components of responding. Moreover, there is evidence reported in the literature that the consequence of an action, rather than the action itself, provides the immediate conceptual basis of responding (e.g., Hommel, 1993). Hence, influences of these factors have to be dissociated and examined separately.

A difficulty of dissociating the influence of response device from that of response mode is that changing one usually alters the other. For instance, though the same computer operation can be performed by using a computer mouse or a touchpad, switching from one device to the other alters the actual actions taken in performing the task, thus confounding two different factors. To exclude the problem, we used two types of response devices that are designed to make key presses, one with a standard keyboard and the other with the response box designed for psychological research (Yamaguchi and Proctor, 2008). The task used physical-location stimuli similar to our preceding experiments for both practice and transfer sessions. The results indicated that nearly perfect transfer was achieved when switching response devices between the two sessions, which excluded this factor from our consideration.

Having excluded the influence of response device, we varied response modes between the practice and transfer sessions (Yamaguchi and Proctor, 2009). Two types of response modes used in our second experiment were key presses on a standard keyboard and deflections of a joystick. This factor exerted a significant influence on transfer of the incompatible mapping. That is, the transfer effect was reduced when response modes varied between the practice and transfer sessions compared to when they were held constant, indicating the importance of response mode (see Figure 2.7). Some researchers suggest, however, that the SRC effect depends on the correspondence of the environmental stimuli with one's intentions (Hommel, Müsseler, Aschersleben, and Prinz, 2001). According to them, taking an action itself is seldom the goal of an actor; instead, people take actions in order to make some changes in the environment. Hence, those researchers believed that a response is represented in terms of an effect that is anticipated to occur as a result of the person's action. Thus, it is the consequences of actions rather than the actions themselves that are central to constitute responses. In the psychological literature, the consequence of responding is called an *action effect*. For example, an action effect can be a light turned on when a switch is pressed. In a task where pressing a left key turned on a light placed on the right and pressing a right key turned on a light placed on the left; the response location and the side of action effect were incompatible, and the Simon effect occurred on the basis of spatial correspondence between locations of stimuli and action effects, rather than between locations of stimuli and response locations, when instructions emphasized the former correspondence (Hommel, 1993).

Therefore, in a third experiment, we assessed the influence of an action effect, in which the two response modes used in the second experiment were employed. The question was whether participants would ignore a change of response mode when the action effects were identical for different response modes. We used a condition in which a left key press or deflection of the joystick triggered the presentation of a filled square on the left side of the screen and a right key press or deflection triggered the presentation of a filled square on the right side. Though the location/direction of response and its effect were compatible in our task, the location of the action effect should have had a stronger influence than that of response location/direction in performing the task if the action effect is the critical factor. If such is the case, a nearly perfect transfer effect would have been observed even when response modes were varied between the practice and transfer sessions.

The results showed that there were significant transfer effects for all conditions, compared to the control conditions. At the same time, there was also a significant reduction of the transfer effect when response modes were varied when participants practiced with the joystick and transferred to the keyboard, compared

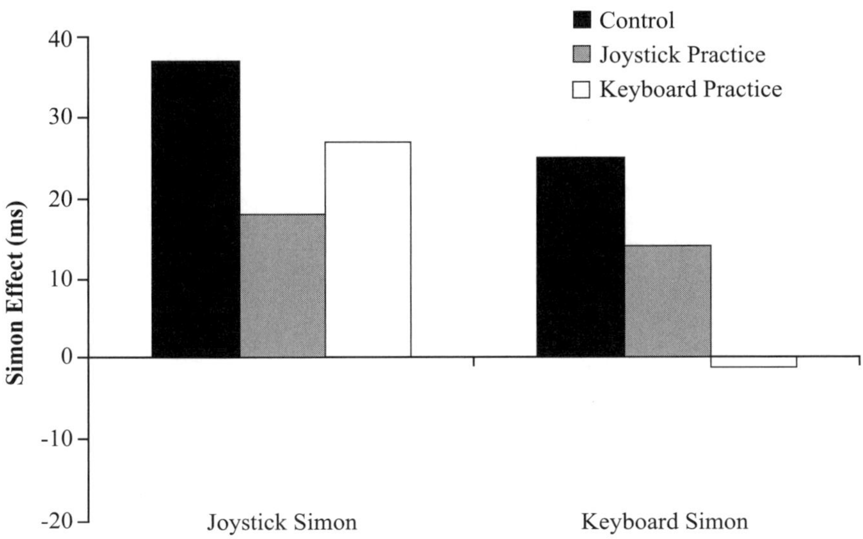

Figure 2.7 The Simon effect for the joystick and keyboard Simon tasks without action effect as a function of practiced response mode in Yamaguchi and Proctor (2009, Experiment 1)

to when they practiced with the keyboard (see Figure 2.8). This reduction was, however, absent when participants practiced with the keyboard and transferred to the joystick, which showed a Simon effect comparable to those who used the keyboard in both the practice and transfer conditions. These observations can be explained if the presence of action effects on practicing the incompatible-mapping task was selectively influenced when the response mode was the keyboard but not when it was the joystick (see Yamaguchi and Proctor, 2009, for more detailed discussions).

The three experiments described in this section collectively point to the importance of response mode in transfer of newly acquired S-R associations. The nearly perfect transfer between the two response devices suggests that the transfer effect is not always reduced by a change in task context. In contrast, it depends on manipulations of environmental factors that are critical to represent performance variables (i.e., stimulus and response). Although these experiments should not be taken to indicate that the response device has no influence on response-selection processes, the specificity of transfer to the trained response mode suggests that construction of a response representation is centered at the actions that are taken to perform the task.

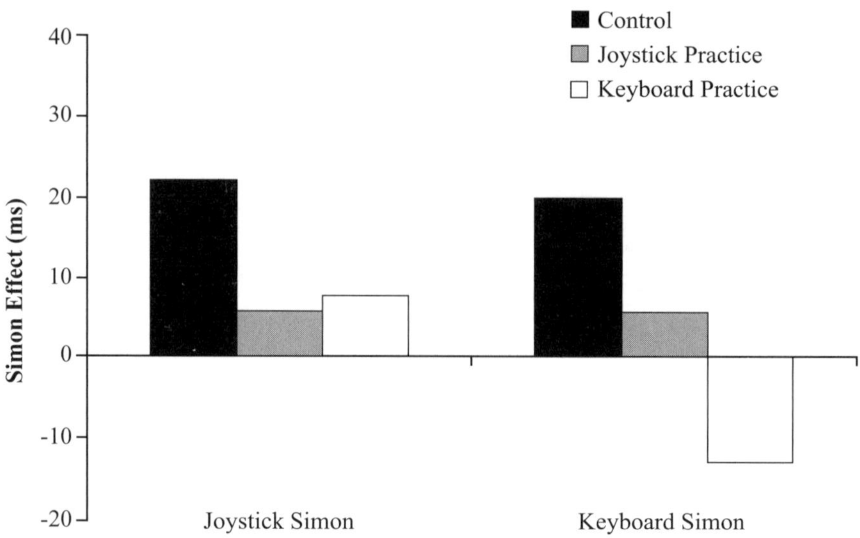

Figure 2.8 The Simon effect for the joystick and keyboard Simon tasks with action effect as a function of practiced response mode in Yamaguchi and Proctor (2009, Experiment 2)

Implications for Interface Design and Skill Training

We have described several lines of research conducted for the MURI project that investigate factors affecting response-selection processes in operational environments, with particular emphasis on basic perceptual-motor skill components. It is widely acknowledged that response selection is the central cognitive process that influences the speed and accuracy of task performance and benefits most from training. Thus, our research provides several results that are suggestive of both designing human interfaces and training efficiency of perceptual-motor skills.

The most important take-home message is, of course, that human interfaces should be designed in a way that maximizes the compatibility of information presentation and actions that are taken in response to the information, if the operational environment is to be made more efficient. This was the starting point of our research project. In the first section, we discussed that mixing the compatible mapping with the incompatible mapping increases both RT and PE, which we called mixing costs. Moreover, the advantage of the compatible mapping is significantly reduced or even eliminated in that condition. It is thus important that great care be taken to eliminate any display-control incompatibility in operational systems. Also, our study suggested that response conflict is a primary cause of the reduction/elimination of the advantage of the compatible mapping. Thus, if it is not possible to exclude incompatibility, operational environments should be carefully designed to separate compatible and incompatible operations to avoid their conflicts. In this way, the advantage of display-control compatibility can be retained in a complex operational environment.

Other experiments showed that the task-irrelevant SRC effect (Simon effect) is robust in a variety of task conditions, across different sensory modalities, stimulus codes and dimensions, and response modes. They indicate that compatibility not only in task-relevant display-control relationships but also in task-irrelevant ones should be considered. Neglect of these factors can lead to human-induced errors, which are often the leading cause of incidents and accidents. Furthermore, these studies suggest that experience with an incompatible task setting can continue to influence performance of subsequent tasks, eliminating or even reversing the compatibility relationships. Surprisingly, the amount of experience does not have to be large to exert its inadvertent influence, especially when these tasks are similar in their operations. This is another reason why the display-control compatibility should be carefully examined in designing human interfaces and task environments.

In addition to the issue of compatibility in interface design, our research suggests several factors that are important for transfer of learning in perceptual-motor skills. In general, transfer can occur with a small amount of practice. However, such transfer is specific to several operational factors, such as sensory modality, stimulus code, and response mode. This is in conformity with the principle of training specificity proposed by Healy, Bourne, and colleagues (e.g., Healy, Wohldmann, Parker, and Bourne, 2005; Healy, Wohldmann, Sutton, and Bourne, 2006). Although generality can be obtained with extended amounts of practice, it is not always the case. Some operational factors have stronger influences on response selection than others so that extended practice cannot overcome the effect, and some cause slow learning processes. Because unique factors of these types exist for different operation skills, it is difficult to generate a general list of such factors. Nevertheless, it is suggested that identification of unique factors would benefit training programs to gain efficiency by either excluding them or developing an intensive battery to train with them.

A general recommendation for skill training derived from the present research is that the context of training should approximate the actual task environments as closely as possible. For instance, although use of virtual reality (VR) and high-fidelity simulator technologies can aid some aspects of training, such technologies are, at their current states, limited in mimicking real contexts. As a consequence, they probably can be used most efficiently to familiarize trainees with specific operations. However, in contrast to the recent popularity of these technologies, a high level of expertise is likely to be achieved more effectively by performing the operations in a simulation that takes place in a physical environment closely related to the actual conditions. Because our research is not specifically designed to investigate this issue, it remains a speculation that future research will need to settle.

Acknowledgements

The research described in the present chapter was conducted for the Multidisciplinary University Research Initiative (MURI) project, *Training Knowledge and Skills for the Networked Battlefield* (http://psych.colorado.edu/~ahealy/MuriFrame.htm), supported by Grant W911NF-051--0153 from the U.S. Army Research Office. We thank Lyle E. Bourne, Jr. and Alice F. Healy for helpful comments on an earlier version of this chapter.

References

Andre, A. D. and Wickens, C. D. (1990). *Display-control Compatibility in the Cockpit: Guidelines for Display Layout Analysis* (Technical Report). University of Illinois Aviation Research Laboratory: Savoy, IL.

Baddeley, A. (1998). Recent developments in working memory. *Current Opinion in Neurobiology*, *8*, 234–38.

Barber, P., and O'Leary, M. (1997). The relevance of salience: Towards an activation account of irrelevant stimulus-response compatibility effects. In B. Hommel and W. Prinz (eds), *Theoretical Issues in Stimulus-response Compatibility* (pp. 135–72). Amsterdam: North-Holland.

Card, S. K., Moran, T. P., and Newell, A. (1983). *The Psychology of Human-Computer Interaction*. Hillsdale, NJ: Lawrence Erlbaum.

De Jong, R. (1995). Strategical determinants of compatibility effects with task uncertainty. *Acta Psychologica*, *88*, 187–207.

Dutta, A., and Proctor, R. W. (1992). Persistence of stimulus-response compatibility effect with extended practice. *Journal of Experimental Psychology: Learning, Memory, and Cognition*, *18*, 801–09.

Fitts, P. M., and Seeger, C. M. (1953). S-R compatibility: Spatial characteristics of stimulus and response codes. *Journal of Experimental Psychology*, *46*, 199–210.

Healy, A. F., Wohldmann, E. L., Parker, J. T., and Bourne, L. E., Jr. (2005). Skill training, retention, and transfer: The effects of a concurrent secondary task. *Memory and Cognition, 33*, 1457–71.

Healy, A. F., Wohldmann, E. L., Sutton, E. M., and Bourne, L. E., Jr. (2006). Specificity effects in training and transfer of speeded responses. *Journal of Experimental Psychology: Learning, Memory, and Cognition*, *32*, 534–46.

Hick, W. E. (1952). On the rate of gain of information. *Quarterly Journal of Experimental Psychology*, *4*, 11–26.

Hommel, B. (1993). Inverting the Simon effect by intention. *Psychological Research*, *55*, 270–79.

Hommel, B., Müsseler, J., Aschersleben, G., and Prinz, W. (2001). The theory of event coding (TEC): A framework for perception and action planning. *Behavioral and Brain Science, 24*, 849–937.

Lu, C.-H., and Proctor, R. W. (1995). The influence of irrelevant location information on performance: A review of the Simon and spatial Stroop effects. *Psychonomic Bulletin and Review, 2*, 174–207.

Monsell, S. (2003). Task switching. *Trends in Cognitive Sciences*, *7*, 1341–40.

Proctor, R. W., and Cho, Y. S. (2006). Polarity correspondence: A general principle for performance of speeded binary classification task. *Psychological Bulletin, 132*, 416–442.

Proctor, R. W., and Lu, C.-H. (1999). Processing irrelevant location information: Practice and transfer effects in choice-reaction tasks. *Memory and Cognition, 27*, 63–77.

Proctor, R. W., and Reeve, T. G. (1985). Compatibility effects in the assignment of symbolic stimuli to discrete finger responses. *Journal of Experimental Psychology: Human Perception and Performance, 11*, 623–39.

Proctor, R. W., and Vu, K.-P. L. (2006a). *Stimulus-response Compatibility Principles: Data, Theory, and Application*. Boca Raton, FL: CRC Press.

Proctor, R. W., and Vu, K.-P. L. (2006b). The cognitive revolution at age 50: Has the promise of the human information-processing approach been fulfilled? *International Journal of Human-Computer Interaction, 21*, 253–84.

Proctor, R. W., and Vu, K.-P. L. (2009). Stimulus-response compatibility for mixed mappings and tasks with unique responses. *Quarterly Journal of Experimental Psychology*. Available at http://dx.doi.org/10.1080/17470210902925270.

Proctor, R. W., Yamaguchi, M., and Vu, K.-P. L. (2007). Transfer of noncorresponding spatial associations to the auditory Simon task. *Journal of Experimental Psychology: Learning, Memory, and Cognition, 33,* 245–53.

Proctor, R. W., Yamaguchi, M., Zhang, Y., and Vu, K. P.-L. (2009). Influence of visual stimulus mode on transfer of acquired spatial associations. *Journal of Experimental Psychology: Learning, Memory, and Cognition, 35*, 434-445.

Simon, J. R. (1990). The effects of an irrelevant directional cue on human information processing. In R. W. Proctor and T. G. Reeve (eds), *Stimulus-response Compatibility: An Integrated Perspective* (pp. 318–6). Amsterdam: North-Holland.

Simon, J. R., Craft, J. L., and Webster, J. B. (1973). Reactions toward the stimulus source: Analysis of correct responses and errors over a five-day period. *Journal of Experimental Psychology. 101*, 175–78.

Smith, M. C. (1967). Theories of the psychological refractory period. *Psychological Bulletin, 67*, 202–13.

Tagliabue, M., Zorzi, M., and Umiltà, C. (2002). Cross-modal re-mapping influences the Simon effect. *Memory and Cognition, 30,* 18–23.

Tagliabue, M., Zorzi, M., Umiltà, C., and Bassignani, F. (2000). The role of LTM links and STM links in the Simon effect. *Journal of Experimental Psychology: Human Perception and Performance, 26*, 648–70.

Vu, K.-P. L. (2007). Influences on the Simon effect of prior practice with spatially incompatible mappings: Transfer within and between horizontal and vertical dimensions. *Memory and Cognition, 35*, 1463–71

Vu, K.-P. L., Proctor, R. W., and Urcuioli, P. (2003). Effects of practice with an incompatible location-relevant mapping on the Simon effect in a transfer task. *Memory and Cognition, 31*, 1461–52.

Wallace, R. J. (1971). S-R compatibility and the idea of a response code. *Journal of Experimental Psychology*, *88*, 354–60.

Welford, A. T. (1960). The measurement of sensory-motor performance: Survey and reappraisal of twelve years' progress. *Ergonomics*, *3*, 189–230.

Yamaguchi, M., and Proctor, R. W. (2006). Stimulus-response compatibility with pure and mixed mappings in a flight task environment. *Journal of Experimental Psychology: Applied*, *12*, 207–22.

Yamaguchi, M, and Proctor, R. W. (2007a, November). *Compatibility for Pure and Mixed Mappings with Discrete and Continuous Stimulus and Response Sets*. Poster presented at the 48th Annual Meeting of the Psychonomic Society, Long Beach, CA.

Yamaguchi, M., and Proctor, R. W. (2007b). Effect of display format on attitude maintenance performance. In R. Jensen (ed.), *Proceedings of the 14th International Symposium of Aviation Psychology* (pp. 7767–81). April 23–26, 2007. Dayton, OH: Write State University.

Yamaguchi, M., and Proctor, R. W. (2008). Transfer of encoding bimanual- and unimanual-key responses. *Manuscript in preparation*.

Yamaguchi, M., and Proctor, R. W. (2009). Transfer of learning in choice reactions: Contributions of specific and general components of manual responses. *Acta Psychologica*, *130*, 11–0.

Zhang, H., Zhang, J., and Kornblum, S. (1999). A parallel distributed processing model of stimulus-stimulus and stimulus-response compatibility. *Cognitive Psychology, 38,* 386–432.

Chapter 3
Measuring Vigilance Abilities to Enhance Combat Identification Performance

Clark Shingledecker
David E. Weldon
Kyle Behymer
Benjamin Simpkins
Elizabeth Lerner
JXT Applications, Incorporated

Joel Warm[1]
Air Force Research Laboratory

Gerald Matthews
Victor Finomore
Tyler Shaw
University of Cincinnati

Jennifer S. Murphy
Army Research Institute

Introduction

Vigilance is the ability to maintain one's focus of attention and remain alert for prolonged periods of time. Tasks that require a high degree of vigilance are an integral part of warfare. In addition to conventional surveillance activities, which base detection on monitoring unprocessed sensory input from the environment, the modern warfighter is likely to engage in computer-generated monitoring tasks associated with the control of missiles, unmanned aerial vehicles, or combat robots, and to perform detection tasks in efforts to counter terrorist activity. Future battlefields are expected to continue to demand that warfighters move and engage rapidly under conditions where friendly forces are intermingled with those of the enemy, and survival depends upon an ability to detect and locate hostile forces quickly (Warm, 1993). In many cases, these activities include the need to monitor sensor-generated signals regarding enemy positions and troop movements, as well as movements of friendly forces in a widely dispersed battlefield scenario where vigilance is the price of survival. In addition to military situations, vigilance is

1 Joel Warm is also a Professor Emeritus from the University of Cincinnati.

a crucial aspect of many civilian activities including air-traffic control, airport baggage inspection, industrial quality control, robotic manufacturing, medical monitoring, and public safety assurance.

Early vigilance research was stimulated by the performance of British airborne radar observers patrolling for enemy submarines. Despite extensive training and motivation, the observers failed with increasing frequency over the course of a watch to detect critical signals. This resulted in submarines going undetected, free to attack allied ships. In efforts to study this problem, Mackworth (1948) developed a simulated radar display called the 'Clock Test.' Using this task, he charted individuals' performance over time and confirmed that the quality of sustained attention is fragile; it wanes over time. This progressive decline in performance over time, labeled the 'decrement function' or the 'vigilance decrement,' has been confirmed in subsequent investigations. Investigations using various tasks indicate that the decline in performance levels off after 20 to 35 minutes from the initiation of the vigil and at least half of the decrement in performance occurs in the first 15 minutes. Under especially demanding circumstances, the decrement can appear within the first few minutes of the watch. As Dember and Warm (1979) have noted, the most striking aspect of this finding is that it seems to result from the necessity of looking or listening for a relatively infrequent signal over a continuous period of time. Since Mackworth began his earliest research, much has been learned about vigilance, its importance, and the characteristics of people with high vigilance tendencies. Much of this research is summarized in Matthews, Davies, Westerman, and Stammers (2000).

Because individuals vary significantly in their ability to perform vigilance tasks effectively, it is important to assign individuals selectively to tasks with high requirements for sustained attention. While the problem of personnel selection for exceptional vigilance and sustained-attention abilities has been researched for years, only limited success has been achieved in predicting individuals' vigilance performance capabilities based on traditional selection tests. More reliable and valid approaches to assessing vigilance aptitudes and capabilities are needed to optimize assignment of personnel and improve overall operational effectiveness in current and future combat environments.

The overarching purpose of the research program described in this paper is to identify a set of individual difference variables that predict vigilance performance, to assemble selection measures to assess these variables, and to validate the measures for use in practical, military personnel-selection settings.

Identifying Individual Differences that Predict Vigilance

Developing a useful solution to the problem of identifying effective individual-difference predictors of vigilance performance requires an examination of past vigilance research findings, their implications for additional research, and the benefits that can be derived from developing a set of predictive vigilance measures that can be easily applied in a routine test environment.

Personality factors Davies and Parasuraman (1982) discussed several personality dimensions related to performance efficiency in vigilance tasks. Included in these are introversion-extraversion, field dependence-independence, internal-external locus of control, and the Type A (coronary-prone) behavior pattern. The findings of these studies indicate that, in general, the performance of introverted observers exceeds that of their extraverted cohorts, field-independent individuals perform better on vigilance tasks than field-dependent observers, individuals with an internal locus of control fare better on a vigilance task than those with an external locus of control, and the vigilance performance of Type-A individuals who are characterized by a rushed, competitive, achievement-oriented life style exceeds that of their more relaxed, Type-B counterparts. Additional research has suggested that boredom-prone individuals may be poorer monitors than those who are less boredom prone (Thackray, Bailey, and Touchstone, 1977). In addition, absent-minded individuals, defined by high scores on the Cognitive Failures Questionnaire, have been found to do more poorly in vigilance tasks than non-absent minded observers and to report higher levels of perceived mental workload than the non-absent minded (e.g., Robertson, Manly, Andrade, Baddeley, and Yiend, 1997). Finally, optimists perform more effectively on vigilance tasks than do pessimists (Helton, Dember, Warm, and Matthews, 1999). Such results indicate that personality profiles should be a part of any approach for developing a vigilance test with reliable predictive features.

Performance sampling as a predictor A second promising source of useful predictors of sustained-attention ability is the objective measurement of an individual's performance on vigilance tasks. Traditional laboratory vigilance tasks require a lengthy watch period that would make them impractical as selection tests for large groups of examinees. However, recent research has shown that brief, highly-demanding vigilance tasks can be constructed that produce performance changes mirroring the vigilance decrements typically observed during long-term vigils (e.g., Matthews, Davies and Lees, 1990; Temple, et al., 2000). These tasks show very rapid perceptual sensitivity decrements over a period of 10 minutes or less, and they demonstrate the key diagnostic indicators of being resource-limited: a sensitivity decrement, high subjective workload, and sensitivity to stress and arousal factors. Thus, a high level of performance on a short task may be a good indicator of aptitude for longer vigilance tasks.

Differences in subjective responses to vigilance task demands Finally, a recent study indicates that the perceived workload of vigilance tasks is quite substantial and that workload grows linearly over time (Warm, Dember, and Hancock 1996) leading researchers (e.g., Johnson and Proctor, 2003) to conclude that, rather than being under-stimulating, vigilance tasks place high information-processing demands upon observers. Thus, resource theories appear to take precedence over Arousal Theory as a model of the factors that control vigilance performance. However, following Kahneman's original model (Kahneman, 1973),

Matthews and Davies (2001) have argued that arousal and resource theories are not necessarily mutually exclusive and that they can be integrated by viewing arousal as the agent responsible for resource production. In support of this view, they have employed a self-report measure of arousal known as energetic arousal (alertness-sluggishness) that correlates with psychophysiological measures of autonomic activity under performance testing conditions. Using this measure, they demonstrated that when confronted with psychophysical challenges designed to increase information-processing load, observers in vigilance tasks who were high on energetic arousal fared better than those who were low on this measure. That is, energetic arousal may provide a marker for resource availability. The finding of an agreement between psychophysiological measures, subjective self-reports, and performance, as predicted by the integrated models, is of significance for selection-test development.

In addition to workload response differences, operators may also differ in the ways that they deploy compensatory effort and coping strategies to adapt to demanding performance environments (Hancock and Warm, 1989). Short tasks are sometimes insensitive to stressor effects, but as time progresses, it becomes increasingly more difficult for the operator to maintain successful coping. Thus, it may be possible to identify useful predictor measures from an operator's reactions to performing a short vigilance task that may offer early warning signs of difficulties in coping. Such difficulties do not necessarily impact performance immediately, but may have adverse effects on a longer, operational task.

There are several indices that may be acquired following performance in order to assess whether the operator is being unduly taxed by task demands. Although workload indices are typically used to discriminate tasks, they may also be used to assess the level of demands perceived by the person. State measures of subjective stress may also indicate the degree of challenge afforded by task performance. As previously mentioned, subjective energetic arousal relates positively to vigilance. Finally, the operator's style of coping with the pressure of the task may facilitate or prevent effective resource deployment. Task-focused coping is preferable to emotion-focused and avoidance coping (Matthews and Campbell, 1998).

A Model for the Development of a Vigilance Selection Test Battery

The challenge presented by the current research problem was to apply the concepts of vigilance and its measurement discussed above to develop a reliable and valid vigilance-prediction toolset. To approach this challenge, a multi-dimensional solution to vigilance prediction was conceived that would sample key constructs related to (1) personality and ability, (2) objective task performance, and (3) stress/workload responses to vigilance tasks. A primary goal was to extract the optimal measurement instruments from these complementary approaches and blend them to produce an efficient, low-cost personnel-selection system capable of predicting vigilance performance. A graphic representation of the 'Vigilant Warrior' personnel-selection battery concept is shown in Figure 3.1.

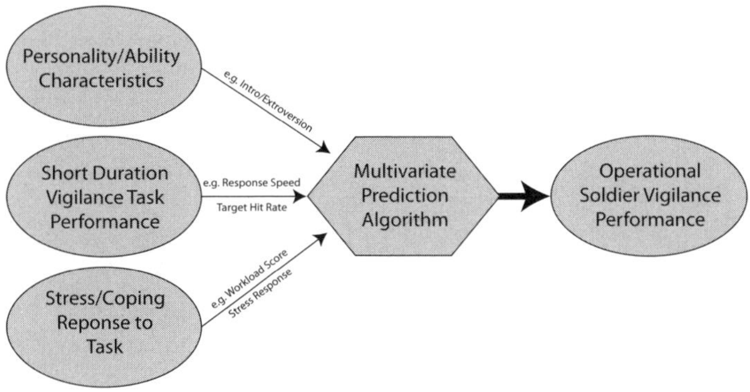

Figure 3.1 Vigilant Warrior model for developing a personnel-selection tool

Preliminary Research

To identify preliminary components for each of the three vigilance-prediction dimensions we evaluated the literature addressing the relationship between various personality and ability variables and vigilance performance. Limitations and strengths of both traditional and more recently explored vigilance predictors were documented, and expert ratings reflecting the degree of research support and projected utility for each dimension were independently assigned by members of the research team. In addition, we examined available brief vigilance tasks that could be included in the battery, subjective rating dimensions and scales that could be used to determine an examinee's perceived workload and coping responses, and attitudes associated with performing a vigilance task. Based on the results of these examinations, we identified a candidate vigilance-prediction battery composed of personality/ability metrics, brief vigilance task performance metrics, and resource depletion and allocation metrics.

The top five rated personality/ability dimensions that have been shown to have some established relationship to vigilance performance were Extraversion–Introversion, Intelligence Quotient, Boredom Proneness, Cognitive Failures, and Conscientiousness. We added the dimensions of Trait Sleepiness, ADHD, Schizotypy, and Propensity to Daydream to this group because recent studies suggest they hold promise for predicting vigilance.

Two versions of a short vigilance task (SVT) were selected for the battery in order to account for the well-known differences in performance and sensitivities to stimulus and environmental variables observed in tasks with comparison stimuli available for classifying an event as a signal or non-signal (simultaneous) and tasks without comparison stimuli (successive). The task is a brief paired-symbol vigilance task. Trials are presentations of letter pairs in any combination drawn from the letters D, O, and backward D. In the simultaneous trials, the signal is any matching pair (e.g., 'DD'). In the successive version, the signal is defined as the occurrence of the pair 'OO.'

The Dundee Stress State Questionnaire (DSSQ), the Coping Inventory for Task Situations (CITS), the NASA TLX workload scale, and the Boles Multiple Resource Questionnaire were selected to assess subject attitudes toward, and responses to, performing the brief vigilance task. Dimensions assessed by these instruments are Task Engagement, Distress, Worry, Task-Focused Coping, Avoidance Coping, Emotion-Focused Coping, Workload, and Multiple Resource Usage.

We conducted the main preliminary investigation in advance of the full-scale battery validation to validate the short vigilance tests, assess the psychometric properties of the candidate personality, intelligence, and stress/attitude/coping measures, and assess their differential ability to predict vigilance performance on the task to be used for the brief vigilance task performance component of the battery. The study was conducted using a sample of 210 participants recruited from psychology classes at the University of Cincinnati.

Method

Participants completed a series of questionnaires and performance-based assessments according to the following sequence: personality tests, intelligence tests, pre-task stress state, 12-minute short vigilance task, and post-task stress state and coping. During the vigilance task, pairs of characters were presented against a masking background at a high event rate. One hundred and five participants performed a simultaneous version of the task requiring a comparative judgment to detect the target. An equal number of participants performed a successive version of the task requiring an absolute judgment to detect the target.

Results

Validity of the short vigilance task One objective of this study was to ensure that the brief vigilance task developed for the battery would show classic performance changes over time that are characteristic of typical longer tasks. Figure 3.2 shows the average number of correct detections made by subjects performing the successive and simultaneous versions of the test over the six continuous 2-minute watch periods. As Figure 3.2 suggests, the short tasks yielded the expected difference between the task conditions ($F_{(1, 208)} = 19.80$, $p < .001$) and a common decrement in performance over the 12-minute watch ($F_{(5,1040)} = 44.74$, $p < .001$).

Factor analysis of the subjective scales A factor analysis was conducted to test whether the initial set of questionnaire scales could be reduced to a smaller number of underlying factors. Analysis of the personality scales showed that these individual difference indicators were intercorrelated. A principal factor analysis was run, followed by an oblique (direct oblimin) rotation. On the basis of the scree test and factor interpretability, a four-factor solution was extracted, explaining 63.7 percent of the variance. Factor loadings are shown in Table 3.1. Factor 1 (labeled Cognitive Disorganization) is defined by various scales linked to disruption of attentional focus, including cognitive failures, mind-wandering and day-dreaming, as well as the O-Life disorganization scale. Factor 2 (Heightened Experience) is defined by O-Life unusual experiences, sensation-seeking, and low internal boredom (i.e., enjoyment of events). It appears to indicate a vivid, excitable mental life. Factor 3 (Sleep Quality) brings together the 3 subscales of the Pittsburgh Sleep Quality Index used in this study. Surprisingly, the other sleepiness index (Epworth) fails to load on this factor. Factor 4 (Impulsivity) contrasts the two impulsivity scales with low premeditation on the UPPS scale (Urgency, [lack of] Premeditation, [lack of] Perseverance, and Sensation Seeking). The factors were intercorrelated, with the highest correlations found between Factors 1 and 4 ($r = .51$) and between 1 and 3 ($r = .44$). Factor 2 was largely uncorrelated with the remaining factors.

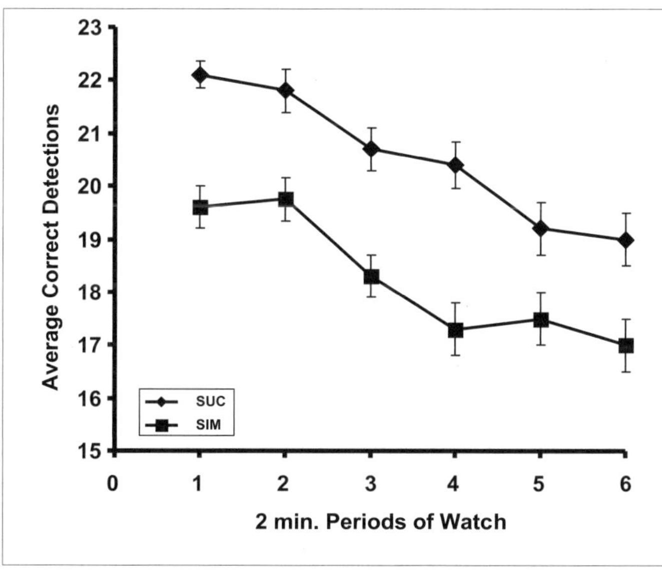

Figure 3.2 **Mean number of correct detections as a function of periods of watch for both simultaneous and successive conditions (error bars are standard errors)**

Table 3.1 Factor analysis of personality traits: Factor pattern matrix (*N* = 210)

	Factor			
Personality Measures	**1**	**2**	**3**	**4**
Cognitive Failures	**.858**	-.091	.123	-.055
O-LIFE Cog. Disorganization	**.839**	-.053	.128	.029
Mind Wandering	**.743**	-.013	-.046	.103
Daydreaming	**.552**	.386	-.070	-.050
Fatigue	**.543**	-.144	.266	.077
ADHD	**.495**	.076	.152	.326
Boredom – External	**.463**	.102	.153	.096
Epworth Sleepiness Scale	**.440**	-.154	-.059	.117
Boredom – Internal	.205	**-.540**	.075	.119
O-LIFE Unusual Experiences	**.404**	**.538**	.091	.012
UPPS Sensation-seeking	-.024	**.485**	.106	.121
Pittsburgh Sleep Quality (2)	-.068	-.001	**.804**	-.051
Pittsburgh Sleep Quality (1)	-.017	.044	**.658**	.031
Pittsburgh Sleep Quality (5)	.116	-.030	**.620**	.003
I7 Impulsivity	.112	.179	-.029	**.845**
UPPS Premeditation	.071	.123	.027	**-.696**
O-LIFE Impulsivity	.157	.184	.162	**.494**

Prediction of performance Satisfied that the SVT possessed the fundamental characteristics of a more classical extended-duration task, we used it as a surrogate to conduct a preliminary exploration of the predictive capabilities of the subjective measures. Pearson correlations were computed to examine the extent to which the personality, intelligence, and stress and coping measures predicted performance on the SVT. Personality was represented by regression-model factor scores computed on the basis of the factor analysis. Detection frequencies within each 2-minute period were highly intercorrelated (alpha = .93), so average target detection frequency was used as the performance measure for this analysis. Table 3.2 provides a summary of the correlations of the various scales with performance, for simultaneous and successive conditions. There were 105 participants in each condition. It can be seen in Table 3.2 that the two measures of intelligence positively correlate with performance on the SVT. The advanced vocabulary test is a better predictor of performance on the successive task, while the letter-sets test correlates with both the simultaneous and successive tasks.

Table 3.2 also indicates that the four personality factors correlate poorly with performance on the SVT.

The best predictors of performance appear to be the subjective stress states and coping style measures. Most notably, pre- and post-task engagement scores correlate highly ($p < .001$) with performance on the simultaneous task, while post-task engagement correlates highly with performance on the successive task. Table 3.2 also suggests that, while simultaneous and successive tasks have some common predictors, the optimal set of predictors for each type of task may differ somewhat.

Table 3.2 Correlations of intelligence and stress variables with performance

Test Type	**Test/Questionnaire**	**Simultaneous**	**Successive**
Intelligence	Advanced Vocabulary	0.084	**0.294**
	Letter Sets	**0.274**	**0.259**
Personality	Cognitive Disorganization	-0.099	-0.089
	Impulsivity	-0.170	-0.132
	Heightened Awareness	0.090	0.048
	Sleep Quality	-0.033	0.077
Stress (pretest)	Engagement	**0.359**	0.122
	Distress	-0.135	-0.089
	Worry	-0.156	-0.152
Stress (posttest)	Engagement	**0.456**	**0.402**
	Distress	*-0.199*	-0.180
	Worry	-0.120	-0.172
Coping	Task Focused	**0.284**	**0.402**
	Emotion Focused	*-0.230*	-0.181
	Avoidance	**-0.429**	**-0.303**

Notes: 1. df = 103; r > 0.19, p < .05, two-tailed test (boldfaced type).
2. *df = 103; r > 0.25, p* < .01, *two-tailed test* (italicized type).

Conclusions

This preliminary study confirmed that the modified versions of the original SVT developed by Temple et al. (2000) show the vigilance decrement characteristic of performance on longer monitoring tasks. Thus, these tasks appear to qualify as the performance-sampling component of the predictive battery. The data also confirm previous findings suggesting that personality traits are, at best, no more than modest predictors of vigilance. However, additional preliminary research has shown that some personality factors have the capacity to predict stress states and coping during vigilance, which may contribute to their utility in the context of performance on a longer, sustained monitoring task. While previous vigilance studies report equivocal findings regarding the intelligence–vigilance association, the present data support inclusion of short intelligence tests in the predictive battery. Finally, consistent with previous findings, both stress states assessed using the DSSQ and coping scales correlated with performance, supporting inclusion of these measures in the vigilance battery. It was found that post-task state measures were correlated more strongly with performance than pre-task measures. This finding may be a consequence of the post-task measures being more representative of states experienced during the task itself. Indeed, the stress response to the SVT resembles that seen with longer-duration tasks. If so, post-task measures taken from the initial SVT may be predictive of a subsequent, longer criterion task. This hypothesis was tested in the validation study for the predictive battery.

The analyses from the preliminary study allowed us to reduce both the number of tests and the number of items on some tests. Table 3.3 shows the tests selected for the validation study and the number of items selected from each test. These tests are being administered along with the SVT and a specially constructed criterion measure in the validation study described below. The DSSQ subscales are administered both as a pre-test and post-test in the validation study.

Criterion Tasks

Two concurrent research efforts are being pursued to validate the ability of the Vigilant Warrior battery to predict vigilance performance using different criterion tasks. In the field validation effort we are using an actual Improvised Explosive Device (IED) detection performance test normally used for the training and operational evaluation of combat threat-detection skills in an outdoor setting. We are collecting data from soldiers and marines as they engage IED-detection problems embedded in several performance lanes implemented by the US Army and Marine Corps. Solvig (2006), for example, provides an unclassified description of these performance lanes, which is available on the Internet.

The primary validation study is being conducted in a controlled laboratory simulation environment. The criterion task developed for this work simulates a real-world, electronic display monitoring environment with high vigilance requirements similar to those that Army operators might use in operational settings. This battlefield situation-monitoring task presents a tactical situation display on a computer monitor, which portrays a two-dimensional plan-view map of a geographical area within which the positions of military combat vehicles are represented. Static components of the display include terrain features and reference grid lines. The dynamic components of the display are moving combat vehicles, the positions of which change with each display update. The symbolic combat vehicles appear in three columns that move from left to right across the screen and return. The movement of all three columns is patterned. Figure 3.3 shows an example of the symbols and background imagery. The center column of the combat vehicles is led by a combat tank with two gun barrels. The display is updated every second with the gun barrels displayed for 50 msec. Participants are asked to respond whenever the gun barrels are of different lengths (simultaneous condition) or are both longer than the standard length (successive condition). In Figure 3.3, the left panel shows the lead tank in the middle column with gun barrels of the standard length (the non-signal event). The right panel in Figure 3.3 shows a 'simultaneous condition' signal event (unequal-length gun barrels). 'Successive condition' signal events occur when the gun barrels are both as long as the long gun barrel in the right panel of Figure 3.3.

Table 3.3 List of tests and items retained in the validation study

Test Name	No. of Items
O-LIFE Unusual Experiences subscale	8
Impulsiveness scale	17
UPPS Sensation Seeking subscale	All
Cognitive Failures Questionnaire Distractibility subscale	All
Pittsburgh Sleep Quality Index	All
ETS Advanced Vocabulary test (Crystallized Intelligence)	All
ETS Letter Sets test (Fluid Intelligence)	All
DSSQ Mood State	29
DSSQ Motivation State Subscale	15
DSSQ Thinking Style Subscale	30
DSSQ Thinking Content Subscale	16
DSSQ General State of Mind Subscale	30
NASA-TLX Workload Scales	6
Opinion of Task questionnaire	20
Task Coping (Dealing With Problems) questionnaire	21

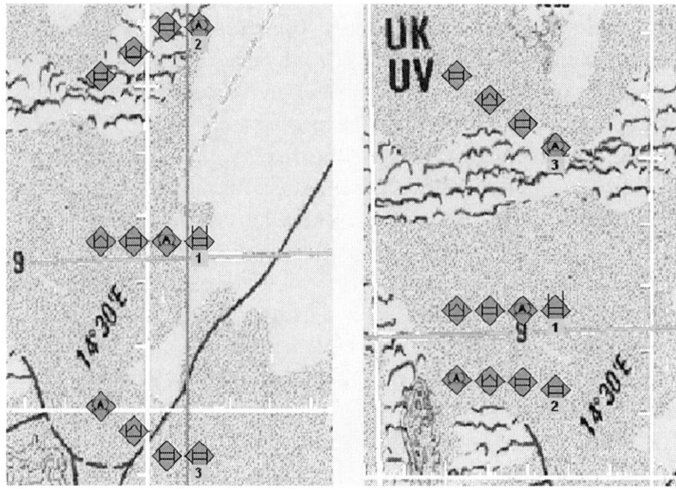

Figure 3.3 Partial screen displays from the battlefield situation-monitoring task showing a non-signal event (left panel) and a simultaneous condition signal event (right panel)

Two versions of the battlefield situation-monitoring task were created to examine the battery's capacity to predict performance under special task conditions and concurrent demands of Army vigilance tasks. The cued version (CUE) captures Army vigilance tasks that are augmented by probabilistic information about potential signal sources. At the heart of the Army's vision for the future soldier is a vast communications and sensor network that will focus on providing both commanders and individual soldiers with near real-time tactical and strategic information to enhance performance and situation awareness. On initial consideration, such a technologically sophisticated information system would appear to resolve fully the classic vigilance problem for the future soldier. Infrequent, perceptually vague signals requiring rapid and accurate responses should be largely eliminated by using combined sensor and intelligence data to predict such events and display them in a salient manner, which should obviate any decrement associated with traditional watch-keeping behavior. However, closer scrutiny of this new perceptual environment reveals, rather than eliminating the vigilance problem, the augmented perception offered by advanced systems may simply shift the problem to a more cognitive domain. The unavoidable problem of enhancing a soldier's information resources is that this information will always be imperfect. Brief (but operationally significant) delays in information arrival, distorted intelligence, data pre-processing that is not always transparent to the receiver, and spoofing of sensors are all probable flaws in any such system that may complicate the detection of significant events. Thus, under some potential circumstances, the soldier's vigilance abilities may be compromised by inaccurate information intended to augment detection. The preceding analysis suggests that the vigilant future soldier will be one who can make effective use of the augmented information to detect significant events, but also maintain an ability to detect rare events that are not spatially or temporally cued by the system. In the cued version of the battlefield situation-monitoring task, display updates include predictors of future signals. These cues are accurate predictions for a majority of signal occurrences. Dependent measures focus on cued vs. uncued signals, and analyses examine the predictive capabilities of the battery for both types of event.

The second version of the battlefield situation-monitoring task (Auditory Secondary Task–AST) represents the common vigilance condition in which the soldier is engaged in tasks in addition to the one of monitoring a visual display for infrequent signals. This condition requires the subject to respond to a secondary auditory display source in order to answer queries about the location of specific vehicles on the map—a task akin to responding to radio messages from collaborators. This additional task increases the mental resource demands imposed upon the subject and permits testing the ability of the battery to predict vigilance performance under multitasking conditions.

Validation Studies

Two validation studies using the Vigilant Warrior battery are completed or nearing completion. We conducted the first at the University of Cincinnati under the direction of Professors Joel Warm and Gerald Matthews. The second validation study was conducted at five military installations located in the United States under the supervision of Dr. Jennifer Murphy of the Army Research Institute (ARI). We describe the two validation studies separately below.

Laboratory Validation at the University of Cincinnati

This study examined the efficacy of the personality variables and the SVT in predicting performance on the simulated battlefield situation-monitoring task described above. Four hundred and sixty-two participants were assigned to one of four criteria conditions as follows: (1) simultaneous or Sim condition (110 participants), (2) successive or Suc condition (122 participants), (3) successive condition with AST or SucAud (108 participants), and (4) successive condition with CUE or SucCue (122 participants). The signal rate was 5 percent in all conditions. Participants in the criterion simultaneous condition received the SVT simultaneous (Sim) task. Participants in the other three criterion groups received the SVT successive (Sim) task. All participants completed the personality tests followed by the assigned SVT condition. They then completed the stress and coping questionnaires. Following a 2-minute practice version of the criterion task, which provided auditory feedback on hits, misses, and false alarms, six 10-minute consecutive periods of watch of the criterion task were presented, and presentation order was fully randomized within each condition.

The analyses showed a strong 'Vigilance Decrement' in the SVT. Figure 3.4 shows the cell means for the two SVT conditions (Sim and Suc); and each 2-minute period using the *d'* measure of perceptual sensitivity (Macmillan and Creelman, 2005). As expected, there were strong main effects for task condition ($F_{(1, 460)} = 62.21$, $p < .001$) and periods of watch ($F_{(5, 2300)} = 36.72$, $p < .001$) in the SVT. As Figure 3.4 clearly shows, the 'Vigilance Decrement' emerged immediately in both task conditions and was reliable by the fourth trial block. The task condition × periods of watch interaction was found to be statistically reliable ($F_{(5, 2300)} = 7.22$, $p < .01$) such that *d'* declined more rapidly in the successive condition.

The analyses also showed a strong 'Vigilance Decrement' in the simulated battlefield situation-monitoring criterion task as well (see Figure 3.5). Figure 3.5 shows the cell means for four conditions (Sim, Suc, SucAud, andSucCue) and each 10-minute trial block, again, using the *d'* measure of perceptual sensitivity. As with the SVT, there were strong main effects for task condition ($F_{(3, 458)} = 14.15$, $p < .001$) and periods of watch ($F_{(5, 2290)} = 51.00$, $p < .001$) in the simulated battlefield situation-monitoring criterion task. As Figure 3.5 clearly shows, the 'Vigilance Decrement' emerged immediately in all four task conditions of the criterion task and was reliable by the second or third period of watch. There was no interaction between the factors. The differing *d'* values suggest that the expected differences in task difficulty were obtained. The cueing manipulation appeared to make the Suc task easier, whereas addition of the auditory secondary task depressed performance (see SucAud means in Figure 3.5).

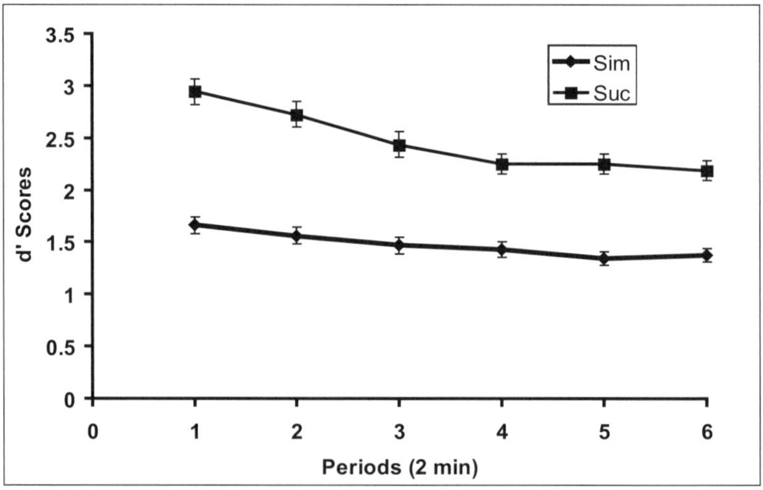

Figure 3.4 Plot of cell means for SVT in validation study

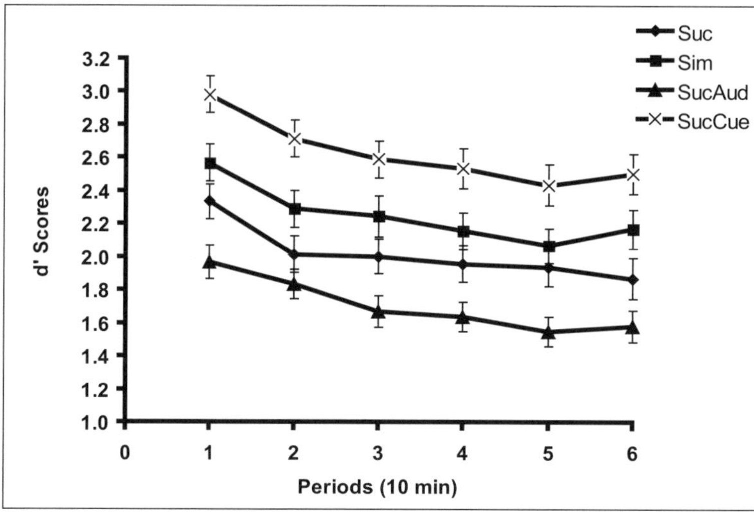

Figure 3.5 Plot of cell means for laboratory-focused criterion task in validation study

In a preliminary test of the efficacy of the SVT in predicting the simulated battlefield situation-monitoring task, we created a derived measure from each. The measure was provisionally called 'Vigilance Skill,' and consisted of the average of the *d'* across all six periods of watch in the SVT and the criterion task. High scores on this measure indicate consistent success on each task. We then tested the hypothesis that high SVT vigilance skill would predict simulated battlefield situation-monitoring task vigilance skill in each of the four conditions of the criterion task. The bivariate correlations between the SVT and simulated battlefield situation-monitoring task vigilance skill measures in each task condition were statistically reliable (Sim: $r = 0.533$, $df = 108$, $p < .01$; Suc: $r = 0.458$, $df = 120$, $p < .01$; SucCue: $r = 0.458$, $df = 120$, $p < .01$; and SucAud: $r = 0.458$, $df = 106$, $p < .01$). Thus, for this derived measure, participants who did well on the SVT were also moderately likely to do well on the simulated battlefield situation-monitoring task.

We next examined the predictive power of the SVT vigilance skill measure in conjunction with the personality, stress, and coping variables in predicting vigilance skill on the simulated battlefield situation-monitoring task using step-wise and multiple regression analysis. The step-wise regression analysis was used to identify the predictors that contributed reliably to vigilance skill scores on the simulated battlefield situation-monitoring task. Multiple regression was then used to establish the regression coefficients for the predictors established by the step-wise regression analysis. Table 3.4 shows the regression coefficients (β weights) for the individual predictors in the final equations for simultaneous and successive vigilance skill on the simulated battlefield situation-monitoring task. SVT vigilance skill and task engagement (a stress reaction) were reliable in both equations, but the contributions of the ability tests differed across the two tasks. The advanced vocabulary test is more predictive of simultaneous performance, whereas Letter Sets is the better predictor of successive performance (perhaps reflecting the role of working memory in both fluid intelligence and successive discriminations in vigilance).

Table 3.4 Regression coefficients (*β*s) for simultaneous and successive performance

Predictor	Simultaneous	Successive
Advanced Vocabulary	.175*	.075
Letter Series	.000	.261**
SVT Vigilance Skill	.407**	.324**
Task Engagement	.216*	.112*

Note: $^*p < .05$, $^{**}p < .01$

These results provide the first concrete evidence to indicate that we have developed a vigilance task with two attractive aspects: (a) the simulated battlefield situation-monitoring task has strong face validity and (b) the findings with this criterion task duplicate those obtained with more abstract laboratory versions. The successive condition manipulation is especially interesting because of the findings for the SucAud and SucCue conditions. Our results indicate that even unreliable cueing in the SucCue condition improved performance while the SucAud reduced performance as expected by increasing the multitasking workload.

Field Validation at Selected Military Installations

To date, we have completed the data analysis for the first of five field studies. Using an abbreviated version of the Vigilant Warrior battery, we collected data from 26 US Army personnel and obtained two measures of IED 'detection lane' performance data on each participant. The first was a 'mounted' condition where participants detected IEDs while riding in a vehicle. The second was a 'dismounted' condition where participants detected IEDs while on foot. Surprisingly, these two criteria measures were unrelated (r = 0.03, df = 24). That is, participants' IED detection scores when observing from a vehicle (mounted condition) were independent of their IED detection scores when walking through a training lane (dismounted condition). The reason for this remains unclear.

Mounted condition We examined the bivariate correlations between the predictors and the mounted IED detection scores (the sample size and the number of predictors precluded regression analysis). The predictors included the SVT vigilance skill measure and post-SVT stress, coping, and workload measures. Two of the predictors, SVT vigilance skill and Task Engagement reached conventional levels of reliability ($r = 0.40$, $df = 24$, $p < .05$ and $r = 0.41$, $df = 24$, $p < .05$, respectively). This result is consistent with the results obtained in the laboratory validation study reported above. The advanced vocabulary and letter sets measures of ability were unrelated to the mounted criterion variable.

Dismounted condition We examined the bivariate correlations between the predictors and the dismounted IED detection scores. The SVT vigilance skill measure was not related to this criterion measure. This result was unexpected, but consistent with the finding that the two criteria measures were unrelated. Surprisingly, some of the post-SVT stress, coping, and workload scales were related to the dismounted criterion. These included the worry stress measure ($r = 0.41$, $df = 24$, $p < .05$), the emotion-focused coping measure ($r = -0.44$, $df = 24$, $p < .05$), the avoidance coping measure ($r = -0.49$, $df = 24$, $p < .05$), and the physical demand workload measure ($r = 0.44$, $df = 24$, $p < .05$). That is, participants who had higher dismounted IED detection scores saw the SVT as physically demanding and were very worried about their SVT performance. On the other hand, they were less likely to get emotional about the SVT task or to avoid it altogether. We are currently exploring the implications of these results.

Future Directions

We continue to analyze the data from the remaining four field studies to see if they are consistent with the laboratory findings. Our next step is to determine the consistent predictors of the field study criterion measures. If so, there is reason to expect that we can provide a reliable and powerful selection tool to the military and to civilian activities, including air-traffic control, airport baggage inspection, industrial quality control, robotic manufacturing, medical monitoring, and public-safety assurance. Because individuals vary significantly in their ability to perform vigilance tasks effectively, the availability of a personnel-selection tool that reliably assigns individuals with exceptional vigilance abilities to tasks requiring sustained attention represents a significant advance in overall operational human effectiveness in current and future combat environments, as well as civilian environments requiring sustained attention.

Furthermore, the development of the Simulated Battlefield Situation Display Monitoring Criterion Task, which duplicates all the characteristics of those obtained with more abstract laboratory versions like the SVT, suggests we can build a selection battery with strong face validity. If it is found that a shortened 12-minute version of the Simulated Battlefield Situation Display Monitoring

Criterion Task exhibits the same characteristics, then such a task may easily substitute for the SVT, which is sensitive to physical display conditions, including ambient light, display brightness, and contrast. If this is the case, we will have developed a selection battery that will be well received by both test administrators and personnel being assessed.

Acknowledgement

This research was supported by a Phase II Small Business Innovation Research grant from the Army Research Institute under contract number W74V8H-06-C-0049. The views, opinions, and/or findings contained in this chapter are those of the authors and should not be construed as an official Department of the Army position, policy, or decision. The authors would also like to thank Robert Shaw for reviewing the original manuscript and Mark Crabtree for his significant contribution as Project Manager during the early stages of the research.

References

Davies, D. R. and Parasuraman, R. (1982). *The Psychology of Vigilance*. London: Academic Press.

Dember, W. N. and Warm, J. S. (1979). *Psychology of Perception* (2nd edn). New York: Holt, Rinehart, and Winston.

Eysenck, H. J. and Eysenck, M. W. (1985). *Personality and Individual Differences: A Natural Science Approach.* New York: Plenum.

Hancock, P. A, and Warm, J. S. (1989). A dynamic model of stress and sustained attention. *Human Factors*, *31*, 519–537.

Helton, W. S., Dember, W. N., Warm, J. S., and Matthews, G. (1999). Optimism, pessimism, and false failure feedback: Effects on vigilance. *Current psychology, 18*, 311–325

Johnson, A., and Proctor, R. W. (2003). *Attention: Theory and Practice.* Thousand Oaks, CA: Sage.

Kahneman, D. (1973). *Attention and effort*. Englewood Cliffs, NJ: Prentice-Hall.

Mackworth, N. M. (1948). The breakdown of vigilance during prolonged search. *Quarterly Journal of Experimental Psychology, 1,* 6–21.

Macmillan, N. A. and Creelman, C. D. (2005). *Detection Theory: A User's Guide* (2nd edn). Mahwah, NJ: Lawrence Erlbaum.

Matthews, G., and Campbell, S. E. (1998). Task-induced stress and individual differences in coping. In *Proceedings of the Human Factors and Ergonomics Society 42nd Annual Meeting* (pp. 821–825). Santa Monica, C A: Human Factors and Ergonomics Society.

Matthews, G., and Davies, D. R. (2001). Individual differences in energetic arousal and sustained attention: A dual-task study. *Personality and Individual Differences, 31,* 575–589.

Matthews, G., Davies, D. R., and Lees, J. L. (1990). Arousal, extraversion, and individual differences in resource availability. *Journal of Personality and Social Psychology, 59*, 150–168.

Matthews, G., Davies, D. R., Westerman, S. J., and Stammers, R. B. (2000). *Human Performance: Cognition, Stress and Individual Differences*. London: Psychology Press.

Robertson, I. H., Manly, T., Andrade, J. Baddeley, B. T., and Yiend, J. (1997). Oops! Performance correlates of everyday attentional failures in traumatic brain injured and normal subjects. *Neuropsychologia, 35,* 747–758.

Solvig, E. (2006, April). Marine base simulates Iraq conditions. *USA Today.* Retrieved from http://www.usatoday.com/news/nation/20060–42–2-marinetraining_x.htm.

Temple, J. G., Warm, J. S., Dember, W. N., Jones, K. S., LaGrange, C. M., and Matthews, G. (2000). The effects of caffeine and signal salience on performance, workload, and stress in an abbreviated vigilance task. *Human Factors, 42*, 183–194.

Chapter 4
Dimensions of Spatial Ability and their Influence on Performance with Unmanned Systems

Thomas Fincannon
A. William Evans III
Florian Jentsch
Joseph Keebler
University of Central Florida

The technology behind unmanned systems has developed over recent years, and this development has increased the potential for impact across a number of different tasks. One set of tasks, such as reconnaissance/surveillance and search/rescue, have a strong identification component and is the primary focus of this chapter. In addition to simply recognizing an observed object, this chapter acknowledges that there are additional factors associated with the operation of unmanned systems that increases the difficulty associated with this task. For example, it is important to remember that operators are physically removed from the search environment, so target identification performance in some tasks is also associated with knowing where a target is located or what objective is associated with that target. Therefore, the research in this chapter focuses on an applied definition of target identification.

As the technology behind unmanned systems develops, more research is needed to advance our understanding of the importance of visual perception, spatial ability, and teamwork in the operation of unmanned systems and combat identification. Presently, field research has been useful in highlighting some perceptual issues associated with identifying targets in remote environments (Casper and Murphy, 2003; Woods, Tittle, Feil, Roesler, 2004). For the first of these, operators have been found to exhibit difficulty with understanding the scale of an environment in relation to the unmanned system under operation, and this lack of situational awareness affects an operator's ability to maneuver through and around obstacles. Second, Woods et al. observed that operators are typically unable to use binocular cues of depth and motion. This hinders the ability to determine the distance between a vehicle camera and an observed object and further increases difficulty associated with navigation. A third issue relates to the fact that vehicle cameras on unmanned systems typically provide a reduced field of view. This limited peripheral vision—often described as looking through a 'soda straw'—affects the operator's understanding of the unmanned system's spatial location and orientation. All of these differences between 'natural' human vision and remote video influence an operator's ability to localize and comprehend visual representations of spatial information for later identification.

Solutions to these perceptual issues can encompass both technological and human-centered approaches, and while technological solutions provide useful tools, this chapter takes a more human-centered approach. One line of research has shown that operator teams identify more targets than individual operators (Rehfeld, Curtis, Fincannon, and Jentsch, 2005) and communication between operator teams is important in the development of situation awareness (Burke and Murphy, 2004). A more recent line of research has found that cognitive abilities, such as spatial ability, are associated with successful target identification (Chen, Durlach, Sloan, and Bowens, 2008; Chen and Terrance, in press). This chapter focuses on the spatial ability of operators and examines how the consideration of cognitive ability can be integrated into existing lines of research.

There are a number of issues that will be discussed regarding spatial ability in this chapter. First, literature surrounding measures and dimensions of spatial ability will be reviewed. Second, there will be a discussion of the effects of spatial ability on individual and team performance. Third, moderating variables will be considered. And finally, a data set will be analyzed and discussed. In reviewing this, the goal is to re-examine spatial ability and how it relates to imagery analysis and performance with unmanned systems.

Spatial Ability

Spatial ability has been found to exhibit both theoretical and practical utility. Measures of spatial ability have not only been found to load highly on general intelligence, but they have also been found to be predictive of performance with pilots, technical trainees, surgeons, mechanical reasoning, and mathematicians (Carroll, 1993; Hegarty and Waller, 2005; Lohman, 1996; McGee, 1979). This practical relationship between spatial ability and performance in technical areas is likely to generalize to performance with unmanned systems.

The most extensive empirical analysis of spatial ability to determine an underlying factor structure and its place in relation to general intelligence was performed by Carroll (1993), who used the term 'visual perception' to describe all spatial and perceptual abilities associated with 'searching the visual field, apprehending the forms, shapes and positions of objects as visually perceived, forming mental representations of those forms, shapes, and positions, and manipulating such representations mentally' (p. 304). Based on Carroll's analysis, sufficient evidence exists to support five primary factors:

- *visualization* (VZ) as the ability to manipulate visually complex patterns;

- *spatial relations* (SR) as the ability to manipulate simple patterns;
- *closure speed* (CS) as the ability to identify obscured patterns that are not known in advance;
- *flexibility of closure* (CF) as the ability to identify obscured patterns that are known in advance;
- *perceptual speed* (P) as the ability to find or compare patterns that are not obscured.

Carroll's analysis provides a useful approach, but it is also important to note that there have been a number of debates within the literature regarding the exact number of factors, how the factors should be defined, and which tests load onto a given factor (Carrol, 1993; Hegarty and Waller, 2005; Lohman, 1996; Lohman, Pellegrino, Alderton, and Regian, 1987; McGee, 1979). One of the major underlying reasons behind this confusion appears to be attributed to the finding that multiple strategies can be used to complete a given test, which hinders the degree to which a given test can represent a specific factor (Carroll, 1993; Hegarty and Waller, 2005; Lohman, et al., 1987). As a result, research in this area continues with attempts to develop more effective tests. One example of this lies in research surrounding the spatial orientation (SO) factor, which has been defined as imagining how a stimulus object will appear from another perspective (Carroll, 1993; Lohman et al., 1987). While earlier research argued for its presence (Lohman et al., 1987; McGee, 1979), Carroll (1993) argued that this factor was difficult to measure and found insufficient evidence to form that distinction with existing tests. Kozhevnikov and Hegarty (2001) responded to this by moving from traditional perspective taking tests (i.e. Guilford-Zimmerman Spatial Orientation) that require a comparison of two 3D pictures of an environment to tests that require a mental understanding of perspective and direction from a two-dimensional array of objects. As hypotheses were supported by these new tests emerging on a separate factor, the domain continues to change, and researchers need to attend to these developments accordingly.

A final consideration relates to how to use tests that load on a single factor of spatial ability. While Lohman (1996) appeared to accept Carroll's (1993) factor structure, he also states that specific variance across tests can be quite large and recommends the use of multiple tests for each factor. In discussing tests that load on the visualization factor, Carroll (1993) classified tests according to the type of task, which may be useful for future research. These included paper foamboard/assembly, block counting, block rotation, paper folding, surface development, perspective, and mechanical movement. Applied research using multiple tests of spatial ability appear to validate this approach. One example of this can be found in research by Pak, Rogers, and Fisk (2006), who examined the relationship between measures of cognitive ability and performance on a computer based information search task. Their results indicated that a block rotation test (Cube Comparison) was predictive of performance, while a paper folding test was not. In a study by Rehfeld (2006), a training manipulation was found to improve performance on measures of block rotation (Guildford-Zimmermann Spatial Visualization and Vandenberg Mental Rotation), but not a traditional perspective test (Guilford-Zimmermann Spatial Orientation; see previous paragraph). As discussed and demonstrated by other researchers (Fincannon, Evans, Jentsch, and Keebler, 2008; Sims and Mayer, 2002; Wiedenbauer, Schmid, and Jansen-Osmann, 2007), this pattern of effects appears to be related to the notion that the applied use of spatial ability tests is highly domain specific, which might also be useful when selecting tests in applied domains.

In summation, there is a long history of research behind spatial ability, and two major trends within this body of research may be useful in predicting relationships. First, there is some degree of consensus on five (Carroll, 1993; Lohman, 1996) or six (Hegarty and Waller, 2005; Kozhevnikov and Hegarty, 2001) major factors within spatial ability that can be used for theoretically driven research. Second, findings around domain specificity add to this by considering how specific tests within a factor relate to a specific area of performance. While both of these should be taken into consideration when examining how spatial ability impacts combat identification with unmanned systems, the analysis presented later in this chapter focuses on the later of these approaches.

Spatial Abilities and Unmanned System Performance

As initially discussed, there is a growing body of research that discusses the influence of spatial ability in the context of unmanned system performance and operations. Biederman (1987) noted that orientation influences ones ability to identify objects, and a defining characteristic of spatial ability is that some are better at mentally resolving differences in orientation than others (Carroll, 1993; Lohman, 1996). Objects that are observed from unmanned systems are typically in unfamiliar orientations, and operators with high spatial ability should best be able to mentally rotate and comprehend these images. As different measures of spatial ability have been found to have different relationships with the use of technology (Pak, et al. 2006), it would be likely to expect this relationship to generalize in some form to target identification with unmanned systems.

Research has indicated that there are differences in the performance of individual operators and operator teams (Rehfeld et al., 2005) and both should be considered separately. One reason for this distinction is that team research has added dynamics, such as communication, that are likely to increase the complexity of the relationships with spatial ability. These added dynamics associated with team research may also offer different approaches for examining the effects of multiple measures of spatial ability. In spite of this potential, questions relating to the relationships between spatial ability and team effectiveness have not received a great deal of attention, and there is more research on the relationships between spatial ability and individual operators.

In addition to the identification of targets, it is also important to consider aspects of localization, such as the learning of and navigation within a new environment. Research has demonstrated that team discussion of an unmanned system's location in an environment is associated with the number of targets that are identified by that team (Fincannon, Keebler, Jentsch, and Evans, 2008). One reason appears to be the separation of an operator from the environment where the vehicle is located, which increases the difficulty associated with localizing targets in that environment. Furthermore, an operator may be able to identify targets correctly, but an inability to assign that target to a location can hinder the operator's ability to identify targets by increasing the chances of confusing the location of friendly and hostile forces.

Individual Operators

One area of performance where a relationship has emerged between spatial ability and individual operator performance is target identification. In a study by Chen et al. (2008), a cube comparison test was used to measure spatial ability of operators who controlled unmanned aerial vehicles (UAVs) and unmanned ground vehicles (UGVs). Whether using a UAV or UGV, Chen and his colleagues found significant correlations between performance on the cube comparison test and the number of hostile forces that were identified and targeted ($r = .46$ and $r = .36$, respectively). A subsequent study by Chen and Terrance (in press) demonstrated that the same test was predictive of the identification and targeting of hostile forces with automation aided target recognition.

A second area where spatial abilities were found to relate to performance involved navigation through remote or virtual environments. Moffat, Hampson, and Hatzipantelis (1998) found that the Vandenberg Mental Rotation test, Guilford-Zimmerman Spatial Orientation test, and Money Road Map test of Direction Sense were all related to the time spent navigating and the number of errors made while attempting to traverse a virtual maze. In a study by Riecke, van Veen, and Bulthoff (2002) participants were tested across multiple measures of homing ability in a virtual environment, and performance on two tests of mental rotation ability (Town and Blobs and Random Triangle tests) were found to have significant relationships with navigation performance. Finally, in a study by Lathan and Tracey (2002), multiple tests of spatial ability were used to create a composite measure that was associated with reduced time on task and navigation errors while using a teleoperated robotic system.

Operator spatial abilities were also related to the degree to which operators learn the environment that a remote system has been navigating. In a study by Fields and Shelton (2006), the participants were asked to judge directional location in an environment while using an aerial or ground perspective of a virtual environment. Tests of spatial ability included the Three Mountain test, Vandenberg Mental Rotation test, Road Map test, and Spatial Perspective test. Not only did each of these tests correlate with performance, but simultaneous regression analyses also indicated that each of these measures contributed uniquely to the prediction of performance with both aerial and ground level perspectives.

In another study by Rehfeld (2006), regression analyses were used to examine the predictive ability of the Card Rotation test, Vandenberg Mental Rotation test, Guilford-Zimmerman Spatial Orientation test, and the Guilford-Zimmerman Spatial Visualization test on the degree to which operators understood their location, while using a UGV. When using an absolute measure of error as distance between a reported location and the actual location, the Guilford-Zimmerman Spatial Visualization test was the only measure to account for unique variance. In contrast, the Vandenberg Mental Rotation test was the only measure to account for unique variance when assessing whether operators could report being on the correct street while operating their UGV.

Moderating Variables

A number of variables have been found to moderate the relationship between an individual operator's spatial ability and performance in remotely operated environments. One moderating variable is the type of remotely operated vehicle, as the relationships of spatial ability and performance have been found to differ between the aerial and ground level perspectives of UAV and UGV operations, respectively. In the study by Chen et al. (2008) described above, participants that performed well on the cube comparison test finished their tasks faster when using a UAV ($r = -.37$). This relationship, however, was not observed for performance with the UGV. Similarly, in the study by Fields and Shelton (2006), while multiple tests of spatial ability correlated with performance, different regression equations and patterns of significance emerged for aerial and ground vehicles. Finally, Diaz and Sims (2003) found an interaction where performance on the Guilford-Zimmerman Spatial Orientation test was associated with subjective ratings of usefulness of an aerial perspective, but not the corresponding ground perspective while navigating through a virtual environment.

Another variable moderating the relationship between spatial ability and performance is the type of visual feedback that operators receive from unmanned systems. In a study by Huk, Steinke, and Floto (2003), an interaction was identified between spatial ability (Tube Figures test) and the visual presentation of educational material, on positive/negative impressions of that educational material. Specifically, the effects of including 3D media in educational material were only observed for participants with high ability had more positive impressions of training material that had 3D media than when 3D media was absent. Similarly, Lathan and Tracey (2002) manipulated operator feedback such that operators received either: (a) video feedback only, (b) video and vibrotactile feedback, (c) video and audio feedback, or (d) video, audio, and vibrotactile feedback. Their composite measure of spatial ability was related to time on task for all conditions, except for the one in which operators received video and vibrotactile feedback. When examining the relationship between the composite measure of spatial ability and the number of task related errors, a significant relationship was only observed for operators in the condition that had both audio and video feedback.

Operator Teams

The introduction of operator teams into unmanned system operation supplies new outlets and research questions to not only improve the understanding of how spatial abilities influence performance, but also to improve the understanding of spatial abilities in general. Fincannon, Evans, et al. (2008) conducted one such study. In this study, UAV and UGV operators collaborated during a reconnaissance task to identify targets. The reconnaissance task was followed by a workload assessment with the NASA TLX. As aerial perspectives had been found to be associated with better understanding of remote environments (Chadwick, 2005; Diaz and Sims, 2003), it was hypothesized that the UAV operator would be better suited to provide navigation support to a UGV operator, and that this relationship should be influenced by the operator's spatial ability. In team relationships, some UAV operators may provide more support than others, and an interaction was hypothesized to exist between the spatial ability of the UAV operator and the degree to which that operator provided navigation support in the prediction of UGV operator workload. For this analysis, both the Guilford-Zimmerman Spatial Visualization and Guilford-Zimmerman Spatial Orientation tests were used as measures of spatial ability. While results indicating that a correlation of .63 existed between these two measures were not surprising, the results of the regression analyses provided different patterns of results. The Guilford-Zimmerman Spatial Visualization score of the UAV operator was only related to UGV teammate workload when the UAV operator provided high amounts of navigation support, where higher scores were associated with higher workload for the UGV operator. With the Guilford-Zimmerman Spatial Orientation score, this pattern was reversed such that navigation support from a UAV operator only decreased UGV operator workload when that UAV operator performed well on the test. Considering the high correlation between these two measures of spatial ability, this pattern of results may seem unexpected, but this does provide support for hypotheses around domain specificity of spatial ability measures.

An Empirical Analysis

Previous research has focused on performance, workload, and situation awareness as outcome measures that are associated with spatial ability. In light of the different questions that can be asked with team research, this analysis extends the existing research by focusing on how spatial ability influenced the actual communication of visual information from one teammate to another.

Similar to the study conducted by Fincannon, Evans, et al. (2008), an assumption in this analysis is that a UAV operator is best suited to provide information of the remote environment to a UGV operator. This analysis focused on the UAV operator's description of the UGV's remote environment to the UGV operator. By highlighting the specific differences in how spatial abilities influence team communication, it may be possible to obtain a better understanding of how different spatial abilities can have different influences on team target identification.

This analysis also considered the effect of visual feedback as one of the moderating variables discussed above. Kraut, Fussel, and Siegel (2003) found that the addition of shared visual information has a main effect on reducing the verbal communication between teammates. In the context of unmanned system operation, a UAV operator that shares a video feed with the UGV operator may be in a better position to understand and anticipate the obstacles, questions, and issues of the teammate, which would increase the location support that is provided by the UAV operator to the UGV operator. If spatial ability is associated with a better understanding of visual information, it is likely that the ability of the UAV operator interacts with a video manipulation. Specifically, the spatial ability of the UAV operator should be associated with more UAV-to-UGV location support when teammates share video feeds.

Method

Participants Data from 82 students from the University of Central Florida who participated in a larger study were used in this analysis. The larger study included more between subject conditions, but for the purposes of this analysis only two conditions of the original study were used. The participants were randomly assigned to the role of being a UAV operator or a UGV operator to create a total of 41 dyad teams.

Design Teams were physically separated in the study in accordance with their team role as a UAV or UGV operator. The primary manipulation was to the interface that teammates used when communicating with each other. In one condition, participants only had access to an instant messaging program for communication, and in the other condition, participants were able to observe the video from their teammate's camera while using the instant messaging software. Spatial ability was measured using the Guilford-Zimmerman Spatial Visualization (block rotation) and Guilford-Zimmerman Spatial Orientation (perspective) tests, which both load on the visualization factor of spatial ability (Carroll, 1993). Regression analyses were used to test main effects of spatial ability and video (shared, not shared), as well as the interaction between the two variables. All predictors in this regression analysis were centered.

Testbed A scaled military operations in urban terrain (MOUT) environment (described by Ososky, Evans, Keebler, and Jentsch, 2007) was used. Participants were able to use an instant messaging program that had satellite representation of the facility in a shared workspace to collaborate for the planning and execution of missions across both manipulations. According to the video manipulation, a UGV operator could see video from the UGV only (not shared) or video from the UGV plus the UAV (shared), and the UAV operator could see video from the UAV (not shared) or the UAV plus UGV (shared).

Process measures Three raters coded communication between the UAV and UGV operators to produce two team process measures, and the two averages of these ratings were used in data analysis. One process measure was the number of statements that a UAV operator used to describe where the UGV was located in the remote environment (ICC = .81; $\alpha = .93$). The second team process measure was the number of times that a UGV operator requested location support from the UAV operator (ICC = .93, $\alpha = .98$).

Results

Descriptive statistics and intercorrelations are reported in Table 4.1, and not surprisingly, an extremely high correlation between measures of spatial visualization and spatial orientation was observed (r = .62). Tables 4.2 and 4.3 report the hierarchical regressions of shared video, spatial ability, and the interaction term with these two variables. The outcome of interest was the number of statements that a UAV operator used to describe the environment of the UGV to the UGV operator. When the regression equations used the block rotation test or the perspective test, the final steps of both analyses were significant, $F(4,36) = 22.08, p < .001$ and $F(4,36) = 18.63, p < .001$, respectively. For the regressions using block rotation, the final step of the analysis revealed an interaction between shared video and block rotation ($sr^2 = .04$) in the prediction of the amount of location support from a UAV operator to a UGV operator.

Table 4.1 Correlation matrix with means and standard deviations ($N = 41$)

	1	2	3	4	5
1. UAV to UGV location support	1.0				
2. UGV requests for location support	.81*	1.0			
3. UAV block rotation test	.27	.26	1.0		
4. UAV perspective test	.22	.23	.62*	1.0	
5. Video manipulation	-.21	-.17	.16	.04	1.0
Mean	.54	.54	16.59	17.63	1.49
Standard deviation	1.02	1.24	7.76	8.88	.51

* $p < .01$

Table 4.2 Block rotation hierarchical regression analysis

Variable	STEP 1 (β)	STEP 2 (β)	STEP 3 (β)
UGV requests for Support	.81**	.78**	.84**
Shared video manipulation		-.09	-.08
UAV block rotation (rotation)		.08	-.05
Rotation x video			.20*
R^2	.66**	.67**	.71**
ΔR^2		.01	.04*

* $p < .05$, ** $p < .01$

Table 4.3 Perspective taking hierarchical regression analysis

Variable	STEP 1 (β)	STEP 2 (β)	STEP 3 (β)
UGV requests for support	.81**	.79**	.80**
Shared video manipulation		-.08	-.08
UAV perspective test		.04	-.03
Perspective x video			.08
R^2	.66**	.67**	.67**
ΔR^2		.01	.01

* $p < .05$, ** $p < .01$

As illustrated in Figure 4.1, the addition of shared video only decreased location support from UAV operators with low spatial ability. When the ability of a UAV operator was high, location support from that UAV operator had a small increase. In spite of the high correlation, an interaction between the perspective measure of spatial ability and shared video was not observed.

Given the discussion of high correlations throughout this chapter, an analysis of multicolinearity, or the correlation between one predictor and the set of remaining predictors, is warranted. One method of measuring multicolinearity in regression (Cohen, Cohen, West, and Aiken, 2003) involves the use of the squared multiple correlation between one predictor and all other predictors in a regression equation to create a Variance Inflation Factor (VIF), which provides an indicator as to how the standard error would change with respect to the complete absence

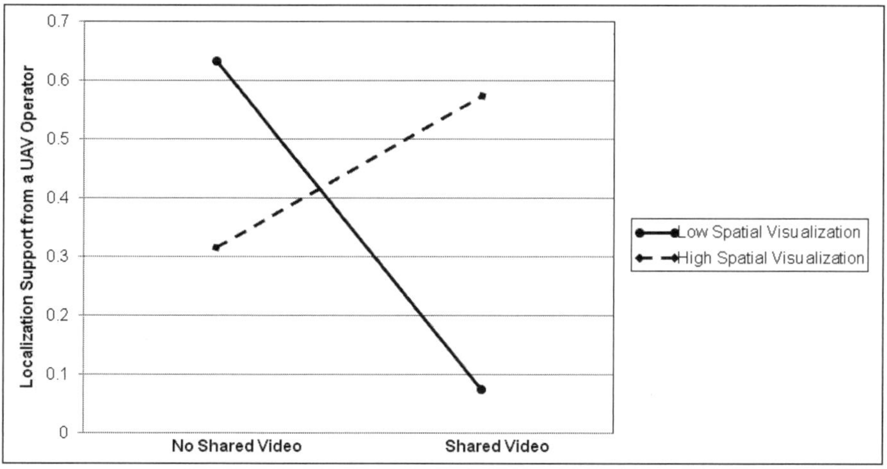

Figure 4.1 Plot of the interaction between shared video manipulation and UAV spatial orientation on UAV

of multicolinearity. A traditional value for the threshold is set at 10, but more conservative thresholds for the VIF have been set at 6. In step 3 of Table 4.2, the VIF for requests for support (VIF = 1.25), video manipulation (VIF = 1.09), block rotation (VIF = 1.16), and the interaction term (VIF = 1.12) were under traditional and conservative thresholds for issues with multicolinearity. For Table 4.3, the VIF for requests for support (VIF = 1.11), video manipulation (VIF = 1.04), perspective (VIF = 1.07), and the interaction term (VIF = 1.01). The inverse of the VIF is a measure of tolerance, which provides an indication of the amount of independent variance for a predictor and ranges from .80 to .99 in this analysis.

Discussion

In this analysis, effects were only found in association with a block rotation test, but not perspective taking test. According to this interaction, UAV operators with lower scores on the block rotation test provided less location support to UGV operators when video feeds were shared. For UAV operators that were high on this measure, the video manipulation had a moderate increase in this type of communication. Had the perspective test been the only measure used, no effect would have been observed. This pattern of effects highlights the need for attending to the measures of spatial ability that a researcher uses in a given analysis. If spatial ability has never been considered in a given context for example, the use of multiple measures to assess multiple dimensions of spatial ability would provide the most accurate assessment as to whether and how spatial ability truly factors into performance. Failure to use of a variety of measures may result in a data analysis that misses important information.

As both of these tests load on the visualization factor, interpretation of these findings might be best guided by understanding differences associated with the domain specificity of applying the specific task of a test to a specific dimension of performance. One question here would involve asking why an effect would emerge around block rotation and the sharing of different visual perspectives. Remembering that the UAV operator primarily uses a top-down view point, the video manipulation adds ground level view point and increases the amount of visual information for this operator to comprehend. If UAV operators are particularly bad at rotating blocks, they might experience difficulty associated with mentally rotating back and forth between perspectives. This difficulty in reconciling differences between different view points would then appear to decrease the degree to which these operators can support their teammates. Operators that are good at three-dimensional rotation are better able to deal with the additional visual imagery. The perspective test involves the comparison of two ground level images. As this would not help a UAV operator interpret top-down imagery, the absence of an effect in this analysis is less surprising.

If one continues with this logic, the perspective test would be expected to relate most to a UGV operator's performance, and by using a one-tailed correlation, the UGV operator's score on this test was significantly related to the team's ability as a whole to correctly report targets on the corresponding reconnaissance task ($r = .27$), and no other measure for the UGV or UAV operator here was significant. This means that only one measure of spatial ability was predictive of one type of performance for a UAV operator, while another measure of spatial ability was predicted another type of performance for a UGV operator. As this highlights how different tasks and abilities are associated with different roles and viewpoints with unmanned systems, care should be given when addressing relevant questions.

Concluding Remarks

As discussed in this chapter, there are a number of differences between different measures of spatial ability. While the research used to differentiate constructs of spatial ability remains somewhat unfinished, the use of multiple measures of spatial ability can provide interesting results. Differences between correlations can be small and require the use of more sophisticated research methodology, but multiple effects can have the potential to be compounded into larger effects. Future research focusing on spatial ability might truly benefit from taking these points into consideration.

Dimensions of spatial ability can also be applied to selection and training. The first step in this process would require researchers to develop an understanding of the measures and dimensions of spatial ability that are most related to target identification with unmanned systems and how this can change from system to system. Once this understanding has been developed, relevant measures can be included or excluded for the selection of personnel that are best suited for that task.

Remembering that training can be used to improve spatial ability (Rehfeld, 2006), relevant measures can also be used to test the efficacy of training interventions.

Even as differences have been identified between measures of spatial ability, this does not necessarily mean that every researcher should try to use measures of every construct of spatial ability. High intercorrelations between spatial ability constructs carries the implication that effects can be found regardless of the construct that the test is intended to capture. Great attention to and selection of these constructs, however, might aid researchers in highlighting specific relationships.

Research with teams of unmanned vehicle operators must also address the way in which individual abilities need to be indexed and combined at the team level. For example, is the spatial ability of the most spatially able team member what determines the performance of the team or is it that of the least spatially able team member? Alternatively, is performance predicted by the total amount of spatial ability in the team (i.e., the sum of the team members' spatial abilities) or by the difference in spatial ability between the most and least able team member? In each case, a different index would have to be used to mark the team's spatial ability. Future research must therefore be mindful not only of the different dimensions of spatial abilities and how to measure them, but also of the ways in which individual-level ability variables must be combined to index the team's ability.

References

Biederman, I. (1987). Recognition-by-components: A theory of human image understanding. *Psychological Review,* 94, 115–147.

Burke, J. L., and Murphy, R. R. (2004). Human-robot interaction in USAR technical search: two heads are better than one. *13th IEEE International Workshop on Robot and Human Interactive Communication,* 13, 307–312.

Carroll, J. (1993). *Human Cognitive Abilities: A Survey of Factor Analytical Studies.* New York: Cambridge University Press.

Casper, J., and Murphy, R. R. (2003). Human-robot interactions during the robot-assisted urban search and rescue response at the World Trade Center. *IEEE Transactions on Systems, Man and Cybernetics, Part B,* 33, 367–385.

Chadwick, R. (2005). The impacts of multiple robots and displays views: An urban search and rescue simulation. *Proceedings of the 49th Annual Meeting of the Human Factors and Ergonomics Society*, 49, 387–391.

Chen, J., Durlach, P., Sloan, J., and Bowens, L. (2008). Human robot interaction in the context of simulated route reconnaissance missions. *Military Psychology*, 20, 135–149.

Chen, J. and Terrance, P. (in press). Effects of imperfect automation and individual differences on concurrent performance of military and robotics tasks in a simulation multitasking environment. *Ergonomics.*

Cohen, J., Cohen, P., West, S., and Aiken, L. (2003). *Applied Multiple Regression Correlation Analysis for the Behavioral Sciences* (3rd edn). Mahwah: Lawrence Erlbaum Associate.

Diaz, D., and Sims, V. (2003). Augmenting virtual environments: The influence of spatial ability on learning from integrated displays. *High Ability Studies,* 14, 191–212.

Eliot, J. and Smith, I. (1983). *An International Directory of Spatial Tests*. Windson: NFER-Nelson.

Fields, A, and Shelton, A. (2006). Individual skill differences and large-scale environment learning. *Journal of Experimental Psychology,* 32, 506–515.

Fincannon, T, Evans, III, A.W., Jentsch, F, and Keebler, J. (2008). Interactive effects of backup behavior and spatial abilities in the prediction of teammate workload using multiple unmanned vehicles. *Proceeding of the 52nd Annual Meeting of the Human Factors and Ergonomics Society,* 52, 995–999.

Fincannon, T., Keebler, J, Jentsch, F., and Evans, III, A.W. (2008). Target identification support and location support among teams of unmanned system operators [Electronic version]. *Proceedings of the 26th Annual Army Science Conference.*

Hegarty, M. and Waller, D. (2005). Individual differences in spatial ability. In P. Shah and A. Miyake (eds), *The Cambridge Handbook of Visuospatial Thinking* (pp. 121–169). New York: Cambridge University Press.

Huk, T., Steinke, M., and Floto, C. (2003). The influence of visual spatial ability on the attitude of users towards high-quality 3D-animations in hypermedia learning environments. *Proceedings of E-LEARN,* 1038–1041.

Kozhevnikov, M., and Hegarty, M. (2001). A dissociation between object manipulation spatial ability and spatial orientation ability. *Memory and Cognition,* 29, 745–756

Kraut, R., Fussel, S., and Siegel, J. (2003). Visual information as a conversational resource in collaborative physical tasks. *Human Computer Interaction,* 18, 13–49.

Lathan, C., and Tracey, M. (2002). The effects of operator spatial perception and sensory feedback on human-robot teleoperation performance. *Presence: Teleoperators and virtual environments,* 11, 368–377.

Lohman, D. (1996). Spatial ability and g. In I. Dennis and P. Tapsfield (eds) *Human Abilities: Their Nature and Measurement* (pp. 97–116). Mahwah: Lawrence Erlbaum.

Lohman, D., Pellegrino, J., Alderton, D., and Regian, J. (1987). Dimensions and components of individual differences in spatial abilities. In S. Irvine and S. Newstead (eds), *Intelligence and Cognition: Contemporary Frames of Reference* (pp. 253–312). Dordrecht: Martinus Nijhoff.

McGee, M. G. (1979). Human spatial abilities: Psychometric studies and environmental, genetic, hormonal, and neurological influences. *Psychological Bulletin*, 86, 889–918.

Moffat, S., Hampson, E., and Hatzipantelis, M. (1998). Navigation in a 'virtual' maze: Sex differences and correlation with psychometric measures of spatial ability in humans. *Evolution and Human Behavior,* 19, 73–87.

Ososky, S., Evans, A. W., III, Keebler, J. R., and Jentsch, F. (2007). Using Scale Simulation and Virtual Environments to Study Human-Robot Teams. *Proceedings of the 4th Annual Conference of Augmented Cognition International,* 4, 183–189.

Pak, R., Rogers, W., and Fisk, D. (2006). Spatial ability subfactors and their influences on a computer based information search task. *Human Factors,* 48, 154–165.

Rehfeld, S. (2006). *The Impact of Mental Transformation Training Across Levels of Automation on Spatial Awareness in Human-robot Interaction.* Unpublished Dissertation, University of Central Florida, Orlando, FL.

Rehfeld, S., Jentsch, F., Curtis, M., and Fincannon, T. (2005). Collaborative Teamwork with Unmanned Vehicles in Military Missions [Electronic version]. *Proceedings of the 11th Annual Human-Computer Interaction International Conference.*

Riecke, B., Van Veen, H., and Bulthoff, H. (2002). Visual homing is possible without landmarks: A path integration study in virtual reality. *Presence: Teleoperators and Virtual Environments,* 11, 443–473.

Sims, V.K. and Mayer, R.E. (2002). Domain specificity of spatial expertise: The case of video game players. *Applied Cognitive Psychology,* 16, 97–115.

Wiedenbauer, G., Schmid, J., and Jansen-Osmann, P. (2007). Manual training of mental rotation. *European Journal of Cognitive Psychology,* 19, 17–36.

Woods, D. D., Tittle, J., Feil, M., and Roesler, A. (2004). Envisioning human-robot coordination in future operations. *IEEE Transactions on Systems, Man and Cybernetics, Part C,* 34, 210–218.

SECTION 2
Visual Discrimination

This section focuses on a specific cognitive process integral to combat identification—visual discrimination. Visual discrimination is important both on the field where the need to identify entities as friend, foe, or neutral arises and in the design of visual displays used to represent the battlespace. Visual discrimination is also relevant to the production of uniforms and military vehicles designed for concealment and the detection of enemies seeking to conceal their locations.

Two types of misidentification errors may occur (Keebler, Sciarini, Jentsch, and Nicholson, Chapter 7). The inaccurate identification of a target as an enemy, a false positive, may lead to fratricide or collateral damage. The inaccurate identification of a target as a friend, a miss or false negative, may lead to a missed opportunity to engage the enemy and could have grave consequences depending on the actions of the enemy. Both types of misidentification errors can have devastating consequences. Understanding the operation of the human visual system and related human factors design principles is central to improving combat identification decision accuracy and reducing misidentification.

In two experiments using the dual-process theory of automaticity, Cleotilde Gonzalez, Rick P. Thomas, and Poornima Madhavan (Chapter 5) test the extent to which a successful visual search of a complex image requires consistency between the representations of targets and methods of response. Based on their findings, the authors offer insights on how to design complex visual displays. Vincent A. Billock, Douglas W. Cunningham, and Brian H. Tsou (Chapter 6) present the results from a study on the discrimination of static and dynamic fractal images displayed under various guises. Their findings have implications for the design of camouflage as well as for detecting camouflaged objects. To explain why misidentification problems occur, Joseph R. Keebler, Lee W. Sciarini, Florian Jentsch, Denise Nicholson and Thomas Fincannon (Chapter 7) integrate three fundamental cognitive theories (Feature Integration Theory, Recognition by Components Theory, and Working Memory Theory) into an explanatory model of combat identification. Next, the researchers describe two studies that compare the utility of using three-dimensional models over two-dimensional cards containing line drawings commonly used by the military for training. Based on a familiarity with how the human visual system processes images, Scott H. Summers (Chapter 8) describes principles behind the design of a visual display that can be quickly and accurately understood. Summers uses the design principles to discuss the development of human machine interfaces used in combat.

Chapter 5
The Effects of Conjunctive Search and Response Mappings on Automatic Performance in a Complex Visual Task

Cleotilde Gonzalez
Carnegie Mellon University

Rick P. Thomas
University of Oklahoma

Poornima Madhavan
Old Dominion University

Complex tasks across fields such as aviation, military, and healthcare require operators to develop highly skilled and automatic levels of performance in response to critical stimuli in the environment. This research extends the findings from the dual-process theory of automaticity by considering the effects of two aspects in a complex visual search task: the stimulus mapping and the response mapping. In realistic visual search tasks, targets are often defined by a combination of cues needed for search (e.g., altitude and speed), and responses are often diverse for the same stimulus (i.e., destroying a target with two different weapons). Results from our experiments indicate that variability of mapping, at both the stimulus and the response sides, results in decreased performance and higher detection time. When all the cues were variably mapped, performance deteriorated compared to situations where at least one of the cues was consistently mapped. These results have implications for designing complex visual systems and training individuals in their use.

Introduction

One of the key contributions to the theoretical development of automaticity is the dual-process theory (Schneider and Shiffrin, 1977; Shiffrin and Schneider, 1977). A classic finding from this theory is that automatic processing develops with extended practice when targets are consistently mapped (they are always targets and never appear as distractors). Such automatic processing is assumed to be fast and parallel, requiring little attention or awareness. Thus, performance under automatic processing is unaffected by workload. On the other hand, under varied mapping conditions wherein stimuli may be targets in one instance but distractors in another, performance occurs under controlled processing, which is voluntary, serial, requires attention, and is significantly affected by workload.

Several complex tasks require participants to develop highly skilled and automatic levels of performance to achieve fast and accurate responses to critical stimuli in the environment. In visual search tasks ranging across fields such as aviation, military, and healthcare, automatic detection processes are relevant to situations such as a pilot detecting the presence of an enemy aircraft among friendly aircrafts, a physician detecting the presence of a tumor in x-rays, or a luggage screener detecting the presence of a hidden weapon among objects in passenger luggage. The development of automaticity is particularly important for these complex visual search tasks, because time pressure and workload may make these tasks more difficult to perform in the absence of practiced skills.

The general guidelines for the design of these tasks drawn from the dual-process theory of automaticity (Schneider and Shiffrin, 1985) are that optimal detection and response times can be obtained by keeping targets consistently mapped. In real-world complex tasks, however, it is hard to determine how to achieve consistent mapping. Real-world tasks are replete with multiple and heterogeneous stimuli, often requiring that we make decisions based on varied or inconsistent conditions. For example, in the field of military aviation, pilots must be able to distinguish the attributes of friendly aircrafts from those of enemy aircrafts in order to select the appropriate response. The nature of visual search is often *conjunctive*: successful detection depends on the combination of multiple cues. For example, the detection of an enemy aircraft may depend on several cues such as size, color, model, altitude, and speed, and it may be classified as belonging to one of multiple categories of targets (e.g., very aggressive, moderately aggressive, minimally aggressive). Also, the responses available for a target in these tasks are varied. For example, an enemy aircraft may be tackled with multiple actions such as 'ignore,' 'attack,' or 'destroy.'

Our goals in this research are to test the dual-process theory of automaticity in a complex visual search task and to determine how the consistency of mapping provides guidelines for the design of visual targets and responses. First, we discuss stimulus and response mappings and the characteristics of complex visual tasks. Next, we present two experiments to test the extent to which consistency of mapping is needed in both the stimulus and the response. Finally, we discuss the results of the experiments and the implications for design and training.

Stimulus and Response Mappings

The complexity of visual search tasks can be determined by mapping the stimulus and mapping the response. Traditionally, experiments demonstrating the dual-process theory of automaticity used simple targets to demonstrate mapping; targets are often defined by one visual cue. For example, Shiffrin and Schneider (1977) used letters or numbers as the targets to be consistently mapped throughout their experiments. In this research we use more complex visual targets defined by more than one visual cue and by the combination of multiple cues. Also, traditional experiments demonstrating the dual-process theory have used one response type to indicate the detection of a target. For example, pressing a space bar upon finding a letter on the screen. Again, in the real world the response modes are not that simple, and in this research we used varied responses after the detection of the visual target.

Thus, the concept of *consistency of mapping* from the dual-process theory of automaticity can be defined in multiple ways in complex visual search tasks: stimulus consisting of multiple cues (*conjunctive search consistency)* and multiple responses (*consistency of responding)*. These two types of consistency have been previously identified in the literature of automaticity in generic task contexts. For example, Fisk and Schneider (1984) investigated whether both consistency of attending and responding (i.e., total task consistency) are necessary for the development of automatic detection in a memory task. They found that performance was better when stimulus mapping was consistent, but consistent response mapping had a negligible processing benefit in their studies. They concluded that response mapping was not necessary for the development of automatic processing and that response mapping was less crucial for skill acquisition than stimulus mapping.

Similarly, Kramer, Strayer, and Buckley (1991) showed a benefit for consistent-attending conditions rather than for varied-attending conditions with practice, but no interaction with response mapping was present. They concluded that consistency of responding was not necessary for the development of automatic detection and that the processing costs of inconsistent responding were small.

Past findings indicate that total task consistency (both consistency of stimulus and response mappings) is not vital for the development of automatic processing (Durso, Cooke, Breen, and Schvaneveldt, 1987; Fisk and Schneider, 1984; Kramer, et al., 1991). However, we believe that consistency in stimulus and response will be important in tasks that are more complex than those used in past research. The reason for this can be illustrated with the following example. In the military field, soldiers often face situations where they need to choose a course of action based on the perceived severity of the threat posed by an enemy aircraft. Specifically, they may have to decide whether to use guns, bombs, or lasers to attack the enemy depending on whether the hostile aircraft needs to be destroyed, captured, or merely 'frightened away.' The challenge in a situation such as this arises from the fact that an enemy aircraft might appear similar to friendly aircrafts on more than one occasion and the required response varies depending on the pilot's assessment of the situation at hand. The need to distinguish targets defined by conjunctive cues and then to respond to a stimulus in multiple possible ways poses a significant challenge to operators in complex systems. These are the characteristics we defined in the Radar task used in two experiments.

The Radar Task

Most automaticity research involves simple and static tasks (items to be identified do not change in time and space), but in the experiments presented in this paper we used a complex visual search task with real-world counterparts called Radar. The Radar task is a single-user visual search and decision-making task in which the goal is to detect and eliminate a hostile enemy aircraft by selecting an appropriate weapon system. Radar is similar to military target visual detection devices, in which a moving target needs to be identified as a potential threat and a decision is made on how to best tackle the target under time constraints. The development of this task was inspired by the theatre defense program (Bolstad and Endsley, 2000) and the task has been used in other research for automaticity (Gonzalez and Thomas, 2008) and training (Young, Healy, Gonzalez, Dutt, and Bourne, 2008).

The Radar task has two components: (1) visual search and (2) decision-making. Due to the nature of the two experiments reported in this chapter, only the visual search component of Radar is described here. In the methods section of each of the experiments, we explain how this component of Radar was adapted to address the two manipulations studied here. A complete description of Radar including the decision-making component can be found in Gonzalez and Thomas (2008).

The visual search component of Radar requires the user to memorize a set of targets and then look for the presence of one or more targets in consecutive radar grids. In each of the radar grids, four blips appear in the four corners of the screen: Northwest, Northeast, Southwest, and Southeast (see Figure 5.1). The blips move from the four corners to the center of the screen, where they collide and disappear. During this time, the player needs to determine if a member of the memory set is present or not on the Radar screen. That is, the detection of an enemy aircraft must occur before the blips collide in the middle of the grid. A target threat may or may not be present among a set of moving blips that represent incoming aircrafts. We manipulated the stimulus mapping, workload (the number of items memorized and the number of items on the radar screen), and the time constraints (called the frame time, measured as the time taken by the blips to collide in the middle of the grid).

Experiment 1: Conjunctive Search Consistency

This experiment tested the dual-process theory of automaticity when targets are formed by the conjunction of two cues in Radar. Rather than using one cue (symbol type), we used two cues (symbol type and color) to characterize the targets. Each of these cues was either consistently mapped (the target was characterized by the same cue throughout practice) or variably mapped (the targets were characterized by some cues in some trials and different cues in other trials). We also used a workload variable–memory-set size (the number of items to remember and search for)–to test whether or not performance under automatic processing was unaffected by workload. We hypothesized that the mapping of both cues together rather than the mapping of only one of the cues would affect performance. Specifically, we expected consistent mapping of both cues to result in the best performance. We also expected the consistent mapping of the conjunction of the two cues would result in automatic performance that would be unaffected by workload.

Method

Participants Five participants (one female and four males) with an average age of 22 years participated in this study. Participants were right-handed, had normal color vision, and had corrected or normal acuity. All participants were recruited from local universities and were paid $11.00 per hour for their participation. Each participant took part in six experimental sessions (each lasting approximately three hours) over six different days for a total of 18 hours of task practice.

Experimental design This experiment was a 2 (color mapping) × 2 (symbol mapping) × 2 (memory set size) within-subjects design. Three variables were manipulated in this experiment: (1) the mapping of the color (consistent or varied), (2) the mapping of a symbol (consistent or varied), and (3) the memory set size (1 or 4 items to remember and search for). The conjunction of two cues, type of symbol and color, determined the target set. The symbols used in the experiments were digits (1, 2, 3, 4, 5, 6, 7, and 8) and letters (K, H, D, G, J, L, M, and C). The colors used in the experiment were gold, gray, green, orange, peach, red, yellow, purple, cyan, and blue; these colors were selected from the Visual Basic 6.0 color palette because they are most distinct from each other. It has been shown that the amount of information that can be gleaned through human color perception is maximized at ten colors (Flavell and Heath, 1992).

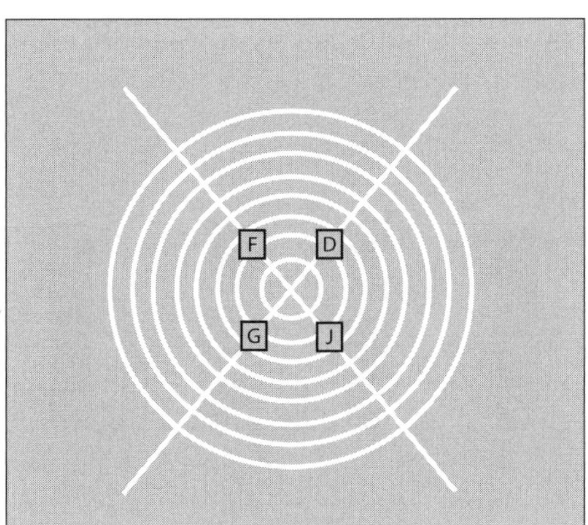

Figure 5.1 An example of the visual search component of Radar

Each of the eight possible conditions was randomly assigned to a block of 16 frames and each frame had a total of 14 trials (detections). The memory set was shown before each frame. The symbols and colors were either consistently mapped (randomly selected symbols and colors were kept as targets within a frame) or variably mapped (the symbols in a target set could appear as distractors in another target set within a block). A target was defined by a conjunction of color and symbol. The memory-set size was set to either 1 or 4 as it was manipulated in the initial experiments of the dual-process theory of automaticity (Schneider and Shiffrin, 1977). When the memory-set size equaled one, there was only one possible target (defined by the conjunction of a single color and a single symbol). When memory-set size was four, there were four targets (out of 16 possible unique conjunctions of the four colors and four symbols).

The dependent variables were the proportion of correct detections and the average detection time for correct detections.

Results

Proportion of correct detections The 2 (color mapping) × 2 (symbol mapping) × 2 (memory set size) repeated measures analysis of variance on the proportion of correct detections showed significant main effects of color mapping, $F(1, 4) = 7.99, p < .05$, symbol mapping, $F(1, 4) = 44.31, p < .01$, and memory set size, $F(1, 4) = 7.82, p < .05$. The proportion of correct detections was higher under consistent color mapping (M = .94, SD = .05) than varied color mapping (M = .86, SD = .09); higher under consistent symbol mapping (M = .96, SD = .05) than varied symbol mapping (M = .84, SD = .08); and higher under memory set size of one (M = .93, SD = .06) rather than four (M = .87, SD = .07).

The two-way interactions were also significant for color mapping and symbol mapping, $F(1, 4) = 7.99, p < .05$, for color mapping and memory set size, $F(1, 4) = 7.99, p < .05$, and for symbol mapping and memory set size, $F(1, 4) = 7.99, p < .05$. These interactions are shown in Figure 5.2.

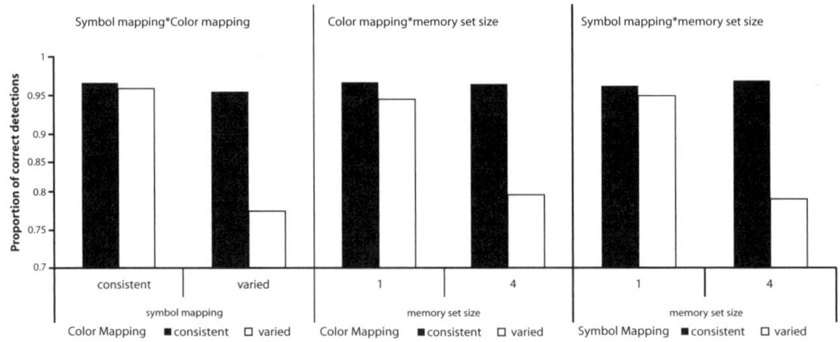

Figure 5.2 Two-way interactions for the proportion of correct detections in Experiment 1

Detection time The 2 × 2 × 2 analyses of variance showed significant main effects of color mapping, $F(1, 4) = 35.75$, $p < .01$, symbol mapping, $F(1, 4) = 16.35$, $p < .05$, and memory set size, $F(1, 4) = 178.87$, $p < .001$. The correct detections' responses were faster under consistent color mapping (M = 756 ms, SD = 177) than varied color mapping (M = 1030 ms, SD = 210); faster under consistent symbol mapping (M = 820 ms, SD = 155) than varied symbol mapping (M = 967 ms, SD = 233); and faster under memory set size of 1 (M = 756 ms, SD = 172) rather than of 4 (M = 1031 ms, SD = 215).

The two-way interactions were also significant: color mapping and symbol mapping, $F(1, 4) = 35.75$, $p < .01$, color mapping and memory set size, $F(1, 4) = 35.75$, $p < .01$; and symbol mapping and memory set size $F(1, 4) = 35.75$, $p < .01$ (see Figure 5.3). In detection time, the triple interaction of color mapping, symbol mapping, and memory set size was also significant, $F(1, 4) = 41.67$, $p < .01$ (see Figure 5.4).

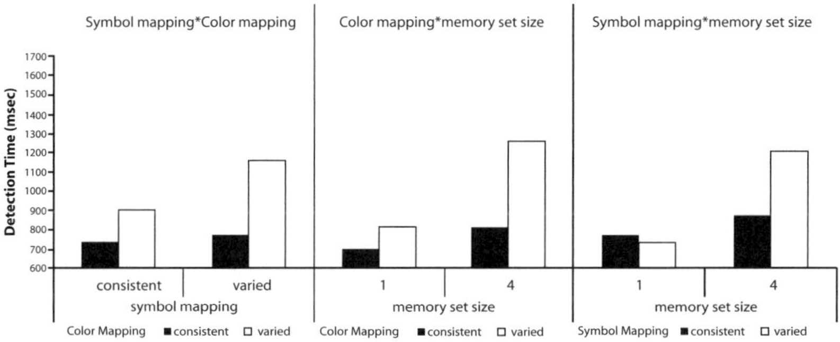

Figure 5.3 Two-way for detection time in Experiment 1

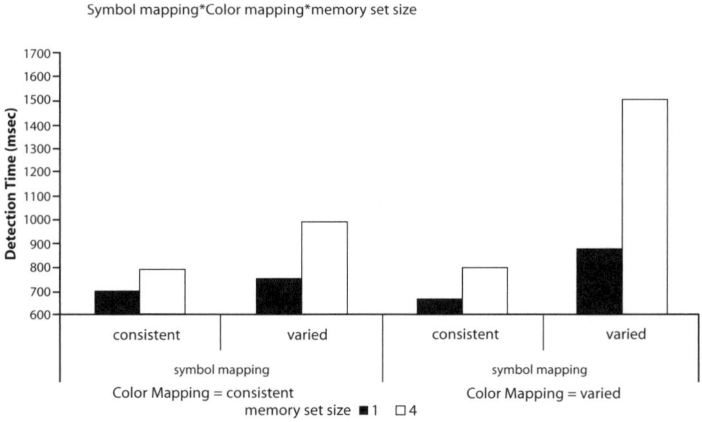

Figure 5.4 Three-way interactions for detection time in Experiment 1

Summary

The results of Experiment 1 revealed that participants were more accurate and faster in detecting targets when each of the cues that defined a target (symbol and color) was consistently mapped than when the cues were variably mapped. More importantly, the mapping of both cues in conjunction also determined performance. Performance was worst when both cues were variably mapped. On the other hand, mapping at least one of the cues consistently had a significant benefit on performance. Furthermore, when at least one of the two cues was consistently mapped, performance was unaffected by workload. In contrast, when at least one of the cues was variably mapped, performance was affected by workload. The significant three-way interaction on detection time shows that worst performance occurred when both cues were variably mapped and individuals worked under conditions of high workload.

Experiment 2: Consistency of Responding

In this experiment, we manipulated the response mapping by varying the number of possible responses (only one or more than one) and the consistency of the responses (the same responses or changing responses) throughout practice. The stimulus was consistently mapped and defined only by one cue (symbol) throughout practice. We also manipulated the time constraint (i.e., the time available for target detection) as a factor of workload. We predicted that response mappings would have a significant effect on performance despite the consistency of stimulus mapping. We also hypothesized frame time would influence performance, specifically in cases where response mapping became more complex (when there were many and inconsistent types of responses).

Method

Participants Eight students (two females and six males) with average age of 20 participated in the study. All participants were right-handed, had normal color vision, and had normal or corrected-to-normal visual acuity. All participants were recruited from local universities and were paid $11.00 per hour for their participation. Each participant completed six experimental sessions of approximately three hours each for a total of 18 hours of task practice.

Experimental Design

We designed a 4 (response mapping) × 2 (frame time) within-subjects experiment. Two independent variables were manipulated in the experiment: (1) response mapping (at 4 levels) and (2) frame time (at 2 levels). Thus, the experiment consisted of eight conditions. Each condition was presented twice during the experiment. Each condition was presented in a block of 30 frames and each frame had 14 trials (detections). A set of targets to remember (memory set) was shown before each frame. Thus, a total of 6,720 trials were distributed in six experimental sessions. The dependent variables were the proportion of correct detections and the average detection time for correct detections.

Response mapping The four levels of response mapping were defined by assigning the four keys corresponding to the four corners of the numeric keypad. Thus, four numbers were the possible responses to the target stimulus: 1, 7, 9, and 3. The keys were color coded with a color sticker on the key (1 = red, 7 = blue, 9 = green, and 3 = yellow). The participants' goal was to respond by pressing the correct key depending on the response mapping condition to which they were assigned to:

1. *Fixed mapping*: participants responded with the same key every time a target was detected. All axes on the radar grid were the same color and this color determined the correct response key. The color of the axis was randomly selected (either red, blue, green, or yellow), shown to the participants, and kept constant in each frame. Thus, any time the participant detected a target he was to respond by pressing the same key, regardless of where the target was detected.
2. *Full mapping*: participants responded with one of the four possible keys according to the color of the axes in which the target was detected. For example, if the target was detected in the bottom-left corner axis, the correct response was to press the 1 key (red). Each of the four axes of the grid were color-coded with the same color as the response keys (bottom-left = red, top-left = blue, top-right = green, and bottom-right = yellow).
3. *Partial mapping*: participants responded with one of the four possible keys. However, in this condition, two of the axes were consistent with the colors of the numeric keys and the other two axes were randomly assigned one of the four colors.
4. *Random mapping*: participants responded with one of the four possible keys, but in this condition, the color of each of the four axes was randomly assigned to one of the four colors. Thus, the consistency between spatial layout and the color coding of the responses and radar axes was random.

Frame time The time between the onset of one frame and the next was manipulated in two ways: the 'slow' condition was 2,050 ms and the 'fast' condition was 1,050 ms. For both frame time conditions, the radar grid was presented to the participants for 1,000 ms before each trial and the inter-trial interval was 1,500 ms.

Procedure

Before the onset of each frame, participants were asked to memorize a set of 4 targets (either digits or letters). Then they were presented with a sequence of 14 trials. Each trial displayed four moving blips. The participant's task was to detect any member of the memory set that appeared in the sequence of trials. Targets could appear in any trial except the first two and last two trials. At the end of each frame, participants received feedback on their performance. In case of an error in detection, a tone sounded at the end of the relevant frame. Participants were also provided with textual feedback at the end of each trial which detailed the type of error made (false alarm, miss, etc.). Stimulus mapping was always consistent and members of the target set never appeared as distractors. For example, when the targets were digits (1–9) the distractors were letters (C, D, F, G, H, J, K, L, and M) and vice versa.

Results

Figure 5.5 shows the proportion of correct detections (Panel A) and the detection times (Panel B) averaged by response mapping condition (stimulus, full, partial, random) and frame time condition (slow, fast).

The 4 (response mapping) × 2 (frame time) repeated measures analysis of variance revealed main effects of response mapping and frame time only for detection time, and no significant effects on the proportion of correct detections. The interactions between response time and frame time were not significant.

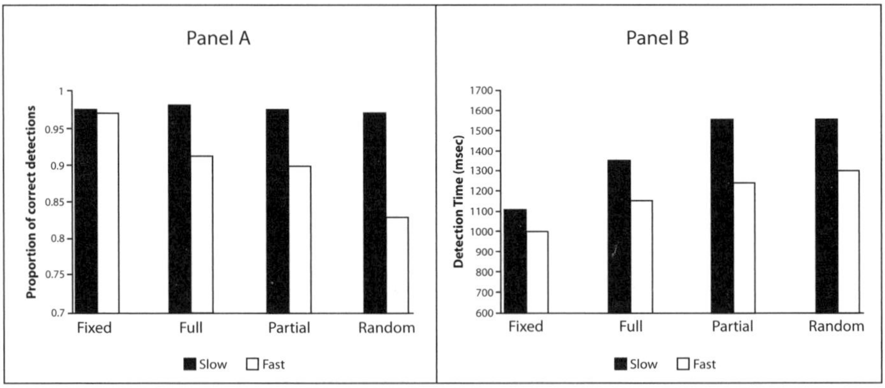

Figure 5.5 Proportion of correct detections and detection time in Experiment 2

The proportion of correct detections remained relatively stable across mapping conditions, particularly in the slow condition. However, although Figure 5.5 (Panel A) shows a different pattern across response mapping conditions for the fast frame times, the interaction between response mapping and frame time was not statistically significant.

On the other hand, detection time of correctly detected targets was clearly influenced by both the response mapping $F(3, 21) = 20.31$, $p < .001$, and the frame time $F(1, 21) = 54.95$, $p < .001$, independently. Detection time increased with the difficulty of the response mapping condition. Detections were fastest in the mapped-to-stimuli condition (M = 1056 ms, SD = 78.14) and slowest in the random-mapping condition (M = 1425 ms, SD = 61.49) as expected. Detection time was also fastest in the fast condition (M = 1173 ms, SD = 73.97) and slowest in the slow condition (M = 1391 ms, SD = 68.15).

Summary

The results indicated that response mapping significantly affected the detection time of correctly detected targets despite consistent stimulus mapping. The results however, also indicated that response mapping did not have an effect on the proportion of correct detections. Furthermore, time constraints also significantly affected detection time, but the effects were constant across response mapping conditions.

General Discussion and Conclusions

The current studies support the prediction of the dual-process theory in that optimal performance after extended practice occurs when targets are consistently mapped, rather than variably mapped. However, this research extends the traditional view of automaticity by testing the effects of conjunctive search consistency and response consistency.

In the first experiment, we demonstrated the importance of conjunctive search consistency. Our results indicated that consistently mapping each of the cues that defined the targets, independently, results in better performance. But, most importantly, our study indicates that the inconsistency of both cues together produced the most detrimental effect on performance. Research on attention indicates that a target defined by multiple cues leads to a serial search (Treisman and Gelade, 1980). That is, search time can increase with the number of cues and the number of levels of cues that are needed to search. Also, research shows that cues are processed in serial order (e.g., colors then symbols or symbols then colors; Fisher and Tanner, 1992). Thus, previous research suggests that an increased number of cues may involve increased serial search. Our results demonstrate that if at least one of the cues is consistently mapped, the serial search is faster and results in more accurate responses, especially when compared with cues that are all variably mapped.

In the second experiment, we demonstrated that response mapping can be critical for skill development above and beyond the stimulus consistency. The more inconsistent the responses are, the longer it takes to detect targets. This extends research that has reported negligible effects of response mapping (Fisk and Schneider, 1984; Kramer, Strayer, and Buckley, 1990). Our results revealed that, on average, the difference in response times between consistent responding (the Fixed response mapping condition) and the Full condition was 184 ms; between the Fixed and the Partial condition was 298 ms; and between the Fixed and Random mapping condition was 421 ms. These processing costs associated with inconsistent responding were considerably higher than in past research by Fisk and Schneider (1984), who found a processing cost of 64 ms for inconsistent responding and concluded that consistency of responding was less important than consistency of attending in the development of automaticity. Also, Kramer et al. (1991) concluded that consistency of responding was unnecessary for the development of automatic detection and that the processing cost of inconsistent responding was small (17 ms). However, as our results demonstrate, when tasks are complex and dynamic, the consistency of responding is equally important. Therefore, we would expect that in real-world, complex search-and-respond tasks, consistency of responding as well as the time constraints in the task can be important factors for effective performance. Although our results did not reveal a statistically significant interaction between response mapping and time constraint, the data suggest that there might be an increasingly detrimental effect of time constraints on performance as the inconsistency of responses (or the complexity of response mapping) increases. Testing this possibility requires further research.

Implications for Design and Training

Consistency of mapping was investigated during the visual search and during the response. We suggest that performance on a visual search task needs to be analyzed on at least these two possible dimensions. As demonstrated by the results of these experiments, analyzing a visual search task in terms of the consistency of responding and the conjunction of search cues helped identify some important impediments in skill acquisition. For example, when the spatial layout of a display and the spatial layout of the response keys are redundant (i.e., location compatibility), it capitalizes on people's natural tendency to move toward stimuli (Simon, 1969; Wickens and Hollands, 2000). When the spatial location of the stimulus and the response location are inconsistent (e.g., a target is preset on the northwest axis, but the appropriate response is the 9 key or the northeast response key), Wickens and Hollands' (2000) collocation and movement compatibility principles are violated. However, even when these principles are not violated (when the spatial location of the stimulus and the response key are redundant), there are still processing costs over inconsistent responding. Thus, when possible, system designers should consistently map stimuli and responses to attenuate the processing costs associated with response selection.

Another implication of this research involves a recommendation for training. If extensive training is provided with a wide range of stimuli in simulated tasks, it is possible to develop automatic detection if at least one of the cues is consistently mapped. Individuals can benefit from automatic detection even when they perform in the presence of other variably mapped cues. Our future research will be directed at exploring the ecological account of the dual-process theory of automaticity in all its dimensions, including the effects of spatial and temporal dynamics and the multiple forms of consistency of attending and responding.

Acknowledgment

This research was partially supported by the Multidisciplinary University Research Initiative Program (MURI; N00014-01-1-0677) and by the National Science Foundation (Human and Social Dynamics: Decision, Risk, and Uncertainty, Award number: 0624228) awards to Cleotilde Gonzalez.

References

Bolstad, C. A., and Endsley, M. R. (2000). The effect of task load and shared displays on team situation awareness. Paper presented at the 14th Triennial Congress of the International Ergonomics Association and the 44th Annual meeting of the Human Factors and Ergonomics Society, Marietta, GA.

Durso, F. T., Cooke, N. M., Breen, T. J., and Schvaneveldt, R. W. (1987). Is consistent mapping necessary for high-speed search? *Journal of Experimental Psychology: Learning, Memory and Cognition, 13*, 223–229.

Fisher, D. L., and Tanner, N. (1992). Optimal symbol set selection: An automated procedure. *Human Factors, 34*, 79–92.

Fisk, A. D., and Schneider, W. (1984). Consistent attending versus consistent responding in visual search: Task versus component consistency in automatic processing development. *Bulletin of the Psychonomic Society, 22*, 330–332.

Flavell, R., and Heath, A. (1992). Further investigations into the use of color coding scales. *Interacting with Computers, 4*, 179–199.

Gonzalez, C., and Thomas, R. P. (2008). Effects of automatic detection on dynamic decision making. *Journal of Cognitive Engineering and Decision Making, 2*, 328–348.

Kramer, A. F., Strayer, D. L., and Buckley, J. (1990). Development and transfer of automatic processing. *Journal of Experimental Psychology: Human Perception and Performance, 16*, 505–522.

Kramer, A. F., Strayer, D. L., and Buckley, J. (1991). Task versus component consistency in the development of automatic processing: A psychophysiological assessment. *Psychophysiology, 28*, 425–437.

Schneider, W., and Shiffrin, R. M. (1977). Controlled and automatic human information processing: I. Detection, search and attention. *Psychological Review, 84*, 1–66.

Schneider, W., and Shiffrin, R. M. (1985). Theoretical note: Categorization (restructuring) and automatization: Two separable factors. *Psychological Review, 92*, 424–428.

Shiffrin, R. M., and Schneider, W. (1977). Controlled and automatic human information processing: II. Perceptual learning, automatic attending, and a general theory. *Psychological Review, 84*, 127–190.

Simon, J. R. (1969). Reactions toward the source of stimulation. *Journal of Experimental Psychology, 81*, 174–176.

Treisman, A. M., and Gelade, G. (1980). A feature-integration theory of attention. *Cognitive Psychology, 12*, 97–136.

Wickens, C. D., and Hollands, J. C. (2000). *Engineering Psychology and Human Performance* (3rd edn). New York: Harper Collins.

Young, M. D., Healy, A. F., Gonzalez, C., Dutt, V., and Bourne, L. E., Jr. (2008). *Effects of Training with Added Relevant Responses on RADAR Detection*. Chicago: The Experimental Psychology Society and the Psychonomic Society.

Chapter 6

What Visual Discrimination of Fractal Textures Can Tell Us about Discrimination of Camouflaged Targets

Vincent A. Billock
General Dynamics Advanced Information Systems

Douglas W. Cunningham
University of Tübingen

Brian H. Tsou
Air Force Research Laboratory

Introduction

Discrimination tasks in combat target identification are legion. For example, operators need to discriminate a target against a background and to discriminate a set of similar targets from one another. The first task is a necessary, but not sufficient condition for targeting, while the second task is essential to solve decoy and friendly-fire problems. Both tasks are complicated by camouflage. If it were necessary to consider the set of all possible targets, backgrounds, and camouflage, the combinatorial problem would be disheartening. However, a consideration of visual psychophysics, image science, and fractal mathematics suggests that a particular optical signature simplifies this problem considerably.

Background: Perceptual Pop-out, Fractals, and Camouflage

It is well known that humans effortlessly (and preattentively) discriminate images which differ significantly in their second-order statistics (the so-called 'pop-out' phenomenon), while images that have similar second-order statistics must usually be compared on a more laborious (attentive) point-by-point basis (Julesz and Caelli, 1979; Caelli, 1981).[1] Most natural (and many artificial) images have surprisingly regular $1/f^{\beta}$ Fourier spatial amplitude spectra (Billock, 2000; Field and Brady, 1997; Ruderman, 1994; Schaaf and Hateren, 1996; Tolhurst, Tadmor, and Chao, 1992; Webster and Miyahara, 1997), a signature which is mimicked by random

1 A second order statistic of an image is any statistic that can be computed from the autocorrelation of the image (e.g., the Fourier amplitude spectra or the fractal dimension).

fractals and which forms the basis for fractal forgeries (Voss, 1985) and digital camouflage.[2] The exponent β is a second-order statistic; natural images typically have β values between 0.8 and 1.3; a review of 11 published studies with 1176 images found a population average of about 1.08 (Billock, 2000). A growing body of evidence suggests that humans are adapted to this statistical regularity in the environment and that this evolutionary/developmental adaptation forms the basis for neural image enhancement and debluring (Billock, 2000; Billock, Cunningham, Havig, and Tsou, 2001; Billock, de Guzman, and Kelso, 2001; Campbell, Howell, and Johnson, 1978; Hammett and Bex, 1996). A hallmark of random fractal images is the presence of statistically similar features at every spatial scale. The lawful relationship between spatial scales is termed self-similarity and is one of the properties of natural images that give rise to $1/f^{\beta}$ spectra. This property is what enables random fractals to mimic natural images and backgrounds. For example, a tree branch gives rise to several smaller branches, which give rise to many twigs–a random fractal that distributes and scales its features similarly can emulate foliage and act as camouflage. The β value (or equivalently, fractal dimension) is often used as a mathematical measure of image texture and its perceptual correlates (Cutting and Garvin, 1987; Kumar, Zhou, and Glaser, 1993; Pentland, 1988; Rogowitz and Voss, 1990; Taylor et al., 2005). It follows that some aspects of fractal image discrimination can emulate natural image discrimination (Hansen and Hess, 2006; Thomson and Foster, 1997; Parraga and Tolhurst, 2000; Tolhurst and Tadmor, 2000).

If natural backgrounds are fractal-like, camouflage should be designed along similar principles. Although some camouflages are designed to mislead, we refer here to camouflage that is intended to conceal an object against its background. Newer camouflage schemes like MARPAT (U.S. Marines) and CADPAT (Canadian Armed Forces) use a two-scale scheme which is noticeably better at blending into terrain and foliage than the older single-scale schemes. For example, detection times for MARPAT camouflaged targets are about 2.5 times longer than detection of NATO single-scale camouflage and recognition times following detection increase by an additional 20 percent (O'Neill, Matthews, and Swiegosz, 2004). To be truly fractal, a camouflage should have feature statistics that are statistically similar at every visible scale. Such camouflages are often created using recursion algorithms on digital computers and are sometimes called digital camouflage. (The square pixels and limited range of scales seen in these textures are cost and technology driven – not a limitation of fractals per se.) More complicated schemes are possible, including the use of multi-fractals, which mimic blends of particular

2 The Fourier spectrum of an image is a recipe for making the image by summing sine and cosine waves of various amplitudes, frequencies, and phases; the amplitude spectrum is the contrast at each frequency (with the phases combined). The spatial frequency is the number of full oscillations that the sinusoid makes across the image. A $1/f^{\beta}$ spectrum means that the contrast of each sinusoid in the image recipe declines (with a slope of β on log-log coordinates) as the frequency increases.

textures that occur in natural images (e.g., plant growth on fractured rock). Here, we study human abilities to discriminate images based on small differences in the β signature and place the results in context with camouflage and with earlier texture discrimination studies.

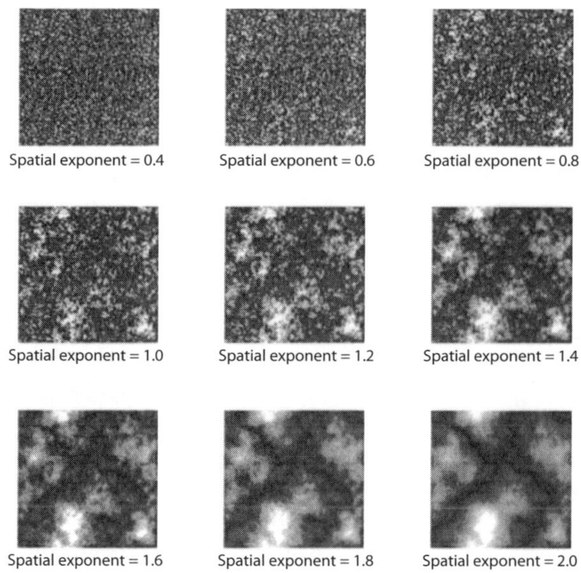

Figure 6.1 Fractal textures like those used in the experiments

Methods

Participants

The four observers were all myopes corrected to at least 20/20 binocular acuity. All are professional psychophysicists and highly experienced observers, with prior work in the psychophysics of 'white' and 'colored' (spatiotemporally non-uniform) visual noise. One subject (SF) was naïve to the purpose of the experiment. Another subject (VB) has a diagnosed mild congenital visual condition – optic nerve hypoplasia (a low density of neurons in the optic nerve). Although his vision is considered normal by standard clinical measures (including acuity), his contrast sensitivity is slightly depressed (about 1 SD) at all spatial frequencies relative to a large sample of age-matched-normals; this depression in sensitivity is worse for higher spatial frequencies. His vision is relevant here because it provides us a gauge of the effects of spatial under-sampling in an otherwise intact visual system.

Apparatus

All stimuli were generated and presented on a Silicon Graphics O_2 graphics workstation with a linearized 30 Hz display. Stimuli were viewed binocularly with natural pupils in a well-lit room (ambient luminance in the plane of the monitor was 3.5 cd/m^2). Subjects were comfortably fixed in place by a chin rest at two viewing distances, 40 cm and 100 cm. The far distance was a limiting case (e.g., each pixel subtends 0.016 deg at 100 cm, matching the subjects' 1 arc min spatial resolution). The stimuli consisted of static, grayscale, random-phase fractals (e.g., see Figure 6.1) whose Fourier amplitude spectra were described by:

$$\text{Eq. 1} \qquad A(f_s) = kf_s^{-\beta}$$

Where k is a constant and f_s is spatial frequency. (In visual psychophysics, amplitude rather than power spectra are used, because amplitude is proportional to perceptual contrast for each spatial frequency component.) For each stimulus, the average luminance was constant at 8.57 cd/m^2 and the Root Mean Square Contrast (a good measure of perceptual contrast for noise-like textures; Moulden, Kingdom, and Gatley, 1990; Peli, 1990, 1997) was 10.98 percent. For consistency with another study, each fractal contained 64×64 pixels (18×18 mm). Thus, at 40 cm, the stimuli subtended 2.58° embedded in a 43.9° horizontal by 36.4° vertical dark surround. At 100 cm, each stimulus subtended 1.03° embedded in a 21.1° horizontal by 16.4° vertical dark surround. Both a reference and a comparison image were generated for each trial. The images were created by filling a 64×64 array with random white noise (256 gray-levels). This white-noise image was Fourier-transformed and the amplitudes of all spatial frequencies were equalized to ensure that the noise was uniformly flat. The resulting amplitude spectra were filtered so that they followed a power law relationship (Eq. 1), and then inverse-Fourier transformed to produce the stimuli.

Procedure

Just noticeable discrimination thresholds (79 percent correct criterion) for fractal spatial exponents were measured using a two-alternative forced-choice adaptive staircase procedure with a 1 db step size (MacMillan and Creelman, 1991). Ten β exponents were used (0.4, 0.6, 0.8, 1.0, 1.2, 1.4, 1.6, 1.8, 2.0, and 2.2) for the reference images. For the comparison image, the fractal exponent was equal to the exponent, β, of its reference image plus a small increment, Δβ. Observers were asked to identify the image with the lower spectral exponent, and were provided with immediate feedback on the accuracy of their response. If the observer correctly identified the reference image three times in a row, the difference between the two images' exponents Δβ was decreased. In contrast, Δβ was increased after each incorrect response. Each staircase continued for 8 reversals, with the mean of the last 6 reversals being used as a measure of the threshold. Two presentation conditions (Sequential and Simultaneous) were used. In the Simultaneous condition, the reference and comparison stimuli were presented side by side (1.1 mm apart) for 2.133 seconds. The location of the reference image (left versus right) was randomized across trials. In the Sequential condition, the two stimulus images were sequentially presented in the center of the screen for 2.133 seconds each. The screen was blanked for 500 ms between the two images to prevent masking effects. The order of presentation of the reference and comparison images was randomized across trials. Combining two viewing distances with two presentation modes yielded four experimental conditions. Two observers were presented with the Near conditions first, and two with the Far conditions first. For all observers, the presentation style (simultaneous vs. sequential) alternated after each threshold. The order of presentation of the ten exponents was randomized for each of the four conditions. Each threshold was measured three times, with the thresholds in all four conditions being completed once before being re-measured. This required approximately 20 hours of data collection per subject, which was generally completed in 2 one-hour sessions each day, over a two-week period.

Results and Analysis

General Findings

Discrimination thresholds (dβ) are generally in the range of 0.05–0.20 for β values of 0.4–2.2 (see Figures 6.2 and 6.3). The discrimination function is not flat; it has higher (worse) discrimination thresholds for both low and high values of β, and lower (better) discrimination thresholds for in-between values of β. The minimum is near β = 1.6, which typifies images with less high spatial frequency content than the vast majority of natural images (β near 1.1). This implies that discrimination between fractal camouflaged objects is somewhat more difficult when the statistics of camouflaged objects are sufficiently similar to the statistics of natural images (as any sensible camouflage scheme should be), compared to the less natural β value of 1.6. This applies regardless of the background's β value, which has implications for fratricide–friendlies and hostiles will be somewhat harder to tell apart for naturalistically camouflaged images, even when friendlies and hostiles are both visible against their backgrounds.

Effect of Viewing Distance

For ideal 1/f images, there should be little effect of viewing distance, because increasing viewing distance would simply shift a lower spatial frequency component into a higher spatial frequency, but the relationship between the spatial frequencies would be preserved. However, all physically obtainable fractals are limited to a range of spatial scales set at the lower end by the size of the image and at the upper end by the size of the pixels. Shifting the viewing distance from

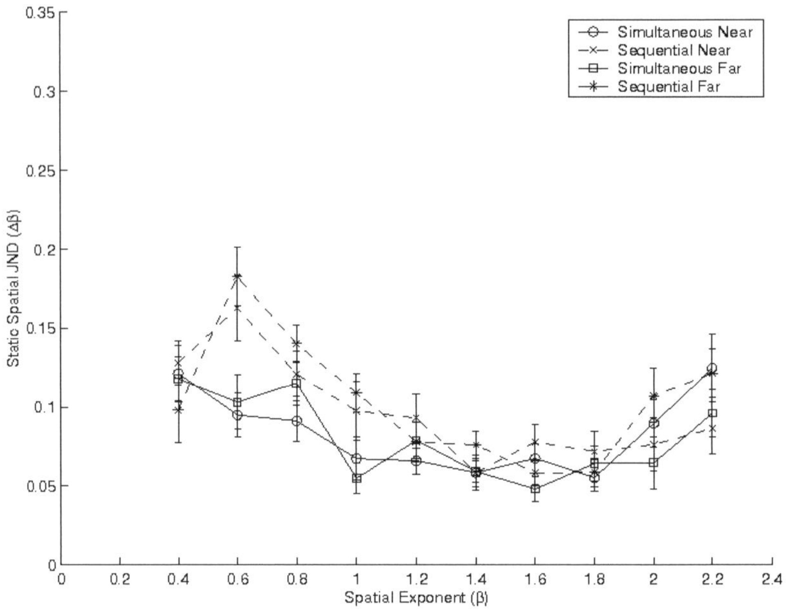

Figure 6.2 Group averages for all four conditions

40 to 100 cm therefore shifts the spatial frequency range of the fractal image by a factor of 2.5, but no information is lost because the stimuli were designed so that the individual pixels were resolvable at the far viewing distance by all observers. Accordingly, for three of the observers (DC, PH, and SF), viewing distance had little effect, although there is a slight trend suggesting lower thresholds in the nearer viewing distances (see Figure 6.3). For VB, however, the Far thresholds (and their variability) were noticeably elevated compared to either his Near thresholds or to the other observers' Far thresholds (see Figure 6.3). VB's anomalous results may be due to sampling problems induced by a mild congenital defect–a developmental paucity of retinal ganglion cells (optic nerve hypoplasia). Electrophysiological studies in VB and other hypoplastic subjects and post-mortem histology in other hypoplastics indicate that both retinal pre-processing and cortical post-processing seem to be normal (Billock et al., 1994) and point to reduction in retinal ganglion cell numbers as the sole cause of abnormal vision in hypoplasia. In the case of VB, perimetric thresholds are flattened relative to normals, suggesting the subject did not gain a full measure of the elevated density of foveal ganglion cells that develops in normals. Since pixel size and stimulus size are fixed, any sampling problems would more likely manifest as a threshold elevation at the far viewing conditions. Moreover, if the reduced sampling is not homogeneous, then this could increase variability (because, from trial to trial, filtered noise features would fall on neighboring retinal locations with different retinal sampling densities).

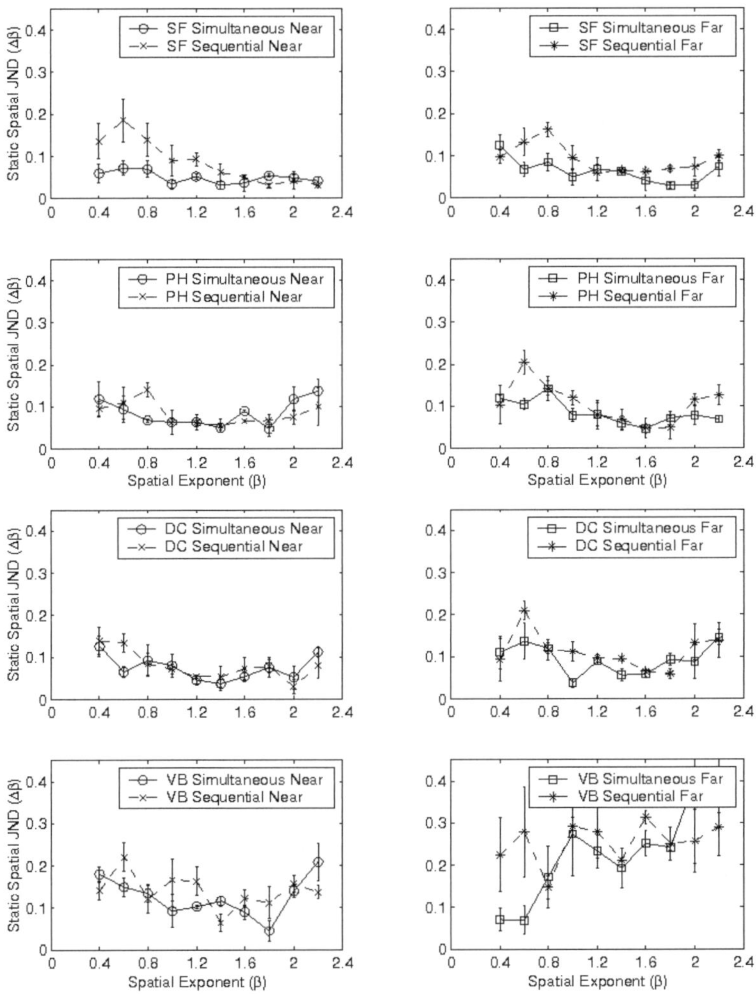

Figure 6.3 Individual data for all four observers for all four conditions

Effect of Presentation Style

Simultaneous viewing simulates the task of making a side-by-side comparison of fractal camouflaged targets, while sequential viewing simulates the task of comparing a target to one that is in memory; in theory and experiment the two paradigms can lead to somewhat different results (Garcia-Perez, Giorgi, Woods, and Peli, 2005; Hansen and Hess, 2006). The discrimination function is similar for both conditions (Figures 6.2 and 6.3) but there is a small advantage for simultaneously-viewed images, relative to sequentially-viewed ones, especially for small values of β. This tendency can be clearly seen when the data from the 4 subjects are pooled (see Figure 6.2). This is contrary to Hansen and Hess (2006), who found an advantage for sequential viewing, and attributed differences between the two conditions to differences in the portions of retina they cover. However, our near-sequential and far-simultaneous stimuli covered very similar regions of central retina (2.6° and 2.3°, respectively), and yet simultaneous viewing yielded lower thresholds for nine of ten exponents (and tied for the tenth). Our data suggest that it would be useful for combat operators to be able to view targets and reference images side-by-side, rather than switching between views.

Discussion

Summary: Fractal Discrimination and Implications for Camouflage in Combat Identification

Based on this and other work we can enumerate some implications for camouflage and combat identification: (1) Natural images have $1/f^{\beta}$ spatial amplitude spectra. The most reasonable value of β for general purpose camouflage is around 1.1. Particular environments will vary in this statistic and in coloration. (2) Keeping the difference between the β_{target} and $\beta_{background}$ less than 0.1–0.2 generally avoids preattentive popout, but discrimination will still be possible using a point-by-point search. (3) Using many spatial scales makes camouflage effectiveness almost independent of distance. (4) For identification purposes, friendly camouflage schemes should have different βs than the unfriendly camouflage patterns, but this may conflict with concealment goals. The best outcome would be for hostile and friendly camouflage statistics to be on opposite sides of the $\beta_{background}$ value, with the friendly scheme not easily discriminable from background but discriminable from the hostile. (5) For identification purposes, side-by-side viewing of sensor and reference images is preferable.

Comparison to Related Studies: Static Fractals

Some prior studies of fractal discrimination overlap with our work. Our discrimination functions resemble those of Knill, Field, and Kersten (1990), particularly their low-contrast near condition (17.5 percent Root Mean Square Contrast at 1 meter with 64x64 pixel images), which is similar to our Far Sequential condition. While the average exponent of natural scenes is around 1.1, the greatest sensitivity to changes in a fractal image's exponent are consistently found to be around 1.6 across a wide range of conditions. We see no evidence for a second minimum at low β (circa 0.6) reported by Tadmor and Tolhurst (1994), even when we used a simultaneous viewing condition similar to Tadmor and Tolhurst. Nor is the discrepancy due to the angular size of the image, as all three studies had stimuli that were similar in size. Hansen and Hess (2006) noted that the spatial presentation task uses two different parafoveal patches of retina, while the temporal task uses the same patch of central fovea; they found that fovea and parafovea yield somewhat different patterns of discriminability as a function of β, but none of their data show a second minimum at low β (rather, they find a maximum at β = 0.8, with better thresholds on either side, similar to our findings). Another possible source of this difference in discriminability functions may be the specific nature of Tadmor and Tolhurst's task. In both the present study and Knill et al. (1990), standard two-Alternative Forced-Choice psychophysical procedures were used. In contrast, Tadmor and Tolhurst (1994) used an odd-one-out task (i.e., three images were presented simultaneously, two of which had identical exponents–the task was to choose the image that was different from the other two). In other words, Tadmor and Tolhurst's task was one of simple discrimination, while our task (and Knill's) requires discrimination and some form of identification (once the two images could be told apart, the subjects had to decide which had a lower exponent). These tasks coincide in difficulty only if all information required for identification is present at the discrimination threshold, which will most often take place when a single channel mediates performance of the task. Indeed, Tolhurst and Tadmor (1997) have shown that simple discrimination data is often consistent with a single channel mediating discrimination. However, since a comparison of channel outputs is required to estimate the spectral exponent of an image, discrimination plus identification would likely require a comparison of channel outputs, perhaps raising the just noticeable difference (JND) for β near 0.4 sufficiently to eliminate the second minimum that Tadmor and Tolhurst (1994) found.

Comparison to Related Studies: Dynamic Fractals

So far we have discussed only perception of static fractals. In general, the effect of motion on the temporal frequencies (f_t) of such fractals is just the spatial frequency multiplied by the velocity. Thus, if the spatial spectrum (f_s) follows a power law ($1/f_s^{\beta}$) for simple movement, the temporal frequency spectrum will be described by a power law as well, which is similar to the rescaling effects of distance in the viewing distance condition discussed above. There are, however, more interesting dynamic manipulations of fractals that are worth study. For example, it is possible to extend our study of discrimination to spatiotemporal fractals–fractals whose individual pixel intensities vary over time in a manner described by fractional Brownian motion. Such images have Fourier amplitude spectra of:

Eq. 2 $$A(f_t,f_s) = kf_t^{-\alpha}f_s^{-\beta}$$

In general, as α becomes larger, the motion of the texture becomes more coherent and can be used to mimic various biological motions. We have published data on human discrimination of spatiotemporal fractals (Billock, Cunnigham, et al., 2001); Figure 6.4 shows the perceptual discrimination space. To define two-dimensional JNDs we measured discrimination in four directions: both increments and decrements for both spatial (β) and temporal (α) exponents. Not surprisingly, the JNDs in this space are ellipsoidal (they resemble color discrimination JNDs). Interestingly, the interior portion of the resulting two-dimensional discrimination space is remarkably flat, a feature that some psychophysicists have gone to great lengths to obtain in nonlinear mappings of other perceptual systems (e.g., color discrimination).

Implications for Practice and Future Research

Based on the performance of the hypoplastic but seemingly normal observer (VB) in Figure 6.3, it might be advisable for sensor operators to be screened for spatial sampling problems (sub-clinical amblyopia) by measuring their contrast sensitivity functions. It is worth noting that humans can become proficient at naming the spectral exponents of images (or equivalently, fractal dimension, which is a linear transform of the exponent; Cutting and Garvin, 1987; Kumar et al., 1993; Pentland, 1988). A neural ability to estimate the spectral drop-off and exploit it has been speculated on and deserves additional attention (Billock, 2000; Billock, de Guzman, et al., 2001; Campbell et al., 1978; Hammett and Bex, 1996; Rogowitz and Voss, 1990). Taken together with the natural image regularities and perceptual pop-out findings discussed earlier, this suggests that β is a key signature, both for images and for the visual systems that evolved to transduce images. Of particular interest is the finding that, under some conditions (e.g., nonlinear systems near threshold), adding noise can facilitate detection and identification of some signals,

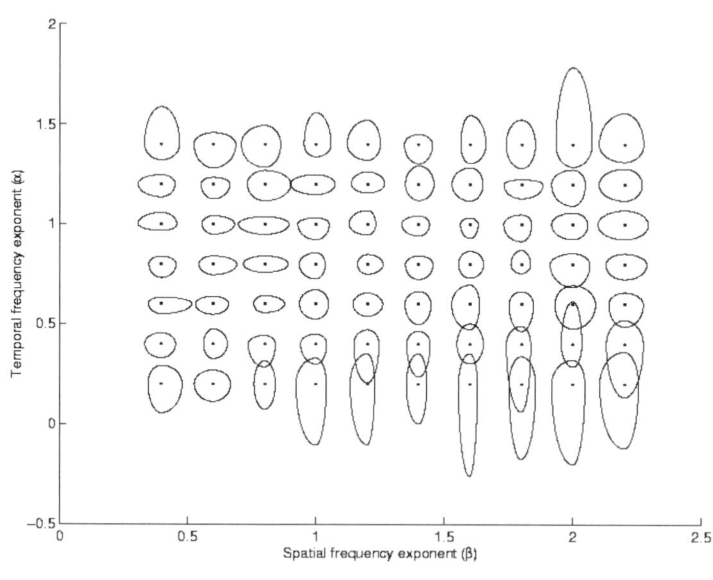

Figure 6.4 Discrimination contours for spatiotemporal fractal textures

including images (Repperger et al., 2001; Simonotto et al., 1997; Yang, 1998)–an example of stochastic resonance as an image enhancement mechanism. Dynamic noise is more effective than static (Simonotto et al, 1997). Because other studies of stochastic resonance show that $1/f^{\beta}$ noise can be more efficient than white noise in inducing stochastic resonance effects (Billock and Tsou, 2007; Hangi, Jung, Zerbe, and Moss, 1993; Nozaki, Collins, Yamamota, 1999), further studies of discrimination in spatiotemporal fractal noise (at various contrast levels) would be warranted and might uncover some practical applications. In particular, it may be possible to break many camouflage schemes by adding filtered noise to the sensor images. This seemingly perverse aspect of stochastic resonance should be exploited if possible. Since stochastic resonance's effectiveness is often dependent on the Fourier spectral qualities of the noise, fractal camouflage may be particularly vulnerable (because the spectral qualities of simple fractals are easily matched by varying one noise parameter). Multi-fractals may be less vulnerable in this regard. It would be ironic if the beautiful mathematical attributes of fractals (which give it so much utility in describing the natural environment and make it such an elegant solution to the problem of designing camouflage) also prove to be its Achilles' heel.

Author Note and Acknowledgements

Vincent A. Billock, General Dynamics, Inc., Suite 200, 5200 Springfield Pike, Dayton, OH 45431; vince.billock@gd-ais.com. Douglas W. Cunningham University of Tübingen, Tübingen, Germany; douglas.cunningham@gris.uni-tuebingen.de. Brian H. Tsou, AFRL/RHCI, Wright-Patterson Air Force Base, OH 45433; brian.tsou@afrl.af.mil.

We thank Jer Sen Chen, Steven Fullenkamp, Paul Havig and Eric Heft for their technical support.

References

Billock, V. A. (2000). Neural acclimation to 1/f spatial frequency spectra in natural images and human vision. *Physica D, 137*, 379–391.

Billock, V. A., Cunningham, D. W., Havig, P., and Tsou, B. H. (2001). Perception of spatiotemporal random fractals: An extension of colorimetric methods to the study of dynamic texture. *Journal of the Optical Society of America A, 18*, 2404–2413.

Billock, V. A. de Guzman, G. C., and Kelso, J. A. S. (2001). Fractal time and 1/f spectra in dynamic images and human vision. *Physica D, 148*, 136–146.

Billock, V. A., and Tsou, B. H. (2007). Neural interactions between flicker-induced self-organized visual hallucinations and physical stimuli. *Proceedings of the National Academy of Sciences of the USA, 104*, 8490–8495.

Billock, V. A., Vingrys, A. J., and King-Smith, P. E. (1994). Opponent-color detection threshold asymmetries may result from reduction of ganglion cell subpopulations. *Visual Neuroscience, 11*, 99–109.

Caelli, T. (1981). *Visual Perception: Theory and Practice*. Oxford: Pergamon Press.

Campbell, F. W., Howell, E. R., and Johnson, J. R. (1978). A comparison of threshold and suprathreshold appearance of gratings with components in the low and high spatial frequency range. *Journal of Physiology, 284*, 193–201.

Cutting, J. E., and Garvin, J. J. (1987). Fractal curves and complexity. *Perception and Psychophysics, 42*, 365–370.

Field, D. J. (1987). Relations between the statistics of natural images and the response properties of cortical cells. *Journal of the Optical Society of America A, 4*, 2379–2394.

Field, D. J., and Brady, N. (1997). Visual sensitivity, blur and the sources of variability in the amplitude spectra of natural images. *Vision Research, 37*, 3367–3383.

Garcia-Perez, M. A., Giorgi, R. G., Woods, R. L., and Peli, E. (2005). Thresholds vary between spatial and temporal forced-choice paradigms: The case of lateral interactions in peripheral vision. *Spatial Vision, 18*, 99–127.

Hammett, S. T., and Bex, P. J. (1996) Motion sharpening: Evidence for addition of high spatial frequencies to the effective neural image. *Vision Research, 36*, 2729–2733.

Hangi, P., Jung, P., Zerbe, C., and Moss, F. (1993). Can colored noise improve stochastic resonance? *Journal of Statistical Physics, 70*, 25–47.

Hansen, B. C., and Hess, R. F. (2006). Discrimination of amplitude spectrum slope in the fovea and parafovea and the local amplitude distributions of natural scene imagery. *Journal of Vision, 6*, 696–711.

Julesz, B. and Caelli, T. (1979). On the limits of Fourier decompositions in visual texture perception. *Perception, 8*, 69–73.

Knill, D. C., Field, D., and Kersten, D. (1990). Human discrimination of fractal images. *Journal of the Optical Society of America A, 7*, 1113–1123.

Kumar, T., Zhou, P., and Glaser, D. A. (1993). A comparison of human performance with algorithms for estimating fractal dimension of fractional Brownian statistics. *Journal of the Optical Society of America A, 10*, 1136–1146.

MacMillan, N. A., and Creelman, C. D. (1991). *Detection Theory: A User's Guide*. Cambridge: Cambridge University Press.

Moulden, B., Kingdom, F., and Gatley, L. F. (1990). The standard deviation of luminance as a metric for contrast in random-dot images. *Perception, 19*, 79–101.

Nozaki, D., Collins, J. J., and Yamamota, Y. (1999). Mechanism of stochastic resonance enhancement in neuronal models driven by 1/f noise. *Physical Review E,* 60, 4637–4644.

O'Neill, T., Matthews, M., and Swiergosz, M. (2004). Marine Corps innovative camouflage. *Midyear meeting of the American Psychological Association, Divisions 19 and 21.* Supplementary data at http://www.hyperstealth.com/digital-design/index.htm

Parraga, C. A., and Tolhurst, D. J. (2000). The effect of contrast randomization on the discrimination of changes in the slopes of the amplitude spectra of natural scenes. *Perception, 29*, 1101–1116.

Peli, E. (1990). Contrast in complex images. *Journal of the Optical Society of America A, 7*, 2032–2040.

Peli, E. (1997). In search of a contrast metric: matching the perceived contrast of Gabor patches at different phases and bandwidths. *Vision Research, 23*, 3217–3224.

Pentland, A. (1988). Fractal-based descriptions of surfaces, in W. Richards (ed.), *Natural Computation* (pp. 279–299). Cambridge: MIT Press.

Repperger, D. W., Phyllips, C. A., Neidhard, A., and Haas, M. (2001). Designing human machine interfaces using principles of stochastic resonance (Report DTIC# ADA412330). *AFRL Technical Report ARRL-HE-WP-TR-2002–0187.*

Rogowitz, B. E., and Voss, R. F. (1990). Shape perception and low-dimension fractal boundaries. *Proceedings of the SPIE, 1249*, 387–394.

Ruderman, D. L. (1994). The statistics of natural images. *Network: Computation in Neural Systems, 5*, 517–548.

van der Schaaf, A., and van Hateren, J. H. (1996). Modeling the power spectra of natural images: statistics and information. *Vision Research, 36*, 2759–2770.

Simonotto, E., Riani, M., Seife, C., Roberts, M., Twitty, J., and Moss, F. (1997). Visual perception of stochastic resonance. *Physical Review Letters, 78*, 1186–1189.

Tadmor, Y., and Tolhurst, D. J. (1994). Discrimination of changes in the second-order statistics of natural and synthetic images. *Vision Research, 34*, 541–554.

Taylor, R. P., Spahar, B., Wise, J. A., Clifford, C. W. G., Newell, B. R., and Martin, T. P. (2005). Perceptual and physiological responses to the visual complexity of fractal patterns. *Nonlinear Dynamics in Psychology and Life Sciences, 9*, 89–114.

Thomson, M. G. A., and Foster, D. H. (1997). Role of second- and third-order statistics in the discriminability of natural images. *Journal of the Optical Society of America A, 14*, 2081–2090.

Tolhurst, D. J., and Tadmor, Y. (2000). Discrimination of spectrally blended natural images: Optimization of the human visual system for encoding natural images. *Perception, 29*, 1087–1100.

Tolhurst, D. J., and Tadmor, Y. (1997). Band-limited contrast in natural images explains the detectability of changes in the amplitude spectra. *Vision Research, 23*, 3203–3215.

Tolhurst, D. J., Tadmor, Y., and Chao, T. (1992). The amplitude spectra of natural images. *Ophthalmic and Physiological Optics, 12*, 229–232.

Webster, M. A., and Miyahara, E. (1997). Contrast adaptation and the spatial structure of natural images. *Journal of the Optical Society of America A, 14*, 2355–2366.

Voss, R.F. (1985). Random fractal forgeries, in R.A. Earnshaw, Ed., *Fundamental Algorithms for Computer Graphics* (pp. 805–835). Berlin: Springer.

Yang, T. (1998). Adaptively optimizing stochastic resonance in visual system. *Physics Letters A, 245*, 79–86.

Chapter 7
A Cognitive Basis for Friend-Foe Misidentification of Vehicles in Combat

Joseph R. Keebler
Lee W. Sciarini
Florian Jentsch
Denise Nicholson
Thomas Fincannon
University of Central Florida

It is important for members of any modern military to quickly and reliably discriminate between friendly and enemy vehicles. Otherwise, friendly fire incidences occur, also known as *fratricide*. During World War II, fratricide rates have been estimated to be as high as 19 percent (U.S. Congress, 1993). Forty-five years later, in Operation 'Desert Storm,' the First Iraq War saw rising rates, with overall casualties due to fratricide near 24 percent (Koehler, 1992). In fact, fratricide accounted for 77 percent of the combat damage sustained by U.S. armored vehicles during this conflict. This clearly demonstrates that, although our technology has advanced significantly over the course of the last century, the limitations of the human perceptual system have remained a constant.

According to Briggs and Goldberg (1995), the potential for fratricide from inaccurate recognition is immense even under favorable environmental conditions. Under a signal detection framework, two types of error may occur. Inaccurate recognition may lead to higher levels of false positives (i.e., fratricide) and the opposite situation (i.e., misses or mistaking an enemy vehicle for a friendly vehicle). Both types of errors in target identification will be labeled under the global term *misidentification* throughout the remainder of this chapter.

Purpose of this Chapter

This chapter aims to integrate fundamental cognitive theories into a usable and explanatory model for combat identification. To help capture why some types of misidentification problems occur, we will look to Feature Integration Theory (Treisman and Gelade, 1980), Recognition By Components Theory (Biederman, 1987) and Working Memory Theory (Baddeley and Wilson, 2002).

Feature Integration Theory (FIT) can lend support towards a better understanding of what is happening to a soldier's visual perception and processing on the battlefield due to attention and distractions, and specifically how or why a soldier would mistake the identity of one vehicle for another when placed under the stress of being in the midst of combat.

To explain the possible advantage of the utility of studying 3D models as a novel, and potentially better method for training, we examine Recognition By Components Theory (RBC; Biederman, 1987). According to RBC Theory, the ability to recognize an object comes from organizing a limited number of constituent 'parts,' referred to as *geons*, into an *object model*, which is a mental representation of the perceived object.

Finally, we will also look at memory, and the model for memory we will be discussing in this chapter will be Baddeley's Working Memory (WM) Model (Baddeley and Wilson, 2002). Baddeley's model treats the brain as if it were a computer. Much like a computer has a hard disk to store information, and a microprocessor to make calculations, we too as human beings have analogous processes in our perceptual systems. Understanding these basic processes of our memory is integral to understanding high stress decision-making, such as that provided by trying to identify targets in a combat setting. Together, the three theories (FIT, WM, and RBC) provide a strong foundation that can be used to conceptualize combat identification and to make predictions about ways in which misidentification can be reduced.

This chapter first discusses three theories, and demonstrates how each theory describes a different facet of the complex task of combat identification. Following our theoretical model, an overview of research conducted by the authors will be reviewed showing how these particular theories may be applied in both experimental and real world settings. We will then discuss implications of our experimental evidence, and future avenues for research and application.

Factors Involved in Misidentification

Misidentification may be a result of many different factors: inadequate training, confusion during battle, decision-making under pressure, or stimulus similarities among friendly and enemy vehicles. There is also the problem of bias on the battlefield. Specifically, it has been shown that during combat, soldiers are more likely to judge an unknown vehicle as a foe (Briggs and Goldberg, 1995). Although humans are capable of recognizing objects of many shapes, forms, and colors under various conditions, battlefields are complex environments that can mask targets and confuse soldiers, making the identification of potential threats much harder (Koehler, 1992). Decisions made under the different pressures exerted on the battlefield can therefore lead to more errors in detecting and identifying targets.

The 'Fog of War'

One factor related to target misidentification is known as the 'fog of war'. The fog of war can be described as a combination of factors: featureless terrain; large complex and fast-moving formations; fighting in precipitation, darkness, or low visibility; and the ability to engage targets from long distances. These factors all contribute to friendly fire incidents (Koehler, 1992). Training, attention, similarity between targets and non-targets, and timed decision-making, all play a vital role in making correct judgments. Although the fog of war will always be present (Koehler, 1992), we believe that through technology and training some of these negative factors can be alleviated and the ability to correctly identify military vehicles can be increased and enhanced (Biederman and Shifrar, 1987).

Stimulus similarities Similarities among vehicles from different militaries is another problem that can lead to misidentification. If a vehicle is similar enough to another vehicle, it can easily be mistaken for a friend or an enemy. An example of this is the Main Battle Tank, which shares its basic architecture throughout the globe (Keebler, Sciarini, Jentsch, Fincannon, and Nicholson, 2008). Frontal views of vehicles are often associated with more mistakes in identification as they show fewer of the necessary and salient cues (Briggs and Goldberg, 1995). Main Battle Tanks share fundamentally the same basic shapes and construction, especially when seen head-on (i.e., a turret on top of a two-track undercarriage; Keebler et al., 2007), and thus provide an example of pan-military vehicle similarity.

Training Generally, having high vehicle image fidelity is not important for proper training, provided that the critical cues (e.g., chassis, turret shape, mortar tubes) are visible (Briggs and Goldberg 1995). Nonetheless, researchers have questioned whether using 2D media (e.g., cards, images in a computer simulation) are effective for familiarizing soldiers with the vehicles that need to be identified and discriminated in a 3D environment (Briggs and Goldberg, 1995). In fact, as critical cues become less salient, the difficulty of the identification task rises drastically. Further, 2D training aids usually show only a limited number of views, whereas vehicles may be seen in the real world from any angle or orientation. This suggests that using 2D training aids may not be effective at facilitating accurate vehicle identification. Later, we will refer to research we have conducted that may lend support to the idea of using 3D objects as superior training media.

Theoretical Framework

Feature Integration Theory (FIT)

Anne Treisman developed a cognitive theory that gives insight into what occurs in the human perceptual system when making an identification decision (Treisman and Gelade, 1980). As an example, imagine a soldier confronting a vehicle that is partially occluded by trees. Certain aspects of the vehicle (e.g., size, shape, color, number of wheels, turret orientation, turret size) may be hidden. Given the small number of cues provided, an observer may try and identify the vehicle using a base of knowledge derived from all of the different types of vehicles that a particular individual has stored in their memory. If the observer is able to provide full attention to the occluded vehicle, then that observer's *attentional spotlight* would interpret this information, and this would lead to a successful description of the vehicle entering that individual's mind. If instead, the observer is unable to focus his or her full attention, some of the important or critical perceptual cues may not make it to the point where they can be used as distinguishing features to recognize specific vehicles. This may lead to errors, and is the main reason that we should consider prescribing to Treisman's model in combat.

According to Treisman's model, *attention* plays an integral role in perception. Other researchers have argued that perception is impossible without attention (Briand and Raymond, 1987). Analogous to a spotlight shining into the darkness to illuminate a specific area, attention can only focus on a small part of the visual field. When focused, attention allows for temporary object representations to form in the observer's mind, which is then compared to stored memories (Treismann and Gormican, 1988). When unfocused, however, these temporary object representations can become chaotic, and information can be mixed up within and between objects. This is known as an *illusory conjunction*, and is an integral finding in Treismann's research. An *illusory conjunction* is an occurrence of misperception, where an object appears to have a different shape, size, or color than it actually has in reality. For example, during one of Treisman's experiments, different colored letters (X's and O's) placed in the visual periphery were misperceived significantly more than chance (a red X would be shown with a yellow O, but participants would report a yellow X and a red O), when participants were involved in another task that consumed most or all of their attention (Treisman and Gelade, 1980).

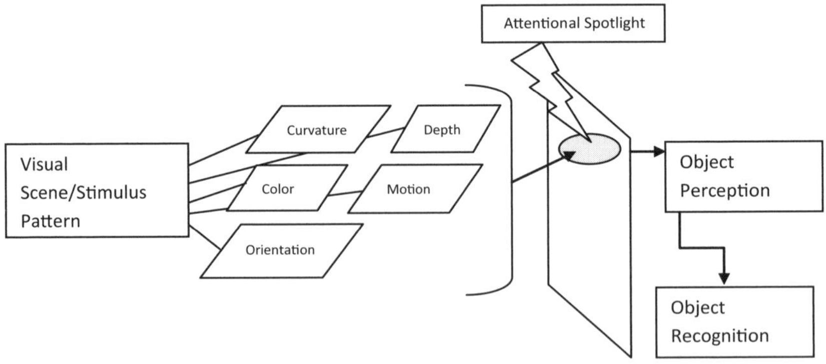

Figure 7.1 Feature Integration Theory Model

If we subscribe to this theory, then it could be argued that a soldier, when under the pressure of combat, must focus most or all of his or her attention to properly make target identification decisions. If the soldier cannot provide adequate attention to a potential threat vehicle to allow for a temporary object representation to form, then the soldier will be unable to make correct judgments as to the identity and alliance of the given vehicle. This leads to a critical need for those soldiers performing identification tasks to attend to the task alone. Any other task involving even minimal amounts of attention could diminish perceptual resources needed to make correct identifications.

FIT demonstrates how integral attention is in the perception of a vehicle's identity. The act of identifying vehicles also requires detailed memories of many friendly and potentially threatening vehicles. Without proper training, soldiers would be unable to recognize different vehicles, even if their attention allowed for perception. We will now discuss Recognition by Components theory, to gain a better understanding of how we should train soldiers to best encode detailed memories of the vehicles they may encounter in battle.

Recognition by Components Theory

Biederman's (1987) Recognition by Components (RBC) theory provides insight into how we perceive all known objects. This cognitive model suggests that objects can be broken down into a limited number of constituent 'parts' referred to as *geons* (geometric icons), the basic volumetric form of object perception. Geons are represented by many objects such as cubes, cones, and spheres. There is an alphabet of approximately 30 of these shapes, and using this alphabet, all known objects can be constructed. We recognize objects by matching the geons in the world to the geons stored in our memories. We first fire 'feature' detectors, and these feature detectors locate basic geons in the objects we perceive. Our ability to recognize an object comes from organizing present geons into an 'object model,' which is a mental representation of the perceived object (Biederman and Cooper, 1991). These geons may be very similar to the object representations explained in FIT theory; however, FIT theory does not include a descriptive alphabet of the actual objects used in perception. This leads to a need to integrate Biederman's model alongside Treisman's.

When perceived geons are properly matched to those in memory, observers can reliably identify a given object. If the object is physically or perceptually obscured; however, the observer may not perceive all of its components. As in FIT, identifying critical components and cues is pertinent to lowering error rates and increasing accurate identification. Targets are often misidentified, as battlefield observers are rarely able to see an entire vehicle because of obscuration by terrain, dust, smoke, camouflage, or poor illumination (Briggs and Goldberg, 1995). This could be interpreted to mean that observers are missing critical cues of the vehicle that allow for proper identification, and therefore make decisions based on the available componential information.

When identifying and discriminating between military vehicles, the RBC model suggests that an observer would first notice the basic outline and structure of the vehicle's geons (e.g., the turret, the main gun and the body). Many tanks and armored vehicles throughout the world have geons that are arranged in the same exact fashion. The similarity in the arrangement of these vehicle geons could make it difficult to discriminate between a friendly and an enemy vehicle. From certain views, especially the frontal view, it is almost impossible to tell many of these vehicles apart from one another. Researchers have even suggested that frontal views of tanks often force rapid decision-making, therefore presenting a potentially dangerous situation (Briggs and Goldberg, 1995).

We believe that RBC theory supports the idea that training media should contain the same geons as the real world object that is being identified. There may be a considerable benefit in using 3D models of military vehicles as training devices. As mentioned earlier, 2D training media may not be the most effective medium for training military vehicle identification (Keebler et al., 2007). Chang, Bowyer, and Flynn (2003) suggested that objects in a 2D space present a problem due to the restriction in the way the object can be viewed or posed. In other words, 2D representations poorly depict the full range of angles that a 3D object presents. According to RBC, 2D representations contain only a few surfaces of the present geons, while 3D models present the geons fully intact.

Pierno, Caria, and Castiello (2004) suggested that 3D objects also provide a perspective-based visual cue that may be easier to utilize than 2D visual cues. The argument for 3D training has a strong foundation and should be introduced in future investigations given that a 3D cue is more easily mapped onto real-world coordinates. If 3D cues are more readily mapped then training specific cues should aid in enhancing correct judgments (Briggs et al., 1995). Further, Pierno et al. (2004) found that experiencing 3D based objects rather than 2D based objects produced faster target acquisition times. Friedman, Spetch, and Ferrey (2005) also argued that the depth cues available in 3D objects may facilitate tank identification. These authors also suggested that 3D objects provide a more detailed account of an object. Accordingly, 3D data should yield better tank identification and recognition performance (see Chang, Bowyer, and Flynn, 2003). Studies have indicated that 3D training of plate tectonics led to significantly better performance than a similar 2D training method (Kim, 2006), and that training on 3D models led to significantly higher scores on a tank identification task, when participants were tested on volumetric, 3D objects (Keebler et al., 2007).

If training can be altered in such a way as to improve the detail of the object model that will be stored in the trainee's memory, we can potentially mitigate some of the problems which contribute to misidentification. When attention can be focused enough to perceive the components of potential threat vehicles, then training must be powerful enough to make sure those components are matched to their proper memory store. If this occurs, there can be an increase in correct identification decisions being made. Biederman's model thus supports potential training enhancements with the introduction of 3D training media. We will now look to Baddeley's Working Memory Model to understand an individual's capacity to store what is learned in training.

Baddeley's Working Memory Model

According to Baddeley's working memory model, the mind functions much like a computer system (Baddeley and Wilson, 2002). There are three main resources in this model: two 'slave memory' areas that allow observers to store (a) visual information (visual-spatial sketchpad) and (b) auditory information (phonological loop), and a third processing area called the Central Executive (CE). The CE takes information from the two perceptual slave-memory areas and processes it, comparing it to old memories, expectancies, and schemata, so as to make decisions about the world. Baddeley's model is integral to this chapter in the sense that an identification task requires comparing the temporary storage of what is being perceived to a long term memory version of the same object. Correctly aligning these two objects should lead to correct judgments in identification. However, to test this notion, it is very important to somehow measure the amount of space or level of detail in the visual-spatial sketchpad. We propose here that the Shape Memory Test (ETS MV-1) may be a measure which quantifies the quality or amount of storage capacity in the sketchpad of an individual.

The Shape Memory Test has been found to be moderately reliable. Waters, Gobert, and Leyden found a split half correlation of .76 ($N = 33$) in their studies involving chess players, and described the Shape Memory Test as having adequate reliability. Dr. Burton of the University of Southern Queensland found a Cronbach's alpha of .68 in her studies using shape memory. In Experiment 2, described below, the test was found to have a Cronbach's alpha of .72 ($N = 57$). The authors believed that these numbers represent stability in the test across different types of studies, albeit a reliability coefficient that tends to be rather low. In the experiments described below, the test demonstrates predictive validity in military vehicle identification tasks.

We theorize that the Shape Memory Test should be predictive of performance on identification tasks, demonstrating that the more information an individual can temporarily store on their sketchpad, the better they will perform when asked to recall and compare that information during a performance task.

Experimental Evidence

The following studies were designed to compare the utility of using 3D models over a current form of training, namely military issued 2D cards containing line drawings. Both studies used the MV-1 Shape Memory Test as a pre-measure of memory capacity. Also, we introduced a simulated observation post to the second study, to investigate effects due to immersion. As previously discussed, 3D models provide a richer training media, and therefore should lead to higher performance scores on identification tasks (Keebler et al., 2007).

Experiment 1

Design

To test our hypotheses, we designed a 2 × (2 × 3) mixed-model factorial design where training (2D vs. 3D) was the between subjects variable, test media (2D vs. 3D) was the first within subjects variable, and test knowledge area (identification, recognition, and friend/foe differentiation) was the other within-subjects variable. To help clarify the terminology for these studies, we used recognition in the sense of whether someone recognized if they had seen the vehicle before or not. Identification was the explicit naming of the test vehicle, while friend/foe differentiation consisted of knowing whether the vehicle was a friendly or an enemy vehicle.

Participants Twenty student volunteers participated in Experiment 1. The participants' ages ranged from 18–32, with a mean age of 24. Fourteen of the participants were female and six were male. There were no participants with prior military experience.

Materials and procedure Typical pre-experimental materials included informed consent and demographics forms. The participants then completed the Shape Memory Test (ETS MV-1). Participants were randomly assigned to one of two conditions: training with cards or training with models. Those in the card training condition viewed six military issue Armored Vehicle Recognition cards. In contrast, those in the model training condition viewed 1:35 scaled models of the same six vehicles. The participants were each given six minutes to study the vehicles.

The order in which the tests were administered was counterbalanced across participants. The identification test required participants to recall the name of the vehicle; the recognition test required the participants to write 'yes' or 'no' to the question of whether they had seen a specific vehicle during the exposure training session. For the friend/foe test, participants were asked to write whether a vehicle was a friend or foe.

Testing Conditions

Card testing For the card test, the cards were all copied onto a sheet of paper. The participants viewed the same graphic cards presented during the exposure training in a different order than was presented in training and with two extra vehicles included as distracters. Participants answered the identification, recognition, and friend/foe questions on blanks provided on the test sheet.

Model testing The same 1:35 scale vehicles were used for the model tests as in the training. The vehicles were reorganized on a table, and two extra vehicles were added as distracters. Participants were asked to answer the identification, recognition, and friend/foe questions.

Results

A repeated-measures multivariate analysis of variance (MANOVA) was used to test whether the participants in the model exposure training group outperformed the participants in the card exposure training group on the three dependent measures of performance (identification, recognition, and friend/foe) using ($p < .05$). Wilk's Lambda multivariate test indicated a statistically significant main effect for testing ($F(5, 14) = 5.09, p < .01$, *partial* $\eta^2 = .64$). Furthermore, Wilk's Lambda indicated a statistically significant interaction for Training x Testing ($F(5,14) = 5.61$, $p < .005$, *partial* $\eta^2 = .67$). A one-tailed Duncan's post hoc analysis was chosen to further analyze the interaction effects.

Card training vs. model training for identification At ($p < .01$), the post hoc analysis indicated that the card training group performed better on the card identification test ($M = .63$, $SD = .37$) than on the model identification test ($M = .35$, $SD = .28$). The model training condition performed better on the model identification test ($M = .57$, $SD = .35$) than the card identification test ($M = .40$, $SD = .34$; $p = .04$).

Card training vs. model training for recognition Participants in the card training condition performed significantly better on the card recognition test ($M = .89$, $SD = .11$) than on the model recognition test ($M = .63$, $SD = .12$; $p < .01$). There was no significant difference on the recognition test for the model training group.

Card training vs. model training for friend/foe The card training condition for both the card friend/foe test and the model friend/foe test had means of $M_{Cards} = .50$ and $M_{Models} = .52$. These means were not compared to the model training condition because they indicated that the participants were doing no better than chance. There was no significant difference between card ($M = .65$, $SD = .19$) and model ($M = .67$, $SD = .34$) friend/foes scores for the model training group.

Shape memory vs. training interaction Using multiple regression analysis, a significant interaction was found between shape memory and training condition. At Step 1, our $R^2 = .141$, $F(2, 17) = 1.392$, $p = .275$. With the addition of the interaction term at Step 2, the amount of variance explained rose to 47 percent, $R^2 = .465$, $F(1,16) = 9.707$, $p < .007$. This suggests that the combination of model training materials and shape memory was predictive of performance on the military vehicle identification task.

Experiment 2

Design

Experiment 2 of this research was designed to increase the realism of our study. The testing modality used for this study was a set of images projected in a desert simulation. In addition to the use of virtual images as the testing materials, participants were brought into a dimly lit immersive foxhole that was decorated with sandbags and other military accoutrements.

Participants Fifty-eight volunteer undergraduate students participated in this study (39 males, 19 females). Participants all received credit in an undergraduate psychology course for participation in this study.

Materials and procedure The pre-experimental materials included the typical informed consent and demographics form, and post-experimental debriefings were also administered. Before viewing the military vehicles, participants were administered the Shape Memory Test. After the Shape Memory Test, participants were randomly assigned to one of two conditions: card training or model training. Those in the card training condition viewed nine military issued Armored Vehicle Recognition cards. Those participants in the model training condition viewed nine 1:35 scaled models of the same tanks. The participants were each given one minute per vehicle to study the tanks (for a total of nine minutes).

The participants were brought into a simulated foxhole for testing. The foxhole consisted of a sandbag emplacement covered by a camouflage net. The foxhole was in a darkened room, and projectors on the ceiling projected images of the surrounding terrain onto a 180-degree wide projection area in front of the foxhole. A vehicle would be placed into the simulation, at which time it would appear as a target on the simulation screen, appearing to be approximately 300 meters away. The participants were asked if they could see the tank in the simulation, and then told to look through simulated binoculars, which had a 'zoomed' image equivalent to a distance of about 50 meters. After ten seconds had passed, the vehicle was removed from sight.

Results

To test the hypotheses, we designed a 2×3 mixed model study where training (cards vs. models) was the between-subjects variable. Test format (identification, recognition, and friend/foe) was the within-subjects variable.

Card training vs. model training for identification An independent samples t-test was executed to investigate if there was a significant difference between the means of the two conditions. The test results showed no significant differences between the two conditions at $p < .05$.

Shape memory and target identification A multiple regression correlation was used to test our hypothesis concerning shape memory. Controlling for knowledge of military vehicles and experimenter ratings of participant interest level, it was found that shape memory was predictive of target identification performance, although an interaction was not present this time. Step 1, which included familiarity with military vehicles and interest level, explained about 35 percent of the variance in performance, i.e., $R^2 = .348$, $F(2,58) = 14.4$, $p < .0005$. Step 2, which added shape memory to the equation, was significant and explained about 12 percentage points more of the variance, $R^2 = .477$, $F(2,57) = 16.06$, $p < .0005$. Overall, the second model explained almost 48 percent of the variance in identification.

General Discussion

Overall, our results suggest that the 1:35 scale models may have a superior training value over the card modality. Below we will discuss our three testing areas of identification, recognition, and friend/foe differentiation, with a discussion of results from both experiments.

2D and 3D Identification

Based on the results of these studies, our hypothesis that 3D training would lead to better performance on target identification tasks was not fully supported when volumetric test objects were being used. When asked to recall the name of a tank, the participants performed better on the tests that matched the media they were exposed to during training. We believe that 3D models have stronger external validity and may provide more realistic training than line drawings on cards. Theoretically, the mental 'object model' referred to in both Biederman's and Treismann's theories was possibly developed more thoroughly by the 3D training media then the 2D media. In the 2D training condition, the participants were only receiving partial views of the vehicles, while in the 3D training condition they were able to see the entire vehicle. This finding is consistent with a study by Hah, Reisweber, Picart, and Zwick (1997) in which participants were trained with whole and partial views. They found training with whole views outperformed partial view training. This lends support to Biederman's theory in that 3D training provides a higher number of complete geons, and therefore, a more enriching training medium.

2D and 3D Recognition

When participants were asked whether they had seen the vehicles that they were previously exposed to, the 2D training group did better on the 2D test than the 3D. This suggests that when viewing a 2D object, such as an x-ray, recognition may be better if the viewer has been trained using 2D training aids rather than 3D training aids. The remaining mean test scores for recognition were consistent. This suggests that the 3D recognition task may be trained using either 2D or 3D training aids. Further investigations should be conducted to find the effects of combining both 2D and 3D media when training for 2D recognition tasks.

2D and 3D Friend/Foe Accuracy

The friend/foe test scores indicated that the participants in the 2D training condition were doing no better than chance when asked to identify a friendly tank from an enemy one. On the other hand, the 3D training group produced consistently higher friend/foe scores for both 2D and 3D testing. 3D training showed higher consistency, and this may have something to do with the amount of information gathered from the training. Specifically, this may be due to the larger amount of geon information provided by the 3D training media. These findings suggest that 3D media may be better for training discrimination tasks.

Based on the results of this study, we concluded that the addition of 3D training media may have important implications. Specifically, the presentation of volumetric forms as a training media seems to impact an individual's ability to later identify vehicles, specifically when those vehicles are also volumetric in nature. Also, individual differences in shape memory capacity has been shown to interact with this training, possibly prescribing that certain people may be inherently better at remembering, and later, identifying targets. We also concluded that improving vehicle recognition tasks may require both approaches to training. For example, 3D training may be useful for naming and discriminating tanks and other vehicles, while 2D training can be used for recognition tasks. As a result, we suggest that both 2D and 3D training media should be incorporated into training programs for 3D object recognition tasks.

Shape Memory by Training Interaction

A statistically significant interaction was found between the Shape Memory Test and training media. Specifically, those trained on the 3D models significantly outperformed the rest of the sample, with improved results as shape memory increased. This finding supports our theoretical framework, as far as Biederman's and Baddeley's theories are concerned. With the addition of an immersive environment in Experiment 2, we found that this interaction disappeared, only to be replaced by a main effect for Shape Memory. This finding shows that visual memory, above and beyond training, can have important implications for identification performance under high stress conditions.

In summary, the findings from Experiment 1 suggested that a more multi-faceted training approach should be considered. Training needs to take into account the various aspects of the tank recognition task, and match the training modality appropriately. In addition, this study approached vehicle recognition from only the perceptual aspect of the task. Other cognitive skills, such as decision-making are required for the task. Both the perceptual aspects and the cognitive aspects of the tasks need to be studied in conjunction.

Conclusions

The main focus of this chapter was to discuss the contribution of three major cognitive theories to the process of identification. Designing training applications and systems around these theoretical frameworks may alleviate some of the issues associated with misidentification.

Whether from the sensor of an unmanned vehicle, a photograph from a reconnaissance plane, or a soldier's direct line of sight in the field; many factors influence the ability to correctly identify vehicles. The first implication we can take away from our discussion is that individuals tasked with vehicle identification must *pay attention to the task*. Whether this necessitates the addition of an entire role designed strictly for target identification is beyond the scope of this chapter, but the need for those soldiers involved in target identification to have their attentional resources fully devoted to that task alone is supported by the cognitive models we have discussed. Without attention, there is no perception.

The next point is the importance of *individual differences in working memory*. Through the experiments discussed above, we have found that having higher scores on tests of visual shape memory can lead to higher performance in identification tasks. Making sure that our soldiers and/or operators have high visual memory is critical for them to properly perceive and identify the vehicles that they may encounter. Screening for this capacity should be an important process in the selection of soldiers and/or operators who will be involved in the identification of vehicles.

The *addition of scale models to training programs* is the final consideration of this chapter. Both studies mentioned above have introduced a novel training method, namely that of utilizing 1:35 scale models as training devices. As discussed in Biederman's theory, these highly detailed, realistic reproductions of actual vehicles can provide a richer representation of the geons needed to perceive, remember, and identify the proper military vehicles. Adding scale models to current forms of training (cards, simulations, etc.), according to our data, should be beneficial. The more realistic and detailed an object is during training, the more realistic and detailed the mental representation. This mental representation may be the only piece of information that a soldier can rely on when they are called upon to make critical decisions. Making sure these mental representations are as reliable, detailed, and useful as possible is the purpose of effective training in identification.

References

Baddeley, A., and Wilson, B. A. (2002). Prose recall and amnesia: Implications for the structure of working memory. *Neuropsychologists*, *40*, 1737–1743.

Biederman, I. (1987). Recognition by components: A theory of human image understanding. *Psychological Review, 94*, 115–147.

Biederman, I., and Cooper, E. E. (1991). Priming contour deleted images: Evidence for intermediate representations. *Cognitive Psychology*, *23*, 393–419.

Biederman, I., and Shiffrar, M. M. (1987). Sexing day-old chicks: A case study and expert systems analysis of a difficult perceptual-learning task. *Learning, Memory and Cognition*, *13*, 640–645.

Briand, K. A., and Raymond, M. K. (1987). Is Posner's 'Beam' the same as Treisman's 'Glue'? On the relation between visual orienting and feature integration theory. *Journal of Experimental Psychology, 13*, 228–241.

Briggs, R. W., and Goldberg, J. H. (1995). Battlefield recognition of armored vehicles. *Human Factors*, *37*, 596–610.

Burton, L. (2003). Examining the relationship between visual imagery and spatial ability tests. *International Journal of Testing*, *3*, 277–291.

Chang, K. I., Bowyer, K. W., and Flynn, P. J. (2003, December). *Face Recognition using 2D and 3D Facial Data*. Paper presented at the Workshop in Multimodal User Authentication, Santa Barbara, CA.

Friedman, A., Spetch, M. L., and Ferrey, A. (2005). Recognition by humans and pigeons of novel views of 3D objects and their photographs. *Journal of Experimental Psychology: General*, *134*, 149–162.

Gauthier, I., Williams, P., Tarr, M. J., and Tanaka, J. (1998). Training 'greeble' experts: A framework for studying expert object recognition processes. *Vision Research*, *38*, 2401–2428.

Green, C., and Hummel, J. E. (2006). Familiar interacting object pairs are perceptually grouped. *Journal of Experimental Psychology: Human Perception and Performance, 32*, 1107–1119.

Hah, S., Reisweber, D., Picart, J., and Zwick, H. (1997). Target recognition performance following whole-views, part-views, and both-views training. *Engineering Psychology and Cognitive Ergonomics: Job Design and Product Design*, *2*, 135–141.

Keebler, J. R., Harper-Sciarini, M. E., Curtis, M. T., Schuster, D., Jentsch, F., and Bell-Carroll, M. (2007). *Effects of 2-dimensional and 3-dimensional Training Media in a Tank Recognition Task*. Proceedings of the 51st Annual Human Factors and Ergonomics Society Meeting. Baltimore, MD.

Keebler, J. R., Sciarini, L. W., Jentsch, F., Fincannon, T., and Nicholson, D. (2008). *Effects of Training Modality on Military Vehicle Identification in a Virtual Environment*. Proceedings of the 52nd Annual Human Factors and Ergonomics Society. New York, NY.

Kent, C., and Lamberts, K. (2006). The time course of perception and retrieval in matching and recognition. *Journal of Experimental Psychology: Human Perception and Performance, 32*, 920–931.

Kim, P. (2006). Effects of 3D virtual reality of plate tectonics on fifth grade students' achievement and attitude toward science. *Interactive Learning Environments, 14*, 25–34.

Koehler, A. C. (1992). *Friendly Fire on Today's Battlefield.* Retrieved October 30, 2008, from http://www.globalsecurity.org/military/library/report/1992/KAC.html.

Pierno, A., Caria, A., and Castiello, U. (2004). Comparing effects of 2-D and 3-D visual cues during aurally aided target acquisition. *Human Factors, 46*, 728–737.

Treisman, A., and Gelade, G. (1980). A feature-integration theory of attention. *Cognitive Psychology, 12*, 97–136.

Treisman, A., and Gormican, S. (1988). Feature analysis in early vision: Evidence from search assymetrics. *Psychological Review, 95,* 15–48.

U.S. Congress, Office of Technology Assessment. (1993). *Who goes there: Friend or foe?* (OTA-ISC-537) Washington, DC: Government Printing Office.

Waters, A. J., Gobet, F. Leyden, G. (2002). Visuo-spatial abilities of chess players. *British Journal of Psychology, 93*, 557–565.

Wolfe, J., Horowitz, T., and Kenner, N. (2005). Rare items often missed in visual searches. *Nature*, *435*, 439–440.

Chapter 8

Preattentive Attributes in Visualization Design: Enhancing Combat Identification

Scott H. Summers
Raytheon Solipsys

The U.S. Joint Chiefs of Staff (JCS) defined combat identification (CID) as '… the process of attaining an accurate characterization of detected objects to the extent that high confidence and timely application of military options and weapon resources can occur.' (2003, p. I-4). While a topic of much debate, the overarching goal of CID is more than avoiding fratricides; the goal is to win conflicts and win them decisively (Dittmer, 2004). As one measure of CID effectiveness however, statistics show that fratricide incidents accounted for at least 10 percent of the total U.S. casualties in World War II, Korea, Viet Nam, and the First Persian Gulf War (Shrader, 1982; Steinweg, 1995). Conflict on the modern battlefield involves a myriad of participants, including joint and multi-national forces, heterogeneous enemy factions, civilians, and non-combatants. In an era characterized by rapidly moving forces and weapons that can strike targets with very high precision at unprecedented ranges, the difficulties of Command and Control (C2) have increased tremendously. This results in increased opportunities for targeting errors, and in fact, it has been asserted that the nature of modern warfare, however technologically advanced, actually raises the risk of fratricide (Defense Update, 2004; Rasmussen, 2007).

As U.S. and coalition forces transform to a network-centric warfare paradigm, increasingly, weapons systems are more dependent on external sources for precise targeting information and firepower is routinely unleashed on distant coordinates provided by remote sensors and network sources (Rasmussen, 2007). The *kill chain* in a network-centric environment can only be as strong as its weakest link; thus it is imperative that the human-machine interface (HMI) at every connected level, from sensor to shooter, be designed to maximize the warfighter's ability to comply with the JCS' dictum to quickly and accurately characterize the detected objects in the dynamic tactical situation that unfolds before them.

History provides well-known examples of instances when flaws in the design of system interfaces were found to be causal in the loss of innocent lives. In 1988, the U.S.S. *Vincennes* shot down an Iranian airliner filled with civilian passengers, killing all aboard. Though the system was found to be operating normally, an overly complex and poorly designed weapons-control interface was blamed for the tragedy, as the weapons system operators were unable to accurately characterize and interpret the aircraft's actions in a timely fashion (Lerner 1989, Cummings, 2006). More recently, during Operation Iraqi Freedom, the confusing and overly complex interface design of the Patriot air defense artillery system, along with inadequate crewmember training and *automation bias* (Mosier and Skitka, 1996), were implicated in two separate fratricide incidents, resulting in the inadvertent destruction of British and U.S. fighter aircraft, killing three aircrew members (Hawley, Mares and Giammanco, 2005; Cummings, 2006). In technology-dense environments, it has often been the case that when errors occur, they are attributed to the human operators. However, some researchers have indicated that very often the fault lies with the design of the systems themselves (Perrow, 1984; Norman, 2002).

Decision Making in High Stress Environments

Following the *Vincennes* incident, the U.S. Navy initiated the Tactical Decision Making Under Stress (TADMUS) project, spending over 10 years researching the problem areas that emerged from the incident (Hawley, 2006). The aims of the program were to analyze decision-making strategies used by people in stressful situations, to determine the ways those strategies sometimes fail, and to prevent these failures (Hair and Pickslay, 1993). The Navy found that Situational Awareness (SA) plays the key role in decision quality in the command and control domain, and that SA is built upon tactical experience and expertise (Cannon-Bowers and Salas, 1998). However, other research indicates that as the primary visualization tool for assessing a tactical situation, the weapons system's HMI is supremely important; indeed, in a command and control context, system HMI usability may be the primary determinant of SA (Bolia, Vidulich, Nelson, and Cook, 2004). The Navy further concluded that operator training must necessarily emphasize the development of adaptive decision-making skills. Anomalous events must be introduced regularly into training regimens, encouraging weapons operators to be able to respond flexibly and effectively to non-routine events (Cannon-Bowers and Salas, 1998). Conversely, however, it has also been asserted that at a certain point, no amount of pre-selection and training of personnel can compensate for a flawed HMI or system design (Crisp, McKneely, Wallace, and Perry, 2001). Clearly, in complex socio-technical systems such as command and control, with decisions that are sometimes made under extreme stress, there are multiple potential points of failure.

Naturally, stress is not exclusive to the military domain; in their study of medical decision-making in level 1 trauma resuscitation wards, Xiao and Mackenzie (1997) determined that stressful environments are characterized by, 'Fast-changing, complex and uncertain situations … in which the performance in decision-making carries high stakes … in which critical decisions have to be made under extreme time pressure … in which decisions are made and carried out collectively by multiple individuals in a team setting' (p. 12). When people are required to make rapid, important decisions in stressful environments, research has found that the effect of stress on the quality of decision-making typically results in two types of outcomes (Janis and Mann, 1977). When stress is moderate, vigilance increases to a constructive level, and search, situational appraisal, and contingency planning are all improved. Under excessive stress, people become hypervigilant, resulting in incomplete search, appraisal, and contingency planning, thus leading to errors of omission or commission (Xiao and MacKenzie, 1997). In the words of U.S. veteran Colonel David Hackworth, 'Fear, nervousness, excitement and exhaustion numb the mind and cause miscommunication and misunderstandings. These circumstances are a recipe for error' (Defense Update, 2004, p. 1). In his book *Emotional Design*, Donald Norman (2004) confirmed that negative affect can inhibit cognitive functioning, and concluded 'Things intended to be used under stressful situations require a lot more care, with much more attention to detail' (p. 26).

Preattentive Processing

With the knowledge that high stress and negative affect severely limit our ability to accurately perceive the world around us, it might be ideal if there were some way to design interfaces that in some measure bypass conscious cognitive processes altogether in environments that can be expected to produce extreme stress. It makes sense that if we can understand the psychophysics of human perception, as well as the information needs of our prospective system users, we can present data in such a way that the most salient points emerge almost effortlessly from the display. Conversely, a failure to account for these needs can render our data in a confusing or even misleading way (Ware, 2000). Perhaps Edward Tufte said it best, 'Confusion and clutter are failures of design, not attributes of information' (1990, p. 53).

One possibility for making systems maximally understandable with minimal mental effort may lie in the exploitation of certain known psychophysical responses of the low-level human visual system known as preattentive processing. Researchers have identified a limited set of basic visual attributes that both real and on-screen objects can possess, which are perceived very accurately and rapidly by the human visual system (within about 200–250 milliseconds), completely outside of conscious thought or reasoning (Treisman and Gelade, 1980). There are several conflicting theories about how and why this phenomenon actually works; for an excellent overview of the primary theories, the interested reader is directed to *Perception in Visualization,* by Christopher Healey (2007).

Current research regarding human visual processing suggests that there are two distinct mechanisms for processing visual information (e.g., Treisman and Gelade, 1980; Theeuwes, 1993). The preattentive mechanism appears to be characterized by a relatively unlimited capacity which works in parallel to process information very quickly and accurately, such that test subjects are able to complete preattentive tasks with little perceived effort, such as detecting targets from non-targets (Healey, 2007).

The other processing mechanism is termed *attentive*, and is characterized by a much slower, limited-capacity serial allocation of attentional resources (Theeuwes, 1993; Healey, 2007). It is hypothesized that utilizing carefully selected preattentive properties in HMI design, particularly for those aspects that characterize primary track attributes on a C2 display, will lead to visual presentations that provide faster, more accurate interpretation, and thus more efficient and effective action with fewer errors.

In order to be detected preattentively, a target object must possess a unique visual property that non-target objects, called distracters, do not have. These unique properties are sometimes referred to as *basic* or *primitive features.* Visual search is most efficient when the target object possesses at least one basic feature that the surrounding distracters do not have (Theeuwes, 1993; Wolfe, 2001). In Figure 8.1, the low-level visual system is able to preattentively detect the targets based upon their possession of a difference in hue relative to the distracters present. In Figure 8.2, the

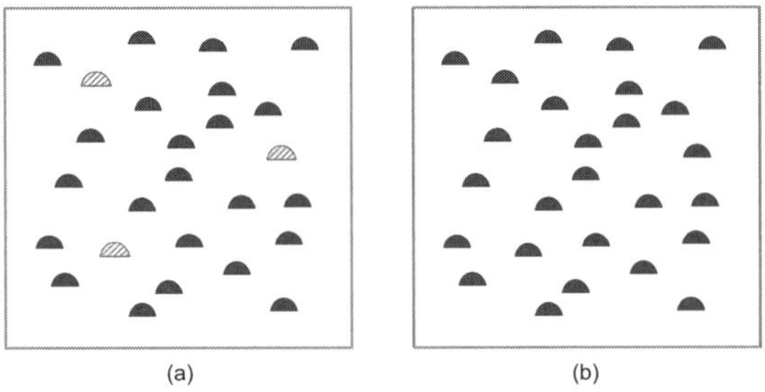

Figure 8.1 Example of a simple search for targets based upon a difference in hue (distinguished in monochrome as a striped fill). (a) Targets are present among distracters of same shape, yet are easily detected preattentively. (b) Targets are absent

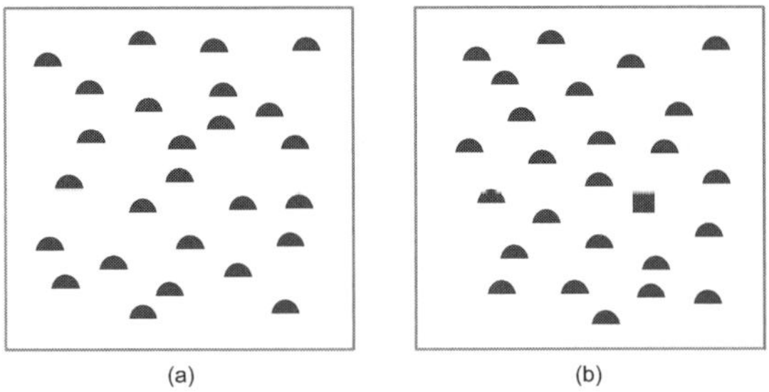

Figure 8.2 Example of a search for a target based upon a difference in shape. (a) Target is absent. (b) Target is present among distracters of the same hue, yet is easily detected preattentively

preattentive visual search is enabled through the target's possession of a difference in the basic feature of shape.

In the real world, a target will possess several basic features, and non-target distracters may share some of those attributes. Targets that possess a combination of basic features that are shared with surrounding distracters or non-targets are known as conjunction targets and require the use of attentive processing, not preattentive processing (Wolfe, 2001; Treisman and Gelade, 1980; Healey, 2007).

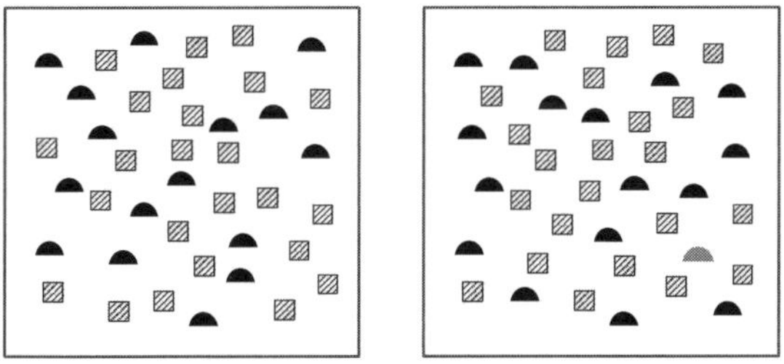

(a) (b)

Figure 8.3 An example of a search for a conjunction target – a light grey half circle. A target which possesses two basic features, one of which is present in each type of distracter, renders this a post-attentive, serial search task. (a) Target is not present. (b) Target is present

Preattentive Variables in Visualization Design

'In fact, what we mean by information – the elementary unit of information – is a difference which makes a difference' (Bateson, 1972, p. 459). With respect to CID and the Joint Chiefs' mandate to accurately characterize detected objects, the question becomes: *what are the differences that make a difference in swiftly and accurately characterizing tracks during the CID process*?

The partial listings of preattentive feature depictions in Table 8.1 are adapted from Healey (2007), and are intended to identify those basic features that may lend themselves most readily to use in C2 displays. There are a number of attributes that C2 display symbols, (representing battlespace entities and referred to as 'tracks'), can possess which will be more or less tactically important at any given time, based upon the context and specific C2 domain involved. The following compilation of attributes and application of basic preattentive features arose out of topical research, as well as informal discussions with several military C2 operations experts regarding those characteristics that they deemed the most important to the topic of CID. Given that the domain these operators were most familiar with is that of the airborne threat, this section may apply primarily to the airborne arena; however, the reader may find applicability across C2 domains. Of note, the airborne community has used a form of cooperative tracking (i.e., Blue Force Tracking-style technology) for over 60 years, via the use of the Identification Friend or Foe/Selective Identity Feature (IFF/SIF) system which the British deployed during WWII. Thus, air domain operators likely have some valuable lessons learned in the use of cooperative systems in the CID process (Rasmussen, 2007; Dittmer, 2004).

The intent of this paper and the practical applications shown are not to prescribe definitive solutions, but to forward the discussion of scientifically-based C2 display design such that the development community is better able to support the process of accurate and timely CID, thus meeting the needs of the warfighter through improved information visualization (Ware, 2000).

Table 8.1 A partial listing of preattentive basic features

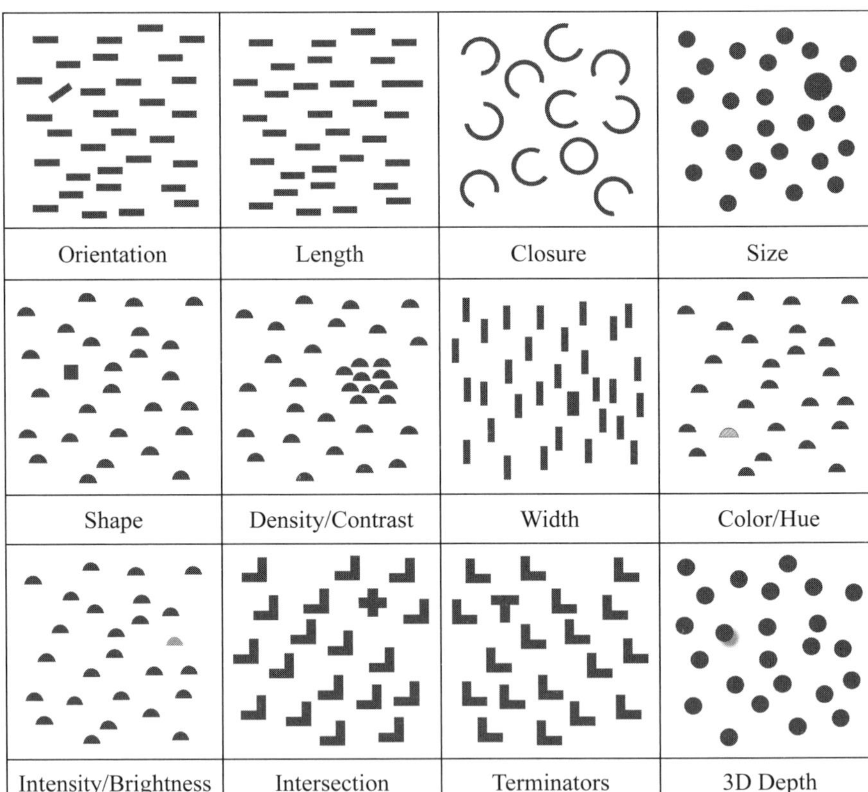

Practical Applications

Track Identification

If an operator had a completely accurate picture, with all tracks correctly identified, this might be considered an optimal situation. This is due to the simple fact that in most or perhaps all military standard symbol sets, each identification type is delineated by use of a unique shape or color, or most likely both. Thus, strictly regarding the display of identity symbols, the possibility of having conjunction targets that confound swift parallel search activities should be minimal. However, with the addition of real-world, potentially cluttering data such as flight routes, airspaces, boundaries, sensor plots, track histories, charts and other imagery, etc., it makes sense to further augment a track's ability to be easily distinguished.

In Figure 8.4, this is accomplished through increasing the target track's size, one of the known preattentive features depicted in Table 8.1. While it is somewhat common to provide an overall adjustment for track symbology size, the design implication here is that *individual categories* and *identities* should be made to be adjustable for size. Thus, for example, one could elect to make Hostile Air tracks larger than any other track symbols, if that suited the needs of the mission.

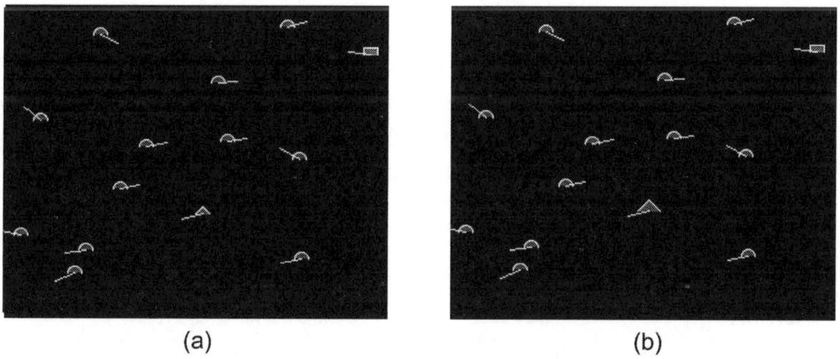

Figure 8.4 An example depicting variable symbology sizing to make a hostile air track more salient. (a) Target is scaled identically to non-target tracks; (b) Target is scaled 200% larger than non-target tracks, improving target saliency[1]

1 The illustrations used for the remaining figures in this section come from fielded C2 systems that utilize the Raytheon Solipsys Tactical Display Framework (TDF) visualization toolset.

Track Heading and Platform

A track's heading communicates a great deal to the trained operator, as it can signal intent. All other factors being equal, target heading alone could induce a decision to employ weapons, and by necessity must be an integral part of any decision to determine hostile intent (Nguyen, 2006). In Figure 8.5, screenshot (a) shows an air picture depicting, among non-targets, three aircraft with an assigned Link 16 platform of *Fighter*, using Military Standard 2525b (MS-2525b) symbols, while (b) shows the identical picture using the Raytheon Solipsys-designed Iconic Naval Tactical Data System (NTDS) symbol set.

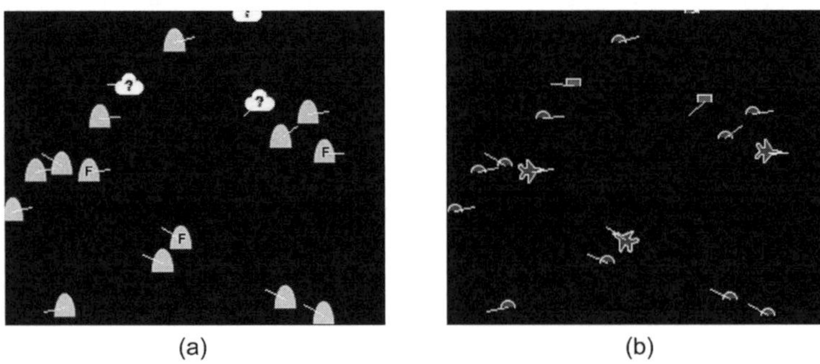

Figure 8.5 An example contrasting two track symbology sets with regard to communicating target heading and platform. (a) Three targets (Fighter aircraft) depicted using the MS-2525b symbol set. (b) Identical air picture as depicted by Raytheon Solipsys Iconic NTDS symbols

Preattentive processing research (e.g., Treisman and Gelade, 1980) supports the notion that since the fighter aircraft depicted in screenshot (b) has an iconic shape that is radically different from the other tracks present on the display, aircraft platform recognition will be more rapid than in screenshot (a), wherein the depiction of the fighter aircraft is less easily distinguished from other friendlies via the relatively subtle addition of the MS-2525b *F* modifier. The less evident benefit of the use of iconic symbols that is relevant here is that track heading is communicated directly through symbol orientation. Because there is a natural nose and tail to the iconic symbols, orientation automatically changes to reflect track heading. Compare this to MS-2525b symbols and indeed, any symbol set where the symbol orientation never changes; heading in these cases is communicated solely via the track 'vector stick.' The iconic NTDS symbol shows an obvious change in orientation, and thus research (e.g., Wolfe, 2001) suggests that iconic symbols communicate track heading, not to mention track platform, much more quickly than standard non-rotating symbols. The use of iconic symbols in military C2 symbology sets is an area that warrants further research.

Track Altitude

Altitude for an individual track, and vertical speed information (whether the track is ascending or descending and the rate over time) is often critical in the characterization of an object in the battlespace (Nguyen, 2006). Regarding vertical speed, the weapons crew aboard the U.S.S. *Vincennes* was apparently unable to accurately characterize an Iranian airliner's actions in the vertical dimension, which appears to have been a major factor in their engagement decision (Cummings, 2006). Figure 8.6 (a) depicts altitude information for two tracks, presented solely with text.

In (b), both tracks augment the standard textual altitude information with vertical speed icons, which communicate trend information. A green arrow indicates a normal rate of descent; a yellow arrow communicates a greater than normal rate of ascent. The user should be able to define the inclusive vertical speed thresholds to use for each category. The categories could include, as examples, descending or ascending at normal, greater than normal, and much greater than normal rates, as well as level flight. Each condition should be represented by unique indicator icons that vary in basic preattentive attributes of shape and hue. Using these preattentive attributes to communicate rapidity and direction of altitude change, research supports the notion that the operator will be able to capture at a glance what would otherwise require precious minutes and cognitive resources to do through diligent observation (Treisman and Gelade, 1980).

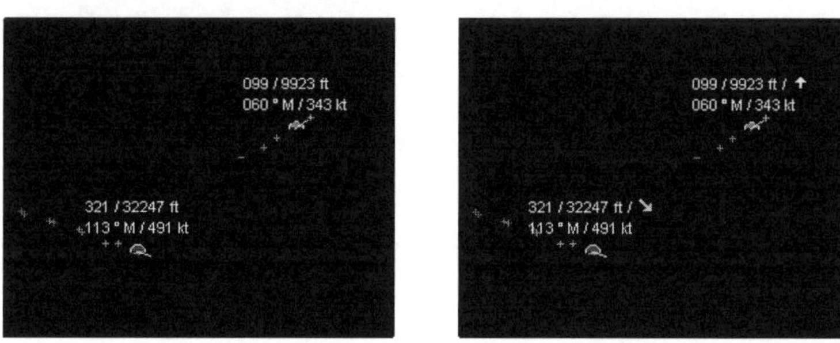

(a) (b)

Figure 8.6 An example contrasting use and non-use of an icon designed to preattentively communicate track vertical speed trend information. (a) Track altitude information presented solely via text; trend information would need to be determined manually, over time. (b) Identical track textual information, but with vertical speed indicator icons shown next to altitude information, providing ascent or descent information at a glance

Track Location

Another primary attribute that can offer tangible evidence of target intent is track location. Of particular interest is the changing location of the track over time, which the author refers to as track history. Having a lengthy track history available provides the operator a way to characterize potential future track actions as a result of understanding what the track's behavior was in the past.

The ultimate goal regarding the display of track history is for the operator to be able to determine a track's point of origin, which can assist tremendously in CID. Every attempt should be made to retain track history for the life of the track in the system, and perhaps beyond, as tracks often drop out of coverage for any number of reasons, only to reappear a short time later. For optimal utility, information about the state of the track in question should be available by interacting with track history points anywhere along the route of travel, providing the ability to retrieve track characteristics at any displayed point in the past.

Figure 8.7 (b) illustrates what may be a subtle distinction with regard to track history display, which is that it is not enough to simply support the display of track history such that it is on or off for all tracks; to utilize track history as a preattentive attribute, support should be provided to display history trails for a single track of interest, or some other limited subset of the total system tracks. This enables the operator to effectively mark tracks of interest using the basic preattentive features of shape, hue, line length, etc. that are inherent in track history depictions.

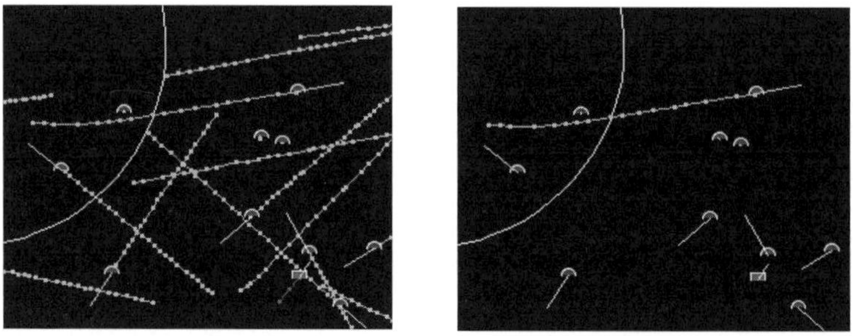

(a) (b)

Figure 8.7 An example illustrating the need to support history trails on not simply all tracks, but for individual tracks as well. (a) Track display showing history trails on all tracks. (b) Identical track picture with a history trail displayed for only a single track of interest

Presence of Friendly

Very early in the CID process, and continually throughout the Find-Fix-Track-Target-Engage-Assess (F2T2EA) cycle, assessment must be made to check potential targets for Presence of Friendly (POF) attributes; at any time the kill chain can be broken to prevent fratricide when POF is detected (Dittmer, 2004; Rasmussen, 2007; Hebert, 2003). Thus, it is critical that attributes which convey POF, such as IFF Mode 1/2/3 returns, Mode IV, Mode V, Blue Force Tracking, Joint Blue Force Situational Awareness, or any other cooperative tracking or ID characteristics are displayed to the operator in an unambiguous, preferably preattentive manner. POF characteristics can manifest themselves in at least two ways–in the IFF/SIF portion of the signal coming from sensors organic to the weapons system being employed or as information attached to system symbology or track via data links, local or remote.

In Figure 8.8, two potential display configurations are contrasted for efficiency of communicating IFF/SIF information. Illustration (a) does not distinguish individual IFF modes by color, nor does it display results of Mode IV interrogations in the plot depictions. Illustration (b), in contrast, depicts radar sensor plots which utilize the preattentive attributes of hue and shape to distinguish a friendly aircraft from the myriad other aircraft in the area. In this case, the radar plots for the military IFF Mode II are configured to appear as orange, and the encrypted military Mode IV return is configured to display with a unique symbol showing the number four, reversed in a small cyan rectangular shape. Research suggests that the judicious use of these preattentive features of shape and color in presenting Presence of Friendly data to the operator will make the CID process much more rapid and accurate (Wolfe, 2001; Treisman and Gelade, 1980; Healey, 2007).

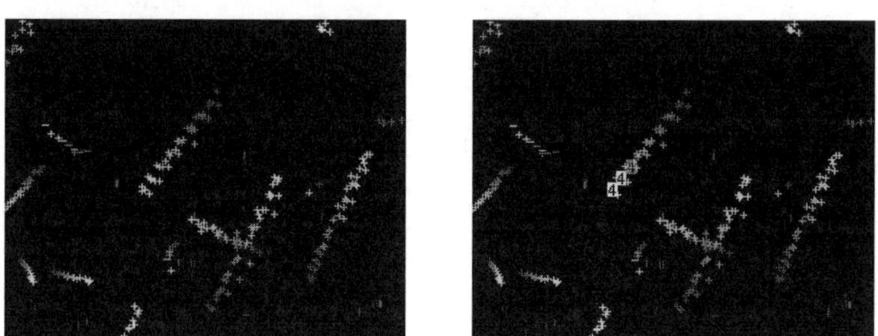

(a) (b)

Figure 8.8 An example depicting effectiveness of communicating friendly IFF/SIF information. (a) Radar sensor plots without distinguishing coloration. (b) Identical track picture using preattentive attributes of hue and shape to distinguish IFF Mode 2 (orange), and unique Mode IV symbol (cyan)

Controlling Display Layers

For the sake of clarity in illustration, the examples used in this paper have been relatively sterile, in that many elements that are required to perform C2 operations in the real world have been omitted. However, highly colored and figured charts and imagery, color-coded weather maps, routes, airspaces and other boundaries, points and markers of varying types, textual information, etc. are all possibilities for display in modern C2 systems. Given this, interface developers need to ensure that the layers of information that they present can be controlled to optimize the display, and assist operators in maintaining their focus on the most important aspects of the picture, which by necessity will vary depending upon mission needs and context. At a minimum, consider having individual layer brightness or transparency controls for maps, lines and areas, points and markers, sensor plots, and track symbology.

In Figure 8.9, the addition of a chart background element reduces foreground to background contrast to an unacceptable level in (a), rendering track elements difficult to interpret. In (b), the map background has been adjusted for brightness such that it is still useable as a secondary or tertiary reference, but the primary subject–the tracks themselves–remain featured and easily distinguished.

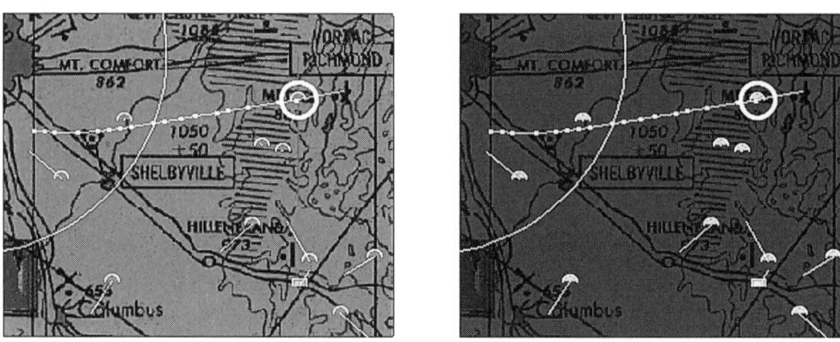

(a) (b)

Figure 8.9 An example illustrating results of layer brightness adjustment to optimize track presentation. (a) Background chart imagery shown in native brightness reduces contrast and visibility of tactical picture. (b) Map layer individually adjusted for brightness to emphasize track presentation, while still allowing map viewing as a secondary visual reference

Conclusion

Efficient and effective combat identification is a problem that must be addressed through more than the development of new technologies. Indeed, as the nature of modern warfare evolves to become increasingly dependent upon rapidly changing technology, it becomes more important than ever to adhere to established principles of user-centered design and development. One of the primary ways we can support this effort is to continue to cultivate a deeper understanding of human perception, so that our increasingly complex systems can be optimally developed and configured for maximally efficient human use. Leveraging rules of perception, we can enable improved CID by helping operations personnel quickly gain situational awareness, and by improving the salience of those *differences that make a difference* in the performance of their duties. History has shown that the penalties for not following the rules of human perception can be quite severe (Lerner, 1989; Hawley, Mares, and Giammanco, 2005); conversely, the rewards for carefully applying the known rules and for discovering new ones will certainly be well worth the effort.

Acknowledgements

The author wishes to gratefully acknowledge the following individuals for their contributions to this chapter. My family, for their understanding and support of all my endeavors; Dr. Margaret McMahon, Tony Jacobs, Vince Bizilj, and Jason Brooks for their encouragement and keen editorial suggestions; and the dedicated engineers of Raytheon Solipsys, whose commitment to serving the needs of those who put themselves in harm's way continues to inspire.

References

Bateson, G. (1972). *Steps to an Ecology of Mind.* Chicago: University of Chicago Press.

Bolia, R. S., Vidulich, M. A., Nelson, W. T., and Cook, M. J. (2004). A history lesson on the use of technology to support military decision-making and command and control. In M. J. Cook (ed.), *Human Factors of Complex Decision Making.* Mahwah, NJ: Lawrence Erlbaum.

Cannon-Bowers, J. A., and Salas, E. (1998). *Making Decisions under Stress: Implications for Individual and Team Training.* Washington, DC: American Psychological Association.

Crisp, H. E., McKneely, J. A., Wallace, D. F., and Perry, A. A. (2001). *The Solution for Future Command and Control: Human-centered Design.* Dahlgren, VA: Naval Surface Warfare Center.

Cummings, M. L. (2006). Automation and accountability in decision support system interface design. *Journal of Technology Studies, XXXII, No. 1*. Retrieved February 12, 2008, http://scholor.lib.ve.edu/ejournals/JOTS/v32/v32n1/pdf/cummings.pdf.

Defense Update (2004). Eliminating fratricide in ground combat. *Defense Update International Online Defense Magazine, Issue 2*. Retrieved March 12, 2008, from http://www.defense-update.com/features/du-2–04/feature-fratricide.htm.

Dittmer, K. (2004). Blue force tracking: a subset of combat identification. *Military Review, 2004, September-October*. Retrieved March 16, 2008, from http://findarticles.com/p/articles/mi_m0PBZ/is_5_84/ai_n7069233/pg_1.

Hair, D. C., and Pickslay, K. (1993). *Explanation-based Reasoning in Decision Support Systems*. Retrieved March 11, 2008, from http://stinet.dtic.mil/oai/oai?andverb=getRecordandmetadataPrefix=htmlandidentifier=ADA270617.

Hawley, J. K. (2006). *Patriot Fratricides: The Human Dimension Lessons of Operation Iraqi Freedom.* Retrieved March 7, 2008, from http://www.trackpads.com/magazine/publish/article_1583.shtml.

Hawley, J. K., Mares, A. L., and Giammanco, C. A. (2005). *The Human Side of Automation: Lessons for Air Defense Command and Control*. Aberdeen Proving Ground, MD: Army Research Laboratory.

Healey, C. G. (2007). *Perception in Visualization.* Retrieved February 10, 2008, from http://www.csc.ncsu.edu/faculty/healey/PP/index.html.

Hebert, A. J. (2003, March). Compressing the kill chain. *Air Force Magazine Online, 86, No. 3*. Retrieved March 17, 2008, from http://www.afa.org/magazine/March2003/0303killchain.asp.

Janis, I. L., and Mann, L. (1977). *Decision making: A Psychological Analysis of Conflict, Choice and Commitment*. New York: Free Press.

Joint Chiefs of Staff (2003). *Joint Tactics, Techniques and Procedures for Close Air Support, Joint Pub 3–09.3.* Retrieved March 2, 2008, from http://www.dtic.mil/doctrine/jel/new_pubs/jp3_09_3.pdf.

Lerner, E. J. (1989, April). 'Lessons of Flight 655.' Aerospace America, 18–26.

Mosier, K. L., and Skitka, L. J. (1996). Automation use and automation bias. *In Proceedings of the Human Factors and Ergonomics Society 40th Annual Meeting* (pp. 204–208). Santa Monica, CA: Human Factors and Ergonomics Society.

Nguyen, X. T. (2006). *Simulating Automated Intent Assessment.* Retrieved March 23, 2008, from http://www.siaa.asn.au/get/2395359717.pdf.

Norman, D. (2002). *The Design of Everyday Things* (2nd edn). New York: Basic Books.

Norman, D. (2004). *Emotional design: Why We Love or Hate Everyday Things*. New York: Basic Books.

Perrow, C. (1984). *Normal Accidents*. New York: Basic Books.

Rasmussen, R. E. (2007). *The Wrong Target: The Problem of Mistargeting Resulting in Fratricide and Civilian Casualties.* Retrieved March 10, 2008, from http://stinet.dtic.mil/cgi-bin/GetTRDoc?AD=A468785andLocation=U2anddoc=GetTRDoc.pdf.

Shrader, C. R. (1982). *Amicicide: The Problem of Friendly Fire in Modern War* (Research survey No. 1 ed.). Fort Leavenworth, KS: Combat Studies Institute, U.S. Army.

Steinweg, K. K. (1995). Dealing realistically with fratricide. *Parameters,* (Spring 1995), 4–29.

Theeuwes, I. J. (1993, November). *Temporal and Spatial Characteristics of Preattentive and Attentive Processing.* Retrieved February 13, 2008, from http://stinet.dtic.mil/cgi-bin/GetTRDoc?AD=A285200andLocation=U2anddoc=GetTRDoc.pdf.

Treisman, A. M., and Gelade, G. (1980). A feature integration theory of attention. *Cognitive Psychology, 12,* 97–136.

Tufte, E. R. (1990). *Envisioning Information* (7th edn). Cheshire, CT: Graphics Press.

Ware, C. (2000). *Information Visualization: Perception for Design.* New York: Academic Press.

Wolfe, J. M. (2001). Asymmetries in visual search: An introduction. *Perception and Psychophysics, 63*, 381–389.

Xiao, Y. and Mackenzie, C. F. (1997). *Stress and Decision Making in Trauma Patient Resuscitation.* Arlington, VA: Office of Naval Research.

SECTION 3
Situation Awareness

Maintaining a high level of situation awareness is a critical component to the decision-making process when prosecuting a target. As presented in this section, Cheryl A. Bolstad, Mica R. Endsley, and Haydee M. Cuevas (Chapter 9) posit that combat identification errors can be mitigated if one is thoroughly aware of what is happening around them. They further suggest that understanding how the information, events, and individual actions impact one's goals and objectives at any given moment or future moment can have decisive impacts to the identification of friend or foe. Further, it is contended that to optimize combat identification and enhance mission effectiveness, greater emphasis must be placed on promoting coherent views of the battlespace through a common operating picture and distinguishing between team and shared situation awareness.

David L. Hall and Stan Aungst (Chapter 10) discuss new sources of information that can be used to improve situation awareness. The authors suggest the use of information gathered by human observers using cell phones, digital cameras, and handheld video cameras that is posted on the Internet through blogs, image and video sharing websites, and self-reported news. This information can be used to augment information obtained by more traditional means (e.g., ground sensors, radar) to enhance situation awareness. Next, the researchers describe the Center for Network Centric Cognition and Information Fusion (NC2IF Center) at the Pennsylvania State University College of Information Sciences and Technology. The chapter concludes with a list of on-going research projects being conducted at two laboratories that are a part of the NC2IF Center: the Extreme Events Laboratory and the User Science and Engineering Laboratory.

Chapter 9
Team Coordination and Shared Situation Awareness in Combat Identification

Cheryl A. Bolstad
Mica R. Endsley
Haydee M. Cuevas
SA Technologies

Introduction

At its core, situation awareness (SA) involves being aware of what is happening around you to understand how information, events, and your own actions will impact your goals and objectives, both now and in the near future. SA is especially crucial in any domain where the effects of ever increasing technological and situational complexity on the human decision-maker are a concern. In particular, combat identification is a function of not only effective target identification capabilities, but also good SA (Greitzer and Andrews, Chapter 11, this volume). Poor SA in combat identification can lead to higher casualties due to combat errors, fratricide, and failing to anticipate enemy actions. Thus, meeting the combat identification requirements across the full range of military operations requires developing and maintaining high levels of SA of the current and future status of an operation to achieve an accurate and timely characterization of entities (both friendly and enemy) in a combatant's area of responsibility. To optimize combat identification processes and enhance combat and mission effectiveness, greater emphasis must be placed on developing technologies that promote a coherent view of the battlespace through a common operating picture. Incoming data must not only be accessible, but filtered, analyzed, and integrated to develop an overall situational picture to support team coordination and shared SA.

To address this issue, this chapter begins with a brief overview of the situation awareness construct, including defining three levels of SA as well as distinguishing between team and shared SA. The next section demonstrates, with examples from relevant research, how a theoretical model of team SA can guide the development of new technologies, and enhances the utility of existing technologies (e.g., Blue Force Tracker) to improve combat identification.

Situation Awareness in Individuals and Teams

Situation awareness can be defined as 'the perception of elements in the environment within a volume of time and space, the comprehension of their meaning, and the projection of their status in the near future' (Endsley, 1995, p. 36). This definition highlights three levels or stages of SA formation: perception, comprehension, and projection. Level 1 SA (perception), which involves the processes of monitoring, cue detection, and simple recognition, leads to an awareness of multiple situational elements (objects, events, people, systems, and environmental factors) and their current states (locations, conditions, modes, and actions). Level 2 SA (comprehension) utilizes the processes of pattern recognition, interpretation, and evaluation to synthesize and integrate disjointed Level 1 SA elements to understand how this information will impact goals and objectives. Level 3 SA (projection) is achieved through knowledge of the status and dynamics of the elements, and comprehension of the situation (Levels 1 and 2 SA), and extrapolation of this information to project future actions and states of the elements in the operational environment. As implied by the definition above, time is also an important concept in SA. Specifically, SA is a dynamic construct, changing at a tempo dictated by the actions of individuals, task characteristics, and the surrounding environment.

As illustrated in Figure 9.1, SA is a key step in decision-making and human performance. However, changes brought about in the workplace due to today's 'information age' have resulted in a huge increase in systems, displays, and technologies, particularly in complex operational environments such as the military. Consequently, development of SA takes up the majority of decision-makers time and effort as they attempt to make sense of incoming information from a multitude of sources with varying degrees of certainty and reliability. Thus, the problem in these environments is no longer a lack of information, but finding, within the volumes of data available, those precise bits of information that are needed to make an informed, reasoned decision.

A widening gap exists between the tons of data being produced and disseminated, and the individual's ability to find the right, disparate bits and process them together to arrive at the actual information sought. Indeed, research on aviation accidents and performance under demanding battlefield conditions has shown that operators have no difficulty physically performing their tasks, and no difficulty choosing the correct action once they understand the situation, but they struggle with developing and maintaining an understanding of the situation (Brezovic, Klein, Calderwood, and Thordsen, 1987; Jones and Endsley, 1996). Thus, a major challenge in combat identification is to overcome this 'information gap' by providing better tools to support warfighters in developing an understanding of the information available and ultimately their SA.

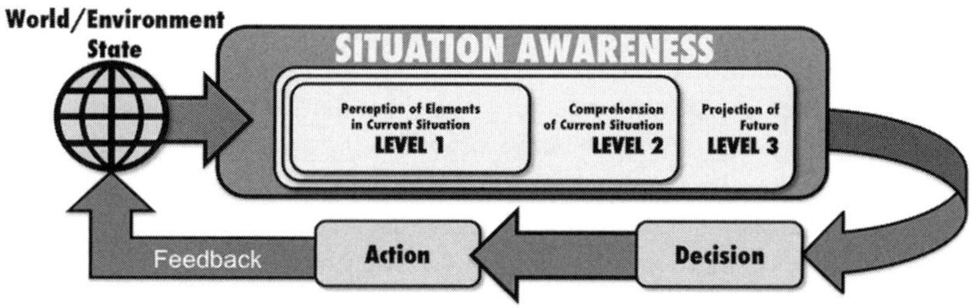

Figure 9.1 Situation awareness forms the basis for decision-making and action

Team and Shared SA

In many systems, people work not just as individuals, but as members of a team. Thus, it is necessary to consider the SA of not just individual team members, but also the SA of the team as a whole, including both team SA and shared SA. Team SA is defined as 'the degree to which every team member possesses the SA required for his or her responsibilities' (Endsley, 1995, p. 39). The success or failure of a team depends on the success or failure of each of its team members. By this definition, each team member needs to have a high level of SA on those factors that are relevant for his or her job. It is not sufficient for one member of the team to be aware of critical information if the team member who needs that information is not aware. For instance, in combat identification, an intelligence officer is concerned with the type and intent of an identified enemy aircraft, while a weapons officer is concerned with its weapons capabilities and trajectory.

Shared SA can be defined as 'the degree to which team members possess the same SA on shared SA requirements' (Endsley and Jones, 2001, p. 48). As implied by this definition, certain information requirements may be relevant to multiple team members. A major part of teamwork involves the area where these SA requirements overlap. Sharing a common understanding of what is happening on those SA elements that are common leads to smoothly functioning teams. However, not all information needs to be shared. Sharing every detail of each person's job would only create a great deal of 'noise' to sort through to get the needed information. In the previous example, information about enemy type, location, numbers, and mission configuration are critical for both officers to perform their jobs effectively and thus, must be shared between these positions.

Model of Team Situation Awareness

The performance of the team as a whole, therefore, is dependent upon both (1) a high level of SA among individual team members for the aspects of the situation necessary for their job and (2) a high level of shared SA between team members, providing an accurate common operating picture of those aspects of the situation common to the needs of each member (Endsley and Jones, 2001). Endsley and Jones described a model of team situation awareness as a means of conceptualizing how teams develop high levels of shared SA and ultimately achieve their team goals. Their model specifies how four factors (SA requirements, SA devices, SA mechanisms, and SA processes) act to help build team and shared SA. The remainder of this chapter will demonstrate, with examples from relevant research, how consideration of these four factors can enhance the design of technologies to effectively promote team coordination and shared SA during combat identification.

SA Requirements

SA Requirements are utilized by team members to identify which information needs to be shared, including their higher level assessments and projections (which are usually not otherwise available to fellow team members) and information on team members' task status and current capabilities. These critical SA requirements can be identified utilizing a Goal Directed Task Analysis (GDTA), a unique form of cognitive task analysis that involves conducting extensive knowledge elicitation sessions with domain subject matter experts (for a detailed description of this methodology; see Endsley, Bolte, and Jones, 2003). The objective of the GDTA is to identify the major goals and decisions that drive performance in a particular job or position as well as to delineate the critical, dynamic information requirements associated with each goal and decision. The GDTA methodology provides a 'technology independent' analysis of operator information requirements at all three levels of SA for each major goal and subgoal (see Figure 9.2). The resultant hierarchy of goals, decisions, and SA information requirements can be utilized for interface design, training development, system evaluation, and job or task analysis.

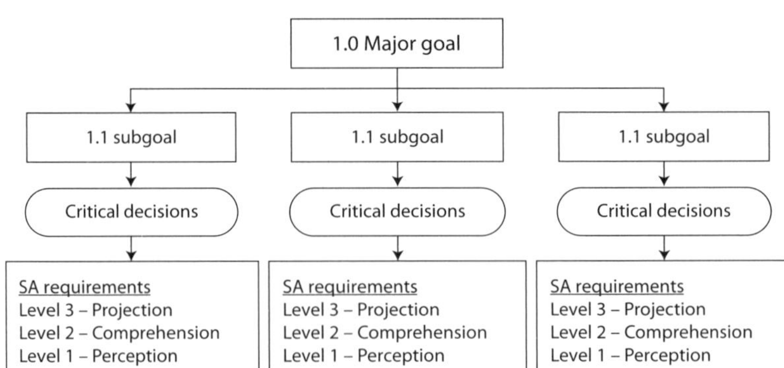

Figure 9.2 Format of GDTA

For example, Bolstad and colleagues created GDTAs for several Army Brigade staff positions (Intelligence [S2], Operations [S3], Logistics [S4], and Engineer) and used these GDTAs to develop a representative command and control (C2) interface, called Synergy, to support SA and collaborative planning and execution processes in military operations (for a detailed description of this research, see Bolstad, Riley, Jones, and Endsley, 2002, Endsley, Bolstad, Jones, and Riley, 2003). Features and information were presented on the displays based on high priority operator goals and critical information requirements, as identified through the GDTAs. As an example, Figure 9.3 shows a portion of the GDTA for the Logistics officer and highlights how these SA information requirements can be satisfied in the design of the Synergy display for this position. In addition, special overlays allowed tailoring of the displays to different positions while the common look and feel of the displays across the different positions supported shared SA.

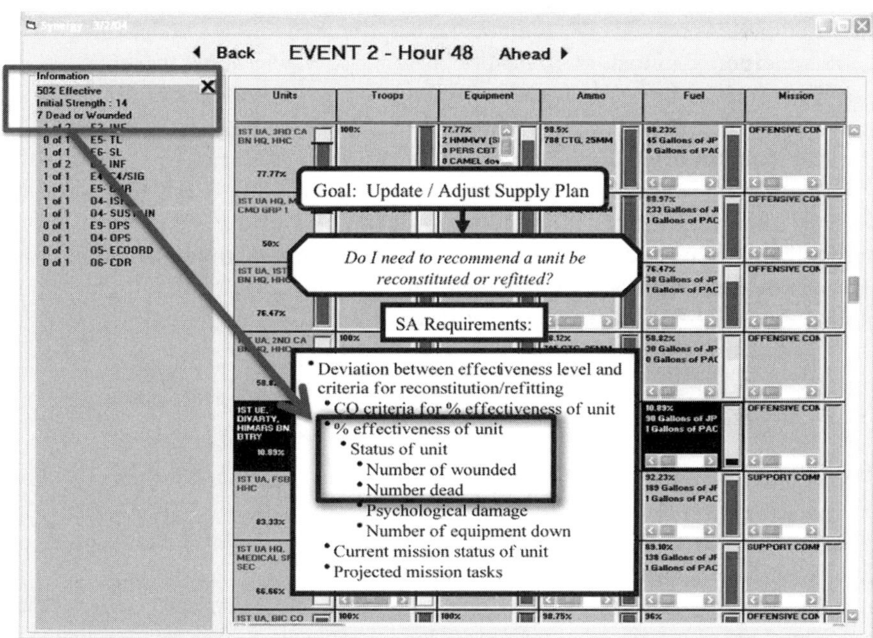

Figure 9.3 Example of mapping of GDTA for Logistics officer onto features of Synergy display

Implications for Combat Identification

A necessary first step to improving SA in combat identification is the creation of a GDTA that delineates the major goals and subgoals, decisions, and information requirements associated with a complex military operation. However, to the best of the authors' knowledge, a GDTA has not yet been created for this domain. Nevertheless, based on available descriptions of combat identification operations (e.g., Greitzer and Andrews, Chapter 11, this volume), one obvious important goal for the warfighter is to 'Maximize combat and mission effectiveness while reducing total casualties due to enemy action and fratricide,' followed by the corresponding subgoal 'Avoid friendly fire accidents.' A key decision for this subgoal is 'Where are friendly forces located?' The associated SA information requirements for this decision include the current and projected location of friendly units and enemy targets. This information would help ensure that no friendly units are within range of ongoing and planned fires and effects operations. Note that this hypothetical example of a GDTA only illustrates one possible goal and subgoal for combat identification. Actual GDTAs are much more comprehensive and detailed.

In addition, combat identification is almost never done individually and instead, involves the coordinated efforts of teams of warfighters. Accordingly, it is not sufficient to simply identify the information requirements needed at the individual warfighter level. Equally important is determining the shared information requirements across team members and how each team member will use this overlapping information. As mentioned earlier, this would include information such as the enemy type, location, numbers, and mission configuration. Thus, as will be discussed next, developing methods for effectively sharing this information is also critical for combat identification.

SA Devices

SA Devices include the different types of devices available for sharing critical SA information requirements, such as direct communication (both verbal and nonverbal), shared displays (e.g., visual and audio displays, tactile devices), and a shared environment. In distributed C2 teams, in particular, nonverbal communication, such as gestures and display of local artifacts, and a shared environment are usually not available. Thus, this places a far greater emphasis on explicit verbal communication and creating effective communication technologies and shared information displays to support distributed team performance.

To illustrate the importance of SA devices, Bolstad and Endsley (1999, 2000) conducted two studies to test if shared displays and shared mental models (an important SA mechanism described in the next section) would assist team members in jointly performing an aviation theater defense task. In essence, the aviation theater defense task involved collaboration between an Air Commander and an Intelligence Officer to detect, identify, and resolve incoming targets. The

Air Commander had to monitor incoming targets that appeared on his radar screen and prioritize these targets based on range and speed. The Intelligence Officer provided the Air Commander with guidance regarding the identity and mission priority of the targets based on information he received from several different sensors. Sensor reliability and the consistency and completeness of the information provided by the sensors were varied to mimic real world conditions.

In the first study, each team member in the shared display condition was provided with a complete replicate of their teammate's display (see Figure 9.4). Additionally, team members in the shared mental model condition were given training on both their task and their team members' tasks. Initially, teams were slower when given the shared display, but at the end of the study the teams with the shared display and shared mental model training had the highest performance. Notably, teams without either shared mental model training or shared displays performed the most poorly.

In the second study, performance using a (fully) shared display (as used in the previous study) was compared to that using an abstracted shared display. The interface for the two displays was identical; however, the abstracted display was enhanced with a small panel that provided only the critical information from their teammate's display, based on an analysis of the shared information requirements, rather than a full duplication of all information. Results indicated that the use of an abstracted shared display improved team performance, while the use of shared displays that completely duplicated the other team member's displays was detrimental.

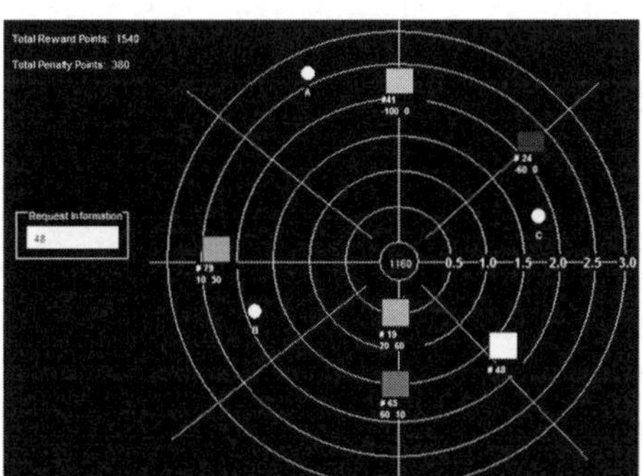

Air Commander

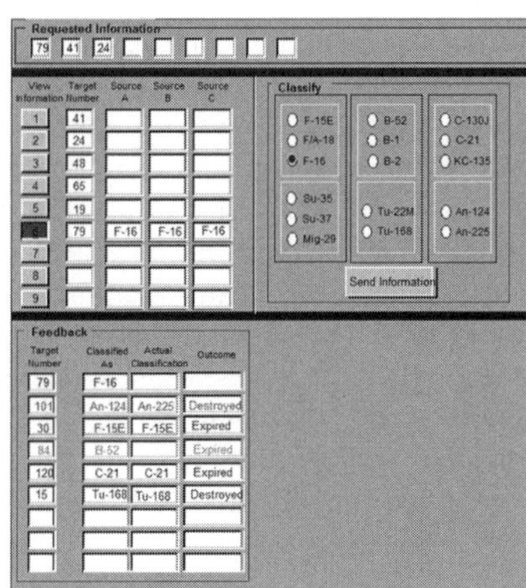

Intelligence Officer

Figure 9.4 Full shared displays for Air Commander and Intelligence Officer

In addition to shared displays, consideration must also be given to the collaboration tools available to distributed team members. For example, Bolstad and Endsley (2006) reviewed current collaborative support tools and organized these into a taxonomy of collaboration to guide the evaluation of these tools in supporting distributed operations. The taxonomy first lists the various categories of collaborative support tools (e.g., video conferencing, chat/instant messaging, file transfer) and then classifies each tool with regard to collaboration characteristics (e.g., degree of interaction), tool characteristics (e.g., recordable/traceable), information types (e.g., verbal, spatial/graphical), and collaboration processes (e.g., tracking information, data gathering). This taxonomy serves as a useful guide for identifying effective tools to support the different processes and conditions of collaboration needed in complex distributed military team operations such as combat identification. Indeed, the taxonomy is currently being used to support Army leaders in promoting 'virtual' team collaboration at higher echelon levels.

Implications for Combat Identification

To support team coordination and shared SA during combat identification operations, technologies need to provide team members with a coherent view of the battlespace through a common operating picture. Systems need to display incoming data in such a way as to assist operators in developing an accurate overall situational picture in a timely manner as well as effectively communicate critical shared SA information requirements across team members. These essential SA user requirements suggest significant implications for how best to enhance the design of existing technologies, such as Blue Force Tracking (BFT), as well as guide the selection of appropriate collaboration tools.

As noted earlier, a critical requirement for successful combat identification is enabling the sharing of a common operational picture and maintaining timely information regarding the location of friendly and enemy targets. Unfortunately, time lags between BFT signal reception and display of icons on the screen often occur, such as due to blindspots, deadspace, weak or disrupted signals, etc., leading to update delays that can sometimes be considerable. These disruptions may result in the display of outdated and inaccurate information regarding the exact location of vehicles. To address this issue, BFT display design could incorporate, for example, a feature that projects 'predicted' vehicle location given current location, trajectory, and rate of movement, as indicated by the most recent update, as well as confidence ratings for the accuracy of the information displayed.

SA Mechanisms

SA Mechanisms involve the degree to which team members possess internal mechanisms, such as shared mental models and shared experiences, which support their ability to interpret information in the same way and make accurate projections regarding each other's actions. Shared mental models can be defined as 'knowledge structures held by members of a team that enable them to form accurate explanations and expectations for the task, and, in turn, to coordinate their actions and adapt their behavior to demands of the task and other team members' (Cannon-Bowers, Salas, and Converse, 1993, p. 228). Research has shown that when team members possess similar mental models their team performance is enhanced (Stout, Cannon-Bowers, Salas, and Milanovich, 1999). Conversely, when teams are not allowed to generate shared mental models, they perform significantly worse (Bolstad and Endsley, 1999). Shared mental models are thought to aid team members in their ability to anticipate information needs of other members, increase coordination between individuals, and reduce the need for explicit communication (McCann, Baranski, Thompson, and Pigeau, 2000). Shared mental models are also crucial to achieving a team's SA (Stout, Cannon-Bowers, and Salas, 1996).

One effective method for fostering shared mental models is cross-training, that is, training each team member on the duties and tasks of the other team members. Several studies have shown that team members who are cross-trained outperform teams that are not (Bolstad and Endsley, 1999; Volpe, Cannon-Bowers, Salas, and Spector, 1996). Cross-trained teams also exhibit more taskwork and teamwork specific knowledge (Cooke, et al., 2003). In addition, Bolstad, Cuevas, Costello, and Rousey (2005) have demonstrated the utility of cross-training for improving team SA. Results from their study, conducted at the Personnel Recovery Education and Training Center, revealed that cross-training, particularly in a leadership role, led to better SA.

Implications for Combat Identification

The previous section discussed how shared displays may improve team performance by promoting shared SA. The research cited in this section suggests that another approach to enhancing shared SA of essential information requirements related to combat identification is to focus on training specifically targeted at improving shared mental models within the team, such as through cross-training on various team member roles. Although cross-training may already be in use for other Air Force operations, cross-training tailored to specifically improve shared SA in combat identification could significantly enhance performance. Taken together, consideration of essential SA Mechanisms, such as shared mental models, coupled with well-designed user-adaptable shared displays, could lead to improved team coordination during combat identification operations.

SA Processes

SA Processes refer to the degree to which team members engage in effective processes for sharing SA information which may include a group norm of questioning assumptions, checking each other for conflicting information or perceptions, setting up coordination and prioritization of tasks, and establishing contingency planning. A series of workshops was conducted within the aviation domain aimed at providing participants with a heightened awareness of SA concepts and their relevance to successful individual and multi-crew operations, together with an introduction to practical skills for use in guarding against the loss of SA during critical periods of operation (for a detailed description, see Henderson, Endsley, and Hayward, 2000; Taylor, Endsley, and Henderson, 1996). Findings from experiential simulation exercises conducted during the workshops highlight several key team processes and behaviors that impact a team's ability to develop sufficient team SA to perform their tasks (see Table 9.1).

Past research has also shown how training programs can be developed specifically aimed at enhancing essential SA processes, such as those cited above. For example, Strater et al. (2004) developed and validated the Infantry Situation Awareness Training (ISAT) Program designed to enhance SA in Infantry Platoon Leaders. The ISAT modules focused on schemata training, SA communications, time management and prioritization, and contingency planning. In general, the results supported the utility of computer-based training programs as a tool to enhance the development of the skills necessary to gain and maintain the higher levels of SA that provide the foundation for decision-making and action.

In another project, Bolstad, Endsley, Howell, and Costello (in press) conducted a series of experiments to evaluate the effectiveness of six training modules designed to teach skills and processes that underlie the development and maintenance of SA in General Aviation pilots. Modules focused on basic skills training (e.g., checklist completion, air traffic control comprehension, psychomotor skills) as well as higher order skills training (e.g., attention sharing, contingency planning) and intensive preflight planning (see illustrative example in Figure 9.5). Overall, the modules were shown to be successful at improving these fundamental skills and also led to improvements in SA.

Table 9.1 Effective and ineffective SA processes

Ineffective	Effective
SA black hole: one member would lead others off	Self-checking: checked against others at each step
Didn't share pertinent info: group norm	Coordinated: to get information from each other
Failure to prioritize: members went in own directions, lost track of main goal	Prioritized: set-up contingencies, re-joining
Relied on expectations: unprepared to deal with false expectations	Questioning: as a group

Figure 9.5 Sample screen from contingency planning module of SA in general aviation training program

Implications for Combat Identification

The research described in this section highlights the value of training programs aimed at developing the skills that support SA processes in combat identification. At this time, the authors are not aware of any existing training programs that specifically focus on SA in combat identification. Targeting the basic skills underlying Level 1 SA (perception) in combat identification, such as monitoring, cue detection, simple recognition, and psycho-perceptual discrimination of situational elements, is a critical first step in the development of such a training program. Subsequently, more advanced training can be developed that builds upon this foundation and emphasizes higher order cognitive skills associated with Levels 2 and 3 SA (comprehension and projection), including, for example, explicit communication of critical shared SA information requirements, training on schemata that supports pattern matching in combat identification operations, and contingency planning to address unforeseen risks to friendly units during combat operations (e.g., delay in update from BFT due to signal disruption). In addition, specific training on the limitations of BFT could be used to help calibrate warfighters' trust in the information displayed, prompting verification and confirmation of potentially outdated data, as needed.

Conclusions

This chapter described how a model of team situation awareness can be applied to meet the team coordination and shared SA challenges associated with successful combat identification in today's complex military battlefield. Consideration of SA requirements, devices, mechanisms, and processes can serve to guide the design, development, and evaluation of the training programs and systems needed to improve warfighters' ability to process and understand the large volumes of oftentimes ambiguous data inherent in combat identification operations. Realizing the benefits of advanced information technology is dependent upon first meeting the challenge of managing this dynamic information base to provide warfighters with the situation awareness they need on a real-time basis. The line of research reported in this chapter offers a promising theoretically-based, empirically-validated approach for addressing this critical requirement.

Acknowledgement

The opinions, views, and conclusions contained herein are those of the authors and should not be interpreted as representing the official policies, either expressed or implied, of the sponsors of the research reported in this chapter, or the organizations with which the authors are affiliated.

References

Bolstad, C. A., Cuevas, H. M., Costello, A. M., and Rousey, J. (2005). Improving situation awareness through cross-training. *Proceedings of the 49th Annual Meeting of the Human Factors and Ergonomics Society* (pp. 2159–2163). Santa Monica, CA: Human Factors and Ergonomics Society.

Bolstad, C. A., and Endsley, M. R. (1999). Shared mental models and shared displays: An empirical evaluation of team performance. *Proceedings of the 43rd Annual Meeting of the Human Factors and Ergonomics Society* (pp. 213–217). Santa Monica, CA: Human Factors and Ergonomics Society.

Bolstad, C. A., and Endsley, M. R. (2000). The effect of task load and shared displays on team situation awareness. *Proceedings of the 14th Triennial Congress of the International Ergonomics Association and the 44th Annual Meeting of the Human Factors and Ergonomics Society* (pp. 189–192). Santa Monica, CA: Human Factors and Ergonomics Society.

Bolstad, C. A. and Endsley, M. R. (2006). Tools for supporting distributed team collaboration. *Ergonomics in Design, 13*, 7–14.

Bolstad, C. A., Endsley, M. R., Costello, A. M., and Howell, C. D. (in press). Evaluation of computer-based situation awareness training for general aviation pilots. *International Journal of Aviation Psychology*.

Bolstad, C. A., Riley, J. M., Jones, D. G., and Endsley, M. R. (2002). Using goal directed task analysis with Army brigade officer teams. *Proceedings of the Human Factors and Ergonomics Society 46th Annual Meeting* (pp. 472–476). Santa Monica, CA: Human Factors and Ergonomics Society.

Brezovic, C. P., Klein, G. A., Calderwood, R., and Thordsen, M. (1987). *Decision Making in Armored Platoon Command* (Prepared under contract MDA903–85-C-0327 for US Army Research Institute, Alexandria, VA [KATR-858(B)-87–05F]). Yellow Springs, OH: Klein Associates.

Cannon-Bowers, J. A., Salas, E., and Converse, S. (1993). Shared mental models in expert team decision making. In N. J. Castellan, Jr. (ed.), *Individual and Group Decision Making* (pp. 221–246). Hillsdale, NJ: Lawrence Erlbaum.

Cooke, N. J., Kiekel, P. A., Salas, E., Stout, R., Bowers, C., and Cannon-Bowers, J. (2003). Measuring team knowledge: A window to the cognitive underpinnings of team performance. *Group Dynamics, Theory, Research and Practice, 7*, 179–199.

Endsley, M. R. (1995). Toward a theory of situation awareness in dynamic systems. *Human Factors, 37*, 32–64.

Endsley, M. R., Bolstad, C. A., Jones, D. G., and Riley, J. M. (2003). Situation Awareness Oriented Design: From user's cognitive requirements to creating effective supporting technologies. *Proceedings of the Human Factors and Ergonomics Society 47th Annual Meeting* (pp. 268–272). Santa Monica, CA: Human Factors and Ergonomics Society.

Endsley, M. R., Bolte, B., and Jones, D. G. (2003). *Designing for Situation Awareness: An Approach to Human-centered Design*. London: Taylor and Francis.

Endsley, M. R., and Jones, W. M. (2001). A model of inter- and intrateam situation awareness: Implications for design, training and measurement. In M. McNeese, E. Salas, and M. R. Endsley (eds), *New Trends in Cooperative Activities: Understanding System Dynamics in Complex Environments* (pp. 46–67). Santa Monica, CA: Human Factors and Ergonomics Society.

Henderson, S., Endsley, M. R., and Hayward, B. (2000). Situational awareness developmental workshop report. In B. J. Hayward and A. R. Lowe (eds), *Aviation Resource Management Vol 2: Proceedings of the Fourth Australian Aviation Psychology Symposium*. Aldershot, UK: Ashgate Publishing.

Jones, D. G., and Endsley, M. R. (1996). Sources of situation awareness errors in aviation. A*viation, Space and Environmental Medicine, 67*, 507–512.

McCann, C., Baranski, J. V., Thompson, M. M., and Pigeau, R. A. (2000). On the utility of experiential cross-training for team decision-making under time stress. *Ergonomics, 43*, 1095–1110.

Stout, R. J., Cannon-Bowers, J. A., and Salas, E. (1996). The role of shared mental models in developing team situational awareness: Implications for training. *Training Research Journal, 2*, 85–116.

Stout, R. J., Cannon-Bowers, J. A., Salas, E., and Milanovich, D. M. (1999). Planning, shared mental models, and coordinated performance: An empirical link is established. *Human Factors, 41*, 61–71.

Strater, L. D., Reynolds, J. P., Faulkner, L. A., Birch, D. K., Hyatt, J., Swetnam, S., and Endsley, M. R. (2004). PC-based tool to improve Infantry situation awareness. *Proceedings of the Human Factors and Ergonomics Society 48th Annual Meeting* (pp. 668–672). Santa Monica, CA: Human Factors and Ergonomics Society.

Taylor, R. M., Endsley, M. R., and Henderson, S. (1996). Situational awareness workshop report. In B. J. Hayward and A. R. Lowe (eds), *Applied aviation psychology: Achievement, Change and Challenge* (pp. 447–454). Aldershot, UK: Ashgate Publishing.

Volpe, C. E., Cannon-Bowers, J. A., Salas, E., and Spector, P. E. (1996). The impact of cross-training on team functioning: An empirical investigation. *Human Factors, 38*, 87–100.

Chapter 10

The Use of Soft Sensors and I-Space for Improved Combat ID

David L. Hall
Stan Aungst
The Pennsylvania State University

Non-traditional warfare in urban environments leads to challenges in combat identification (Combat ID) and motivates observing and understanding the physical terrain as well as understanding the human terrain (individuals, organizations, and culture and non-physical concepts such as intent). New sources of information for improved situational awareness and threat refinement include the use of human observers (humans as 'soft sensors') and data available on the internet from blogs, news reports, YouTube™, and self-reported news (e.g., CNN's *News to You*). This chapter explores these concepts and describes a test-bed developed at the Pennsylvania State University College of Information Sciences and Technology (IST) to conduct human-in-the-loop experiments to evaluate visualization tools, collaboration techniques, decision aids, and risk models.

Introduction

Combat ID seeks to determine identities of friendly, neutral, and enemy targets and threats in a dynamic environment. Failure of accurate combat ID leads to fratricide and undue losses of friendly forces when soldiers are unable to accurately assess and understand a rapidly evolving situation. These challenges are ever increasing due to non-standard and urban warfare operations in which U.S. forces must operate in, around, and among native populations and non-government organizations such as relief organizations. Moreover, the increasing complexity of targets (moving from traditional weapon systems to individuals and groups of people planning and executing asymmetric warfare) further complicates identification and situational awareness. Traditional information and data fusion has sought to address these challenges by integrating and fusing data from heterogeneous sources and sensors to improve situational awareness. In addition, recent advances in hand-held sensing and communications devices provide the opportunity to leverage the observational ability of humans (viz., using humans as 'soft sensors'). This chapter describes some concepts, challenges, and on-going research to use a combination of traditional sensors, human 'soft' sensors, and data available on the internet to improve situational awareness and combat ID.

The Data Fusion Tradition

The tradition of data and information fusion systems for applications such as military situational awareness and combat ID has focused primarily on the use of physical sensors such as radar, LIDAR, and acoustic and seismic sensors to monitor physical objects (e.g., Waltz and Llinas, 1990; Hall and McMullen, 2004; Hall and Llinas, 1997; Hall and Llinas, 2001). Military fusion systems have traditionally sought to observe, characterize, and identify targets such as tanks, trucks, aircraft, weapon systems, and sensors. The input data has included observations from physical sensors with limited inputs from human observers. The goal of information and data fusion (using techniques such as signal and image processing, statistical estimation, pattern recognition, and limited automated reasoning) has been to transform physics-based observations about physical objects into knowledge about those physical objects. The main output of this fusion process is the creation of state vectors providing information about target location, identification, and characteristics. Extensive research has focused on key aspects of this transformation from detected energy about a target or entity to creation or estimation of a state vector which represents knowledge about the target. Examples of relevant research include the development of increasingly more sophisticated sensors, advanced signal and image processing methods for representing the sensor data, new target tracking and state estimation methods, and techniques for sensor resource allocation (Liggins, Hall, and Llinas, 2008). With few exceptions, the focus of data fusion research has been data and observation driven–that is, the development of new and improved physical sensors, new methods for processing data, and architectures that 'served the data' (i.e., process input data to result in a common operational picture, situation display, or database of tracks and state vectors). In this approach, the human user has been viewed primarily as an interpreter of the processing results (via interactions with a situation display or databases) and decision-maker who made tactical decisions based on the evolving situation presented via a common operational picture.

New Trends in Information Fusion

There are two major trends that impact this traditional view of data fusion. First, the types of targets or entities that we are interested in are no longer primarily physical. Instead of specific vehicles, sensors and weapon systems, we are becoming interested in the location, identity, and interactions of individuals and groups (social networks). Addressing a military threat such as improvised explosive devices (IEDs) involves not only the identification, location, and characterization of physical explosive devices and delivery vehicles, but also networks of people who plan, design, manufacture, and deploy these devices (e.g., Burgeon and Varadan, 2007). Thus, there is a hierarchy of physical to non-physical targets sought from physical devices, vehicles, and communications devices to human networks with various hierarchies of authority, intents, belief systems, cyber-connectivity, policies, and procedures. Waltz (2003) described this in detail and provided an information hierarchy with layers of information from physical entities to human thoughts and intentions. This is a data rich but model poor environment. While physics based models exist for relating the observations of physical sensors to physical targets, very limited models exist for non-physical targets and for social networks. Beyond military applications, analysis of threats such as health hazards and cyber-attacks on national infrastructure also involve trying to identify and characterize human networks including physical communications and virtual relationships.

The second major trend in information fusion is the emergence of two new major sources of information that have previously been relatively neglected: human observations and web-based information. With the advent of ubiquitous cell phones (with associated GPS, image sensors, and on-board computing), we can consider formal and informal 'communities of observers' that provide information about an evolving situation. Over four billion cell phones are currently used throughout the world. New web sites that allow sharing of data (e.g., YouTube™, Flickr), ad hoc reporting to national news networks (e.g., Yahoo's You Witness News, http://news.yahoo.com/you-witness-news), Blogs, FaceBook™, and MySpace™ provide huge sources of data. While this data collection is ad hoc and uncoordinated, it provides a potential source of information that we term 'soft sensing' or the 'S-space.' Robert Lucky (2007), for example, described the concept of Internet based information. Lucky stated, 'Meanwhile, those billion amateurs are taking pictures of everything on the planet and placing the images on Flickr and other sites. There are thousands upon thousands of pictures of every known place, taken from all angles and under all lighting conditions. Researchers are now using those pictures to create three-dimensional images and panoramic vistas' (p.96). This information can significantly augment data obtained from traditional sensors such as unattended ground sensors, radar, airborne vehicles, and others. Similarly, Burke et al. (2006) described the concept of participatory sensing, in which a community of observers might be tasked to provide information for applications such as urban planning and public health. This concept is one example of crowd-sourcing described by Howe (2006) in which distributed groups of people are tasked to use their spare time, knowledge, and energy to create content for documents (such as Wikipedia), solve problems, or conduct research and development.

There are numerous challenges in accessing and utilizing such data. Examples of research challenges include:

- Soft-sensor tasking – how to effectively task civilian observers and direct human attention (akin to directing a hard sensor's 'look angle'),

- data and knowledge elicitation—how to solicit information about a target's activity, location, characteristics, and identity,
- representation of uncertainty and second order uncertainty—how to quantify the uncertainty of human reports and the uncertainty of the uncertainty,
- reporting and observational biases—how to account for and mitigate the effects of biases such as the confirmation bias, framing effects, and problems with subjective probabilities (see Pohl, 2005; Piattelli-Palmarini, 1994),
- deliberate information corruption—how to account for human spoofing, dissemination of rumors, and information warfare; and
- hard/soft information fusion—how to effectively fuse heterogeneous sources that include real-time physical sensors such as video and acoustic data with human text reports.

While challenges exist, this new information space becomes a very important part of the new information fusion concepts. The U.S. Army has embraced some of these concepts via their 'Every Soldier a Sensor' program, (see http://www.ausa.org/pdfdocs/IP_Sensor08_04.pdf). In addition to soldiers acting as sensors, civilians may also act in this manner, providing information about local conditions, activities, events, and other context-based information to improve our understanding about a situation or threat. Increasingly, the first reported observations of disasters, accidents, and adverse weather conditions are reported by local observers posting pictures taken via cell-phone.

A Research Framework

Research at the Pennsylvania State University College of Information Sciences and Technology (in the Center for Network Centric Cognition and Information Fusion [NC2IF]; see http://ist.psu.edu/facultyresearch/facilities/nc2if/) is focusing on a new view of multi-sensor data fusion to address these new trends and explicitly consider the active role of a human user/analyst. We view the inputs to the data fusion process as comprised of three pillars: (1) traditional sensing resources ('H-space'), (2) dynamic communities of human observers ('S-space'), and (3) resources such as archived sensor data, blogs, reports, and dynamic news reports from citizen reporters via the Internet ('I-space'). The sensors in all three of these pillars need to be characterized and calibrated. In H-space and I-space, calibration issues related to motivation, truthfulness, etc. must be considered in addition to the standard physical characterization and calibration issues that need to be considered in S-space. Our research explicitly considers the role of human observers as a major source of input that augments traditional sensor systems (Hall, Llinas, McNeese, and Mullen, 2008). Figure 10.1 provides an overview of the functions that support this concept.

The bottom of Figure 10.1 shows the concept of an observed real-world environment consisting of a physical terrain including land, hills, vegetation, roads, buildings, and other physical objects such as vehicles, sensor systems, and other objects. This physical landscape is often observed by networked sensors. In addition, the real-world environment consists of the human terrain involving individuals, groups, organizations, activities, and events. This human terrain is typically observed and reported on by ad hoc human observers providing direct information via cell phone calls and reports as well as information placed on the internet via blogs, chat rooms, shared photographs, shared videos, and related information. In order to transform this diverse information into a useable form for fusion processing requires functions such as signal and image processing (for the physical sensor data), knowledge elicitation (to obtain information from the human observers), and software agents or search engines to obtain information from the web. In turn, these data are processed by a variety of meta-data generation methods that range from automated semantic labeling of images, to extraction of information from text, to feature extraction and interpretation (Liggins, Hall, and Llinas, 2008). The original data and meta-data may be fused via traditional level-1 fusion methods including association/correlation, state vector estimation, and pattern recognition methods. At higher levels of abstraction, situational awareness tools may be applied to understand an evolving situation, with still other tools such as cognitive aids and collaboration tools used to support human

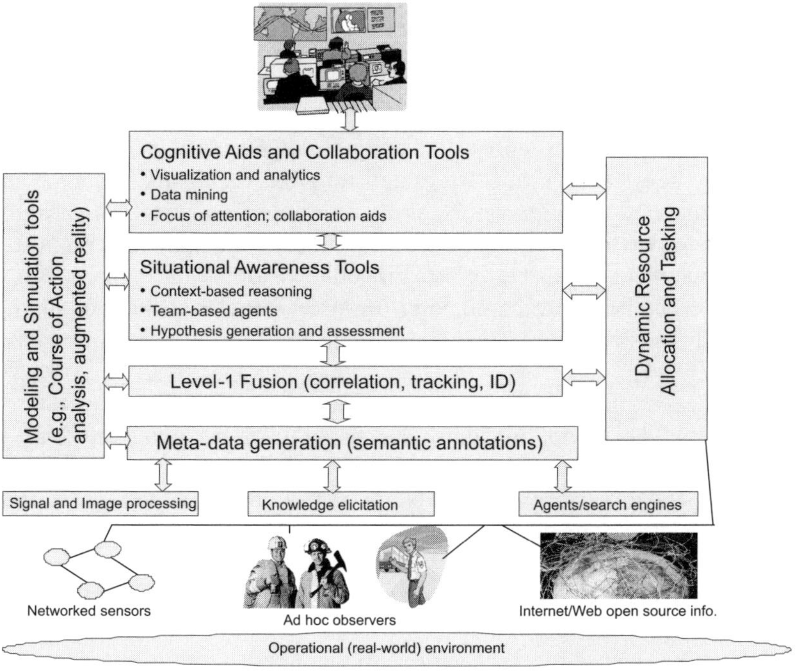

Figure 10.1 Concept of human centered information fusion

analysts. All of these functions are supported by modeling and simulation methods such as course of action (COA) analysis techniques.

In addition to using humans as soft sensors, we consider a new role for the analyst-in-the-loop in data fusion. In this new role, the human analyst augments the traditional automated reasoning of computer-based fusion systems by explicitly using human cognition for pattern recognition (via visual and aural processing) as well as using semantic reasoning for context-based interpretation of evolving situations. The concept is to develop computer displays, use sonification (transformation of data into sounds), and generate semantic meta-data from signals and images (Li and Wang, 2003, 2006) to allow the human user/analyst to become cognitively engaged in the inference process. Humans have excellent context-based pattern recognition and reasoning abilities that far out-strip automated computing. For example, if asked to identify objects in a room that could contain liquids, it is an easy matter for a human to identify bottles, water glasses, coffee cups, vases for plants, sinks, and other items that could hold liquids. By contrast, this would be a very challenging pattern recognition task for computer processing in part due to the wide variety of objects that fit the requested definition. Similarly, humans use language easily to characterize context–but such processing is still challenging to automate. Hence, we seek to explicitly use the human as part of a reasoning/pattern recognition system and develop aids to enhance this hybrid machine/human cognition.

The use of novel displays and sonification are aimed at improving the ability of analysts to interpret complex data and to identify anomalous conditions that require further attention. This can be different than providing a realistic situation display. Figure 10.2 shows an example of a display of a 3 dimensional cube of 27,000 data points or sensor inputs represented as pyramids. Background or ordinary data are shown as being translucent. By establishing thresholds and dynamic search criteria the analyst's attention can be easily focused on those data or sensors that require attention by transforming the translucent data into opaque pyramids.

In our approach humans participate on both ends of the fusion process: on the input side as members of a community of observers, and on the output side as engaged analysts supporting pattern recognition and semantic-based analysis (see Figure 10.1).

We are also exploring an emerging concept in which an ad hoc community of analysts could support dynamic analysis of evolving situations (see Hall, Hall, McMullen, McMullen, and Pursel, 2008). That is, just as a community of observers may be tasked (or volunteer) to provide input data of value regarding an evolving event, crisis, emergency, or another situation, we consider the possibility of a future in which a community of analysts could collaborate to analyze evolving situations using the media of virtual worlds such as *Second Life* or *Olivia*. In this concept, civilian or amateur analysts may seek the solution of a complex problem by collaborating over the web analogous to the creation and maintenance of the Wikipedia. Thus, just as the national news media have begun to use civilian

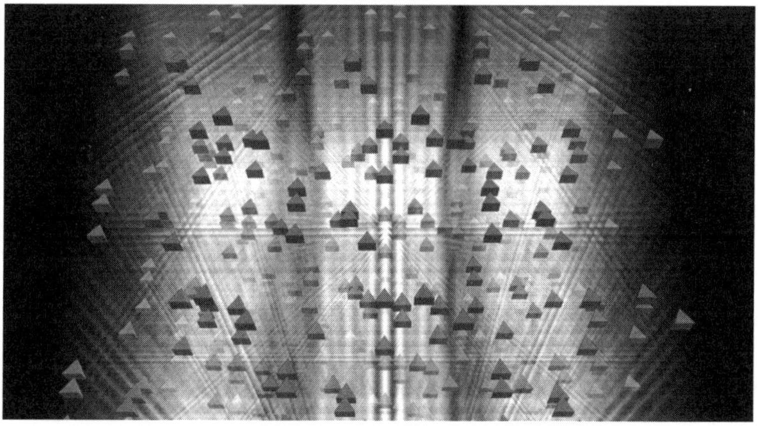

Figure 10.2 Example of 3-D display for identifying key data among a large data set (W. Shumaker, personal communication, October 14, 2007)

reporters (see, for example, CNN's *News to Me* [CNN I-reports]) as observers, fusion systems could be established to solicit analytical results.

An Extreme Events Laboratory

In order to conduct human in the loop experiments to investigate these issues, we have developed an extreme events laboratory shown conceptually in Figure 10.3. In effect, we have turned the entire Pennsylvania State University Park campus into a laboratory environment. The extended laboratory consists of:

- A 3-D visualization and analysis facility (upper right hand side of Figure 10.3) located in the Penn State Information Sciences and Technology building (upper left hand side of Figure 10.3). This facility provides the capability for 3-D visualization of data including situation displays and advanced displays of complex data. It also includes the capability for experiments with sonification (use of sound to assist in interpreting data).
- A wireless mesh network has been established on campus, which is connected to the Penn State information technology backbone. Meraki line-of-sight relay nodes provide campus wide wireless connectivity (see http://meraki.com/oursolution/mesh/).
- Mobile handheld devices include Nokia N810 Internet Tablets and Neo1973s cell phones (http://nokia.us/A4626058) to support solicitation and creation of human incident reports.

- A mobile command and analysis facility (pictured via a van) allows information to be processed and displayed in laptop computers.
- We have demonstrated public key encrypted communications among all of these components including streaming video from the Nokia N10 device to the mobile laptops and to the 3-D visualization facility in the IST building.

The Extreme Events Laboratory provides the capability to support human in the loop experiments involving humans as observers, analysts, decision-makers, and collaborators. Our User Science and Engineering Laboratory (http://minds.ist.psu.edu/) also supports experiments involving team-based decision-making and collaboration.

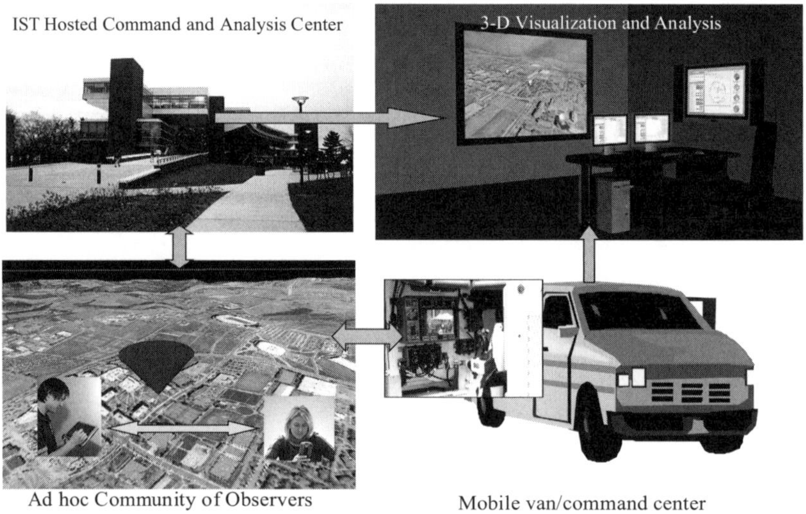

Figure 10.3 NC2IF Extreme Events Laboratory

On-going Experiments

Currently a number of experiments are being conducted in the NC2IF Center. These include the following:

- Characterization of human reports—We are developing mathematical models to characterize human reports including models for first and second order uncertainty.

- Computer assisted knowledge elicitation—Experiments are being conducted with variations in computer mediated knowledge elicitation for human observers (ranging from no mediation–simply allowing users to input data free form, to computer based mediation via templates and guiding advisory agents, to human assisted elicitation analogous to a 911 center; see Jones, Jefferson Jr., Connors, McNeese, and Obeita, 2005).
- Role of stress and emotion—Experiments have been conducted to understand the role of human emotion and stress in effectiveness in team based decision-making for a situational awareness exercise (Pfaff, 2008).
- Role of peer dynamics in reporting—Experiments are being designed to understand how observers are affected by the actions of their crowd peers in accuracy of reporting.
- Advanced visualization and sonification for data understanding—Development of 3-D and 2 ½ D displays and sonification for understanding complex data sets and focusing attention (Hall et al., 2008).
- Exploration of military/NGO interactions—Development of models and experiments to improve understanding and cooperation between military personnel and non-government organizations (NGO) personnel for applications such as disaster relief.

While these are only a beginning, these experiments are anticipated to assist in addressing the new role of humans as observers, analysts, and collaborators for addressing problems such as situational awareness and combat identification.

References

Burke, J., Estrin, D., Hansen, M., Parker, A., Ramanathan, N., Reddy, S., and Srivastava, M. B. (2006). Participatory sensing. *Proceedings of WSW'06 at SenSys'06*, Boulder, CO: ACM.

Burgoon, J., and Varadan, V. V. (2007). *Detecting and Countering IEDs and Related Threats* (NSF Workshop Report). Washington, DC: National Science Foundation.

Hall, C. M., Hall, D. L., McMullen, S. A., McMullen, M., and Pursel, B. (2008). Perspectives on visualization and virtual world technologies for multisensor data fusion. *Proceedings of Fusion 2008, the 11th International Conference on Information Fusion*, Cologne, Germany.

Hall, D. L., and Llinas, J. (1997). An introduction to multi-sensor data fusion. *Proceedings of the IEEE, 85*, 6–23

Hall, D. L., and Llinas, J. (eds). (2001). *Handbook of Multisensor Data Fusion.* Boca Raton, FL: CRC Press.

Hall, D. L., Llinas, J., McNeese, M., and Mullen, T. (2008). A framework for dynamic hard/soft fusion, *Proceedings of FUSION 2008, the 11th International Conference on Information Fusion*, Cologne, Germany.

Hall, D., and McMullen, S. (2004). *Mathematical Techniques in Multisensor Data Fusion*. Norwood, MA: Artech House.

Howe. J., (2006). The rise of crowdsourcing, *Wired, 14*.

Jones, R. E., Jefferson Jr., T., Connors, E. S., McNeese, M. D., and Obeita, J. F. (2005). Exploring cognitive maps for use in a crisis-management simulation. *Proceedings of the Conference on Behavior Representation in Modeling and Simulation*. Universal City, CA.

Li, J., and Wang, J. Z. (2003). Automatic linguistic indexing of pictures by a statistical modeling approach. *IEEE Transactions on Pattern Analysis and Machine Intelligence, 25*, 1075–1088.

Li, J., and Wang, J. Z. (2006). Real-time computerized annotation of pictures. *Proceedings of the ACM Multimedia Conference* (pp. 911–920). Santa Barbara, CA: ACM.

Liggins, M., Hall, D. L., and Llinas, J. (eds). (2008). *Handbook of Multisensor Data Fusion: Theory and Practice* (2nd edn). Boca Raton, FL: CRC Press.

Lucky, R. (2007). A billion amateurs. *IEEE Spectrum, November*, 96.

Pfaff, M. (2008). *Effects of Mood and Stress on Group Communication and Performance in a Simulated Task Environment.* Unpublished doctoral dissertation, The Pennsylvania State University College of Information Sciences and Technology, University Park, PA.

Piattelli-Palmarini, M. (1994). *Inevitable Illusions: How Mistakes of Reason Rule our Minds*. Hoboken, NJ: John Wiley and Sons.

Pohl, R. (2005). *Cognitive Illusions: A Handbook on Fallacies and Biases in Thinking, Judgment and Memory*. Hove, East Sussex: Psychology Press.

Waltz, E. (2003). *Knowledge Management in the Intelligence Enterprise.* Norwood, MA: Artech House.

Waltz, E., and Llinas, J. (1990). *Multisensor Data Fusion.* Norwood, MA: Artech House.

SECTION 4
Teams

Teams of various sizes and functions are an integral part of combat operations. The modern battlefield's dangers demand that teammates work closely together to have the best chance of making accurate and reliable combat identification decisions. The dynamics can either enhance or degrade a teams' capability to make the correct CID decisions. This section examines the important cross of team functioning and decision-making with the CID challenge.

Frank Greitzer and Dee Andrews (Chapter 11) examine the most human of all human dimensions that impact the friendly fire problem–emotion, stress, and response bias. Combat is a tremendously stressful activity that can raise emotions to their highest state and then plunge them into despair. The stress that comes with combat can color perceptions and cognitive decision-making in deleterious ways. This chapter explores some of what is known about emotion and stress, and attempts to explain how reasoned training strategies might mitigate the effect of those emotions. Emotion and stress can be contributing factors to a warfighter's tendency to respond inappropriately to cues received before and during battle. For example, if a platoon commander tells the troops that the enemy is over the next hill, the stress of the situation and their emotional state may cause the troops to respond in a biased manner to any humans they see over the next hill and they may shoot first before making a good combat identification. The chapter explores training strategies that might help to reduce response bias and thus reduce the chance of friendly fire.

Verlin Hinz and Dana Wallace (Chapter 12) use the Airborne Warning and Control Station (AWACS) teams to 'apply models of information processing in teams to show how teams can use and remember more information and thus make more accurate judgments than individuals' (p. 191). Their experiment allowed them to 'examine how teams and individuals compare in their identification judgments about targets when they are presented with a cognitively demanding situation involving a number of targets' (p. 195).

Wayne Shebilske, Georgiy Levchuk, Jared Freeman, and Kevin Gildea (Chapter 13) describe an experiment they conducted as part of their effort to develop a paradigm to train experienced teams for combat identification. In their study, they simulated Air Force Intelligence, Surveillance and Reconnaissance (ISR) and Dynamic Targeting Cell (DTC) teams because these teams are an integral part of CID operations. They were interested in how teams that 'are experienced in defending against one enemy strategy will have to learn to defend against new strategies' (p. 211).

Accurately defining the individual and team tasks that must be undertaken in combat settings is crucial to the development of good CID practices and tools. Beejal Mistry, Gareth Croft, David Dean, Julie Gadsden, Gareth Conway, and Katherine Cornes (Chapter 14) describe how this might be done with the INCIDER (Integrative Combat Identification Entity Relationship) model. The INCIDER model integrates physical representations of sensors and identification friend or foe (IFF) systems with human cognitive and behavioral characteristics to provide a simplified representation of detection and classification processes set within an operational context…' The model contains representations of a number of human factors, such as fatigue, stress, experience, expectation, and personality (p. 225). They used task analysis to add team tasks to the INCIDER model which had been focused only on individual tasks. Their chapter offers insight into how to best characterize team CID functions and tasks.

Teams can have great difficulty making appropriate CID decisions in the heat of battle if they do not share the same mental model. Measuring a team's mental model is fraught with difficulties. Jerzy Jarmasz, Richard Zobarich, Lora Bruyn-Martin, and Tab Lamoureux (Chapter 15) developed a behaviorally anchored rating scale (BARS) to measure behavior in a simulation-based exercise for assessing team cognition in tasks that required CID.

Chapter 11
Training Strategies to Mitigate Expectancy-Induced Response Bias in Combat Identification: A Research Agenda

Frank L. Greitzer
Pacific Northwest National Laboratory

Dee H. Andrews
Air Force Research Laboratory

Introduction

Combat Identification (CID) is the process of attaining an accurate characterization of detected objects (friendly, enemy, or neutral) throughout the Joint battlespace (DoD, 2000). Combat Identification is a function of Situation Awareness (SA) and Target Identification (TI) capabilities; effective CID requires adherence to doctrine, unit tactics, techniques and procedures, and approved rules of engagement. The goal of CID is to improve unit combat effectiveness while preventing fratricide (friendly fire) and minimizing collateral damage. CID is the process that human shooters or sensors go through to identify entities on the battlefield prior to making shoot/don't shoot decisions. To perform CID, the warfighter uses all available means at his disposal to sort the entities on the battlefield prior to applying combat power. The focus of this chapter is on exploring ideas for training mitigations that address stress-induced emotional and cognitive factors that introduce biases and expectancies that undermine CID.

Fratricide, as defined by the U.S. Army's Training and Doctrine Command (TRADOC) Fratricide Action Plan, is 'the employment of friendly weapons and munitions with the intent to kill the enemy or destroy his equipment or facilities, which results in unforeseen and unintentional death or injury to friendly personnel' (U.S. Department of the Army, 1993, p. 1). Fratricide has been a concern since humans first engaged in combat operations, although it gained much emphasis in the Persian Gulf War (U.S. Department of Defense, 1992). The percentage of deaths attributed to fratricide has ranged from 21 percent during World War II (American War Library, 1996) to 17 percent in the Persian Gulf War (Garamone, 1999). During recent major combat operations in support of Operation Iraqi Freedom, fratricide studies have reported a 25 percent increase in platform-to-soldier incidents and an increase in soldier-to-soldier incidents of 10 percent. It is difficult to know with certainty what the actual fratricide rate is because of the fog of war and the negative stigma that fratricide brings.

Rates of fratricide are increasing in part due to the increased accuracy and lethality of weapons, and despite the introduction of advanced technologies designed to increase target identification performance. Indeed, as TI is only part of the equation underlying CID, it is clear that enhancing SA is a continuing and critical need. Reliance on technology alone is a flawed strategy because technology is not infallible; technology may fail or be unavailable, and it may be undermined by technology developed by an adversary. Human SA will always be part of the equation because, ultimately, the human gives the order and pulls the trigger. Because of the background of human error in the equation, there is a sense of inevitability associated with the fratricide problem. It has been argued that fratricide is one of the inescapable costs of war (Marine Corps University Command and Staff College, 1995). But just as causal analysis studies of human error have produced insights and effected design/organizational improvements to reduce accidents, studies of the human factors underlying CID errors can reduce friendly fire incidents. The challenge is to minimize this unwanted companion to war that has been shown to produce devastating effects on troops in addition to the tragic loss of life. Data collected through the U.S. Army's Center for Army Lessons Learned (CALL) suggest 10 potential effects of friendly fire incidents (U.S. Department of the Army, 1992), including disrupted operations, a loss of initiative, loss of team cohesion, and loss of confidence in the team leader.

Reported Causes and Contributing Factors

A report produced by the U.S. Army's CALL center cited primary causes of fratricide (U.S. Department of the Army, 1992) as poor SA, combat identification failures, and weapons errors; with contributing factors including anxiety, confusion, bad weather, inadequate preparation, and leader fatigue. The report stated that these contributing factors are a critical dimension of realistic training conditions. Inadequate training is often cited as a contributing factor by studies of fratricide; other factors that have been cited include poor leadership, inappropriate procedures, language barriers, lack of appreciation of own platform position and heading, an inability to communicate changing plans or situations, and disorientation, confusion, and carelessness of aircraft crews (BBC News, 2004a, 2004b; Marine Corps University Command and Staff College, 1995; Penny, 2002).[1] While these studies provide some insight into contributing factors, identifying these factors as contributing does not by itself illuminate diagnostic factors underlying these failures. Wilson, Salas, Priest, and Andrews (2007) examined human factors literature for underlying human factors causes of friendly fire incidents. As

1 Over reliance on technology should also be included in any list of factors contributing to fratricide.

argued by Wilson et al., to accomplish tasks on the battlefield requires cognitive processes, performed as a collective effort that requires shared cognition. Using a human-centered approach, they concluded that in the absence of adequate shared cognition, warfighters can have problems interpreting cues, making decisions, and taking correct action. They concluded that when shared cognition 'fails,' the incidence of fratricide increases. They derived a taxonomy of behavioral markers that may help military leaders reduce the consequences of fratricide in war (see Table 11.1) and they identified factors (based on the individual, task, organization, technology, and environment) that influence shared cognition (see Figure 11.1). Addressing CID and fratricide requires mitigation strategies to reduce human errors and better prepare warfighters for factors that undermine SA.

The taxonomy presented in Table 11.1 has the potential to be a useful tool in diagnosing the contribution of shared cognition breakdowns in fratricide, and in identifying possible training strategies to prevent or overcome such breakdowns. Identifying portions of the taxonomy that are most influenced by combat stress, and more particularly, by stress-induced emotional and cognitive factors, can further define a training roadmap and strategies for reducing cognitive biases that undermine CID.

Table 11.1 Behavioral markers of teamwork breakdowns (from Wilson et al., 2007)

Communication	
Information exchange	Did team members seek information from all available resources? Did team members pass information within a timely manner before being asked? Did team members provide 'big picture' situation updates?
Phraseology	Did team members use proper terminology and communication procedures? Did team members communicate concisely? Did team members pass complete information? Did team members communicate audibly and ungarbled?
Closed-loop communication	Did team members acknowledge requests from others? Did team members acknowledge receipt of information? Did team members verify information sent is interpreted as intended?
Coordination	
Shared mental models	Did team members have a common understanding of the mission, task, team and resources available to them? Did team members share common expectations of the task and team member roles and responsibilities? Did team members share a clear and common purpose? Did team members implicitly coordinate in an effective manner?

Table 11.1 ***Concluded***

Mutual performance monitoring	Did team members observe the behaviors and actions of other team members? Did team members recognize mistakes made by others? Were team members aware of their own and others surroundings?
Back-up behavior	Did team members correct other team member errors? Did team members provide and request assistance when needed? Did team members recognize each other when one performs exceptionally well?
adaptability	Did team members reallocate workload dynamically? Did team members compensate for others? Did team members adjust strategies to situation demands?
Cooperation	
Team orientation	Did team members put group goals ahead of individual goals? Were team members collectively motivated and did they show an ability to coordinate? Did team members evaluate each other, while using inputs from other team members? Did team members exhibit 'give-and-take' behaviors?
Collective efficacy	Did team members exhibit confidence in fellow team members? Did team members exhibit trust in others and themselves to accomplish their goals? Did team members follow team objectives without opting for independence? Did team members show more and quicker adjustment of strategies across the team when under stress based on their belief in their collective abilities?
Mutual trust	Did team members confront each other in an effective manner? Did team members depend on others to complete their own tasks without 'checking up' on them? Did team members exchange information freely across team members?
Team cohesion	Did team members remain united in pursuit of mission goals? Did team members exhibit strong bonds and desires to want to remain a part of the team? Did team members resolve conflict effectively? Did team members exhibit less stress when performing team tasks?

Source: Reproduced with permission from *Human Factors*, 49(2), 243–256.

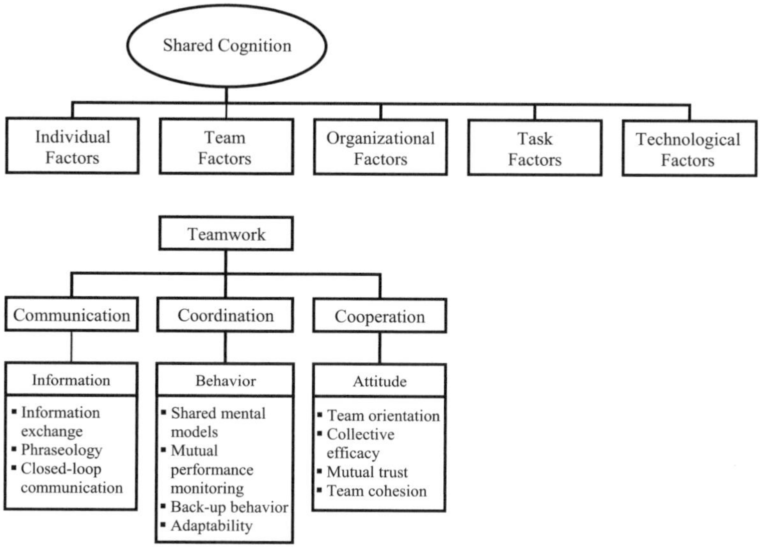

Figure 11.1 Framework for classifying teamwork breakdowns (after Wilson et al., 2007)

Source: Reproduced with permission from *Human Factors*, 49(2), 243–256. Copyright 2007 by the Human Factors and Ergonomics Society. All rights reserved.

The Role of Emotion, Stress, and Cognition

Emotions play a powerful role in everyday life and in military planning and operations, as well as military training. For a comprehensive examination of the psychology and performance effects of emotion and stress, the reader is referred to excellent reviews of the effects of stress (Staal, 2004, Kavanagh, 2005) and emotions (Blascovich and Hartel, 2007) on cognition. Emotions influence our perceptions and they bias our beliefs; they influence our decisions and in large measure guide how people adapt their behavior to the physical and social environment (Musch and Klauer, 2003; Judge and Larsen, 2001). Because emotions can impair decisions, military training developers are well advised to incorporate an emotional element into training to elicit the strong emotions soldiers will feel on the battlefield.

Effects of Emotion and Affective State

Emotion has effects at all levels of cognitive processing; many of them are directly relevant to military contexts. Military situations are fraught with uncertainty, and understanding the role of emotion in arriving at accurate SA may prove useful in optimizing decision processes.

A person's affective state is primarily influenced by a largely automatic process termed evaluation (Bargh and Ferguson, 2000; Barrett, 2006a). Evaluation is a fast analysis, often unconscious (Moors and De Houwer, 2006), in which something is judged 'good for me' or 'bad for me'–in other words, an analysis of whether or not properties of a situation are important to one's survival, well-being, and goals (Ellsworth and Scherer, 2003). Thus, affective states influence what people attend to and how they interpret what they see. MacLeod (1996) suggested that anxiety impairs cognitive performance by diverting mental resources toward task-irrelevant information that relates to the perceived threat. Emotions also influence what people remember about an event, or details just before or after an event that elicit strong emotions (such as intense fear). Research has also shown that emotions can bring about self-deception (e.g., Mele, 2000) or overwhelm reason (Shiv and Fedorikhin, 1999) in making decisions.

Effects of Stress

Stress has strong effects on every aspect of cognition from attention to memory to judgment and decision-making. A general framework describing performance effects of stress is shown in Figure 11.2. In general, under stress, attention appears to channel or tunnel, reducing focus on peripheral information and centralizing focus on main tasks (Kavanagh, 2005). Originally observed by Kohn (1954), this finding has been replicated often, first by seminal work from Easterbrook (1959) demonstrating a restriction in the range of cues attended to under stress conditions (tunneling) and many other studies (see Staal, 2004). Peripheral stimuli are likely to be the first to be screened out or ignored. Decision-making models proposed by Janis and Mann (1977) support this hypothesis and suggest that under stress, individuals may make decisions based on incomplete information. Friedman and Mann (1993) suggested that when under conditions of stress, individuals may fail to consider the full range of alternatives available, ignore long-term consequences, and make decisions based on oversimplifying assumptions—often referred to as heuristics.[2] There is also literature on the effects of stress on vigilance and sustained attention, with a particular focus on stress caused by fatigue and sleep deprivation. A review by Davies and Tune (1970) concluded that vigilance tends to be enhanced by moderate levels of arousal (stress), but sustained attention appears to decrease with fatigue and loss of sleep. In the cognitive domain, a study by Wickens, Stokes, Barnett and Hyman (1991) found that under time pressure, noise, and financial risk, individuals performed more poorly on vigilance and attention tasks, but declarative knowledge tasks were not affected.

2 While researchers who argue that perceptual narrowing reduces the quality of individual decisions, Klein (1996) observed that the use of heuristics may allow individuals to respond more quickly to external demands while under stress or when provided only partial information.

Stressor →	***Stress*** →	***Performance***
An external demand or event	A response to the external event	Response affects performance / behavior
		Perceptual narrowing *Reduced cognitive processing* *Use of heuristics* *Longer task completion time*

Figure 11.2 Performance effects of stress (from Kavanagh, 2005, p. 3)

Several investigations have shown that tasks that are well-learned tend to be more resistant to the effects of stress than those that are less-well-learned. Extended practice leads to commitment of the knowledge to long term memory and easier retrieval, as well as automaticity and the proceduralization of tasks. These over-learned behaviors tend to require less attentional control and fewer mental resources (Leavitt, 1979; Smith and Chamberlin, 1992), which further results in enhanced performance and greater resistance to the negative effects of stress—i.e., they are less likely to be forgotten and more easily recalled under stress. Van Overschelde and Healy (2001) found that linking new facts learned under stress with preexisting knowledge sets helps to diminish the negative effect of stress. On the other hand, there is also a tendency for people under stress to 'fall-back' to early-learned behavior (Allnut, 1982; Barthol and Ku, 1959; Zajonc, 1965)—even less efficient or more error prone behavior than more recently-learned strategies—possibly because the previously learned strategies or knowledge are more well-learned and more available than recently acquired knowledge.

Research suggests that high stress during learning tends to degrade an individual's ability to learn—perhaps due to interference or disruption in the encoding and/or maintenance phases of working memory. An implication for instructional strategies is that a phased approach should be used, with an initial learning phase under minimum stress, followed by gradual increasing exposure to stress more consistent with real-world conditions. Stress inoculation training attempts to immunize an individual from reacting negatively to stress exposure. The method provides increasingly realistic pre-exposure to stress through training simulation; through successive approximations, the learner builds a sense of positive expectancy and outcome and a greater sense of mastery and confidence. This approach also helps to habituate the individual to anxiety-producing stimuli.

Finally, it is important to consider group processes in this context. Historically, research has focused on individuals, but there is a growing literature on team decision-making. Effective teams are able to adapt and shift strategies under stress; therefore, team training procedures should teach teams to adapt to high stress conditions by improving their coordination strategies. Driskell, Salas, and Johnston (1999) observed the common finding of Easterbrook's attentional narrowing is a phenomenon also applicable to group processes. They demonstrated that stress can reduce group focus necessary to maintain proper coordination and

SA—i.e., team members were more likely to shift to individualistic focus than maintaining a team focus.

Cognitive Biases

Gestalt psychology tells us that we tend to see what we expect to see. Expectancy effects can lead to such selective perception as well as biased decisions or responses to situations in the form of other cognitive biases like confirmation bias (the tendency to search for or interpret information in a way that confirms one's preconceptions) or irrational escalation (the tendency to make irrational decisions based upon rational decisions in the past). The impact of cognitive biases on decision performance—particularly response selection—is to foster decisions by individuals and teams that are based on prejudices or expectations that they have gained from information learned before they are in the response situation. For example, if a combat pilot is told that only the enemy is on the north side of a river, the pilot may be biased to fire prematurely at the first potential target seen on the north side of the river. The pilot has an expectancy that this action will lead to a successful first level outcome, namely the enemy will be destroyed or disrupted.

The disruption of rational decision-making processes by cognitive biases is only exacerbated by the stress experienced in life and death situations. Stressful, emotionally-charged combat situations thus provide a stimulus for the effects of cognitive biases that overcome the effects of prior training.

Following are two well-publicized incidents of friendly fire where expectancy and response bias appear to have played a major role.

'Pa. Guard pilots cleared in "friendly fire" incident; 10 Marines died, 4 hurt when A-10 jets and Iraqis struck U.S. force last year' (*The Sun*, 2004). The Central Command placed sole blame on an unidentified Marine captain who called in the two Air Force A-10 attack jets without realizing that dozens of Marines were in the area. Because the Marines were attacked by both friendly and enemy fire, the exact source of their wounds could not be determined, investigators said. The pilots, who used binoculars, said they could pinpoint only white pickup trucks and not the Marines' armored vehicles, two of which were attacked by the jets, according to the investigative report. Investigators said the Marine captain gave the pilots blanket approval to attack an area on the outskirts of Nasiriyah. The Marine captain faces possible disciplinary or administrative action. Col. Gregory Marston, vice-commander of the 111th Fighter Wing, said the pilots were 'miles' from the Marines when they began their bombing and strafing runs, and not as close and 'low' as some Marines reported after the incident. The report said the pilots circled at 15,000 feet before descending and beginning their attack. 'People on the ground were shooting' at the two pilots, Marston said. 'They staged this at the prescribed altitude and the prescribed distance from the target.' Marston said he doubted that the pilots had flown directly over the Marines, because 'that's not how they train.' Marston said he could not recall the last time the squadron trained with Marine units. The pilots could see the white pickup trucks near Nasiriyah because the vehicles stood out against the desert background, said Marston, unlike the Marines' green armored vehicles, which the pilots said they did not see.

Friendly fire: A recent history – CBC News On Sept. 4, 2006, two U.S. A–10 Thunderbolts mistakenly attacked Canadian troops in Afghanistan during Operation Medusa, a major operation aimed at retaking control of two dangerous districts west of the city of Kandahar. In April 2002, [An] American fighter pilot ... killed four Canadian soldiers when he dropped a 225-kilogram bomb on a unit conducting military exercises near Kandahar. [He] saw gunfire on the ground, which he mistook for surface-to-air fire. [He] attacked, killing Sgt. Marc D. Leger, Cpl. Ainsworth Dyer, Pte. Richard Green and Pte. Nathan Smith. Eight other Canadians were wounded in the bloodiest friendly fire incident to hit this country since the Korean War.

Implications for CID Training

Based on the foregoing discussion, the challenges and needs for more effective CID training in general terms as well as more specifically can be summarized as: addressing deficiencies in scenarios, addressing needs for incorporating stress and stress management techniques, and addressing challenges in preparing warfighters to overcome cognitive biases. The following factors should be included in CID training:

- Training should provide extended practice, promoting more persistent memory and easier retrieval, and to encourage automaticity and the proceduralization of tasks to make them more resistant to the effects of stress.
- Team training should focus on strategies for maintaining group cohesion and coordination, mitigating the tendency for team members to revert to an individual perspective and lose shared SA.
- Training should exercise the execution of cognitive tasks by both individuals and groups.

Deficiencies in Typical Combat Training Scenarios

Warfighters are trained based on threat scenarios, but deficiencies in the characteristics of such scenarios may prevent the learning of strategies to overcome cognitive biases while under stress. CID training should provide sufficiently complex scenarios that induce stress by forcing warfighters into 'uncomfortable territory.' Complex or dynamic changes (threats) must be injected into scenarios that induce trainees to experience uncertainties of the real world, rather than simply exercising previously-learned skills and 'recipes' learned to face typical or expected threats. In other words, to ensure that the trainee is forced to operate without perfect information and in the face of 'surprises' that challenge preconceptions or assumptions. Without such complex and dynamic threats, training can cause the warfighter (and battle planners) to overestimate their capabilities. As argued by Sawyer and Pfeifer (2005) in a homeland security training context, '…organizations must recognize that the threat is dynamic and is characterized by extensive uncertainty. To move beyond preparing for the last war, our training must challenge and test our assumptions about operating in complex environments, examine our operational and strategic constraints, and evaluate our capabilities to respond effectively to challenging, changing events.' [p. 250]. Thus, the following suggestions apply to scenario construction and management in CID training:

- CID training scenarios should include complex/dynamic threats that reflect the uncertainties of the real world—scenarios that force trainees to operate without perfect information and that incorporate surprises that challenge preconceptions or assumptions.
- CID training scenarios should be designed to encourage the habit of testing one's assumptions to produce more adaptive, resilient CID performance in the face of uncertainty.

Need for More Realistic Stress and Stress Management Training

All training involves instruction and practice in exercising knowledge, skills, and abilities necessary to accomplish a task. Thus, in the context of CID, warfighters must be trained on how to accurately perceive stimuli that will inform the trainee's decision-making process as to whether the stimulus of interest is friend, foe, or neutral, and how to recognize a failure in TI technology (see Cannon-Bowers and Salas, 1998). Additionally, because of the intense nature of the battlefield (also referred to as the 'fog of war'), warfighters experience extreme pressures that they must overcome to apply the knowledge, skills, and abilities that they acquire during training, and to control intense emotions associated with battle. While they receive extensive training on strict rules of engagement, procedures and requirements to pursue the commander's intent, none of the training experiences can match the real battle where they must make life-or-death decisions quickly. Therefore, a major challenge for CID training—and one that is distinctly different from training on knowledge, skills, and abilities—is training to enhance awareness of the effects of stress on cognitive performance and to exercise the execution of cognitive tasks, individual, and group decision-making (maintaining shared SA) under conditions of stress that are comparable to operational environments, with the ultimate aim of reducing human errors associated with CID.

Training (and development of effective combat ID training) becomes more complicated when potential affective issues that might produce cognitive biases are emphasized. However, it is vital that warfighters be able to recognize and then be trained to overcome, if possible, those biases. It is not a case of the warfighters not knowing what cues they should be looking or listening for. Their initial skill-based training taught them those cues. Rather, it's a case of warfighters being taught to recognize their proclivity to bias their interpretation of the cues when they are emotionally charged. As was discussed previously, emotional stress can cause warfighters to narrow their attentional focus and to start looking for reasons to fire instead of reasons not to fire. Can warfighters be taught to recognize and overcome these cognitive biases? A large part of the answer rests on whether warfighters can be taught to recognize their emotionally charged state and then to manage that state. It is not realistic to expect warfighters to overcome their emotions in a combat situation; however, it should be possible to train them to recognize and undertake the management of those emotions in unfamiliar situations.

The following suggestion should be included in a prescription for improving CID training through the use of more realistic accommodation of stress factors:

- Training strategies should incorporate an emotional element into training to elicit the strong emotions soldiers will feel on the battlefield.
- Because high stress during training tends to impair learning, a phased approach should be used, beginning with minimum stress and building up to stress levels more consistent with real-world conditions.

Need to Address Cognitive Biases

CID training must be designed to more effectively address cognitive biases. Cognitive biases such as confirmation bias and irrational escalation can cause experienced warfighters to spend critical time searching for familiar cues or indicators associated with situations with which they have had experience or training, to the detriment of their ability to think outside the box and observe cues and stimuli that are most relevant to the novel situation that they face. Therefore, training on combat ID should attempt to teach warfighters to identify and assess the relevant indicators in a new environment, without automatically resorting to preconceived lists of indicators.

A training approach to address the effects of stress on cognitive biases, and management of such biases, may include detailed 'after-action reviews' to raise trainee awareness about the ways they gathered information to help them recognize threats, identify problems, and make correct decisions (or incorrect ones). The focus of this type of training, which occurs after the traditional skill-based training, is to help the warfighters learn to keep their eyes and minds open to crucial elements in situations they have not experienced before. They must be able to weigh all information—even unexpected information—and keep an open mind to overcome cognitive biases that restrict their perception, attention, and decision-making performance. The ability to imagine or anticipate unexpected outcomes is critical to effective decision-making under stress.

Training requirements to better meet the objectives of addressing stress-induced cognitive biases in CID should:

- enhance awareness of the effects of stress on cognitive performance—such as tunneling and flawed decision-making strategies that ignore information—and coping strategies to moderate these effects. The training should be designed to make as explicit as possible what might happen to skill and knowledge under stress on the battlefield.
- Train awareness of cognitive biases and practices for managing these biases;
- Emphasize habits of testing assumptions and moving beyond traditional reactive behaviors to train techniques for more adaptive, resilient CID performance in the face of uncertainty.

Conclusions: Summary of New CID Training Requirements and a Preliminary Research Agenda

In conclusion, this survey of relevant literature on warfighter affective conditions applicable to CID performance with the objective of describing new CID training requirements addresses stress-induced cognitive limitations and biases. The suggestions for enhancing traditional CID training emphasize the need to expose warfighter trainees to high-stress training in completely unfamiliar scenarios, and to provide meaningful cognitive feedback to help them cope with and manage their limitations and biases. A summary of these requirements is listed below, followed by a discussion on the need for a research agenda to further define CID training challenges.

Summary List of CID Training Effectiveness Requirements

- CID training must move beyond core competency training by training warfighters to cope with increased stress and cognitive biases in unfamiliar situations.
- Training strategies should incorporate an emotional element into training to elicit the strong emotions soldiers will feel on the battlefield.
- Training should provide extended practice, promoting more persistent memory and easier retrieval, and encourage automaticity and the proceduralization of tasks to make them more resistant to the effects of stress. Because high stress during training tends to impair learning, a phased approach should be used, beginning with minimum stress and building up to stress levels more consistent with real-world conditions.
- CID training should enhance awareness of cognitive biases and the effects of stress on cognitive performance—i.e., to train warfighters to recognize and avoid, or at least manage, their emotional state so that effects of cognitive biases are reduced.

- Team training should focus on strategies for maintaining group cohesion and coordination, mitigating the tendency for team members to revert to an individual perspective and lose shared SA.
- Training should exercise the execution of cognitive tasks by both individuals and groups.
- CID training scenarios should include complex/dynamic threats that reflect the uncertainties of the real world—scenarios that force trainees to operate without perfect information and that incorporate surprises that challenge preconceptions or assumptions.
- CID training should emphasize habits of testing assumptions and moving beyond traditional reactive behaviors to train techniques for more adaptive, resilient CID performance in the face of uncertainty.

CID Training Research Challenges

Clearly, performing research on warfighter affective conditions is very difficult. It is not clear that the created simulated conditions would adequately replicate battlefield conditions in such a way as to bring about the kind of expectancy and response bias described. In addition, there are ethical issues that would have to be considered, a research agenda is needed to properly explore this topic.

A necessary first exercise is to extend the training taxonomy developed by Wilson et al. (2007). First, the taxonomy must be extended in the area of stress-induced response bias and deleterious effects on expectancies. Second, it would be useful to describe and speculate on the additions to the taxonomy for training individual warfighters to avoid response bias induced fratricides. Finally, a research agenda should be established and executed with the main elements summarized below.

- Research is needed to examine possible effects on decision-making performance while warfighters are expending limited cognitive resources trying to 'manage' their emotions.
- Research is needed to assess whether systems like Blue Force Tracker can improve the warfighter's expectancy of the stimuli they are likely to see. How can Blue Force Tracker displays be improved to reduce cognitive biases?
- Research is needed to further understand the effects of cognitive bias in combat settings.
- Define stress factors that exacerbate cognitive bias. There is substantial evidence that stress is a key factor. What roles are played by; fatigue, illness, fear of fratricide, past experience, lack of skill and/or knowledge of the weapon system or environment, poor team communication, trust in intelligence, trust in superiors?
- Define aspects of cognitive bias that most strongly apply to combat settings. It is not clear that every combat situation could be covered by the same biases.

- Determine whether it is possible to mitigate stress-related cognitive bias through better and/or more training.
- Our assumption is that cognitive bias will be less of a problem if a warfighter is better trained, but is that assumption correct? The training community's general feeling is that the better trained the warfighter, the better they will be able to overcome the negative effects of stress. Is that assumption correct? Because of the difficulty of conducting empirical studies where valid stressors are introduced and measured, this question may remain difficult to answer, but it should receive more research attention than it has.
- Continuing on the training theme, what training methods and technologies can best be used to mitigate cognitive bias? Do these methods and technologies need to be used differently for different warfighters?
- Determine whether anecdotal reports of friendly fire incidents should be trusted by researchers who are investigating cognitive bias in combat. First- and second-person reports of actual incidents are notoriously prone to bias. Are we taking the right lessons from actual incidents, and can these lessons be reliably relied upon to shape training programs to reduce cognitive bias?

These research challenges form a minimal set of requirements for developing a valid research agenda. A review of present literature indicates that there are few valid answers for many of these research questions. In addition, there may be value in doing research to identify training approaches to help warfighters overcome the terrible effects of being involved in a friendly fire incident—undoubtedly related to training and psychological interventions aimed at alleviating the effects of post-traumatic stress syndrome. Without such intervention, warfighters suffering these effects bring great danger upon their unit and themselves if their emotional grief overwhelms their training and desire to continue the fight. It is possible that no training could help a warfighter overcome those situations and the warfighter might have to be quickly removed from the combat situation. There are, of course, many situations where such removal would not be possible.

The potential gains from the proposed training approaches that emphasize cognitive/affective management skills are evident. The use of dynamic threat scenarios to promote coping with uncertainties in unfamiliar situations will build warfighter abilities to think ahead rather than merely react, and to be better equipped to perform in the midst of the 'fog of war.' If, on the other hand, we continue to limit our CID training objectives to core competency/skill development issues, then it seems that we will continue to run 'limited' exercises well, build false confidence in our abilities, and fail to meet our most critical challenges in protecting our forces from friendly fire incidents.

Acknowledgement

This work was supported by the Pacific Northwest National Laboratory operated for the U.S. Department of Energy by Battelle under Contract DE-AC05–76RL01830. [Information Release No.: PNNL-SA-64673].

References

Allnut, M. (1982). Human factors: Basic principles. In R. Hurst, and L. R. Hurst (eds), *Pilot Error* (pp. 1–22). New York: Aronson.

American War Library. (1996). *The Michael Eugene Mullen American Friendly Fire Notebook.* Retrieved from http://members.aol.com/veterans.

Bargh, J. A., and Ferguson M. J. (2000). Beyond behaviorism: On the automaticity of higher mental processes. *Psychological Bulletin, 126*, 925–945.

Barrett, L. F. (2006). Valence as a basic building block of emotional life. *Journal of Research in Personality, 40*, 35–55.

Barthol, R. P., and Ku, N. D. (1959). Regression under stress to first learned behavior. *Journal of Abnormal and Social Psychology, 59*, 134–136.

Blascovich, J. J., and Hartel, C. R. (eds). (2007). *Human Behavior in Military Contexts.* Washington, DC: Board on Behavioral, Cognitive, and Sensory Sciences, Division of Behavioral and Social Sciences and Education, the Report of the National Research cCouncil of the National Academies. Retrieved February 12, 2008, from http://www.nap.edu/catalog/12023.html.

Bowman, T. (2004, March 31) Pa. Guard pilots cleared in 'friendly fire' incident. *The Sun,* p. A3.

British Broadcasting Corporation News. (2004a, March 13). *Marine Killed in 'Friendly Fire.'* Retrieved February 12, 2008, from http://news.bbc.co.uk/2/hi/uk_news/3507514.stm.

British Broadcasting Corporation News. (2004b, May 14). *'System Error' Link to RAF Deaths.* Retrieved February 12, 2008, from http://news.bbc.co.uk/2/low/uk_news/england/norfolk/3714251.stm.

Cannon-Bowers, J. A., and Salas, E. (eds). (1998). *Making Decisions Under Stress: Implications for Individual and Team Training*. Washington, DC: American Psychological Association.

CBC News Online. (2006, September 5). *Indepth: Friendly Fire. Friendly Fire: A Recent History.* Retrieved September 12, 2008, from http://www.cbc.ca/news/background/friendlyfire/friendlyfire-2006.html.

Davies, D. R., and Tune, G. S. (1970). *Human Vigilance Performance*. London: Staples Press.

Driskell, J. E., Salas, E., and Johnston, J. (1999). Does stress lead to a loss of team perspective? *Group Dynamics: Theory, Research and Practice, 3*, 291–302.

Duncan, S., and Barrett, L. F. (2007). Affect as a form of cognition: A neurobiological analysis. *Cognition and Emotion, 21,* 1184–1211.

Easterbrook, J. A. (1959). The effect of emotion on cue utilization and the organization of behavior. *Psychological Review, 66*, 1873–201.

Ellsworth, P. C., and Scherer, K. R. (2003). Appraisal processes in emotion. In R. J. Davidson, K. R. Scherer, and H. H. Goldsmith (eds), *Handbook of Affective Sciences,* (pp. 572–595). New York: Oxford University Press.

Friedman, I. A., and Mann, L. (1993). Coping Patterns in Adolescent Decision-Making: An Israeli-Australian Comparison, *Journal of Adolescence, 16*, 187–199.

Garamone, J. (1999, February 2). *Fixes Touted to Combat Friendly Fire Casualties*. American Forces Information Service. Retrieved August 28, 2004, from http://www.defenselink.mil/news/Feb1999/n02021999_9902027.html.

Janis, I. L., and Mann, L. (1977). *Decision Making,* New York: The Free Press.

Judge, T. A., and Larsen, R J. (2001). Dispositional sources of job satisfaction: A review and theoretical extension. *Organizational Behavior and Human Decision Processes, 865,* 67–98.

Kavanagh, J. (2005). *Stress and Performance: A Review of the Literature and its Applicability to the Military.* (RAND TR-192, ADA439046). Santa Monica CA: Rand Corp. Retrieved February 12, 2008, from http://www.rand.org/pubs/technical_reports/2005/RAND_TR192.pdf.

Klein, G. (1996). The Effects of Acute Stressors on Decision-Making. In J. Driskell, and E. Salas (eds), *Stress and Human Performance* (pp. 49–88). Hillsdale, NJ: Lawrence Erlbaum.

Kohn, H. (1954). Effects of variations of intensity of experimentally induced stress situations upon certain aspects of perception and performance. *Journal of Genetic Psychology, 85*, 289–304.

Leavitt, J. (1979). Cognitive demands of skating and stick handling in ice hockey. *Canadian Journal of Applied Sport Sciences, 4*, 46–55.

MacLeod, C. (1996). Anxiety and cognitive processes. In I. G. Sarason, G. R. Pierce, and B. R. Sarason (eds), *Cognitive Interference: Theories, Methods, and Findings* (pp. 47–76). Mahwah, NJ: Lawrence Erlbaum.

Marine Corps University Command, and Staff College. (1995). *Fratricide: Avoiding the Silver Bullet*. Unpublished student paper. Retrieved October 28, 2008, from http://www.globalsecurity.org/military/library/report/1995/DJ.htm http://12.1.239.226/isysquery/irl34aa/1/doc.

Mele, A. R. (2000). Self-deception and emotion. *Consciousness and Emotion, 1,* 115–137.

Moors, A., and De Houwer, J. (2006). Automaticity: A theoretical and conceptual analysis. *Psychological Bulletin, 132*: 297–326.

Musch, J., and Klauer, K. C. (2003). *The Psychology of Evaluation: Affective Processes in Cognition and Emotion*. Mahwah, NJ: Lawrence Erlbaum.

Penny, P. H. G. (2002). Combat identification: Aspirations and reality. *Military Technology, 26*, 50–54.

Rao, P. S. (2000). *Personnel and Human Resource Management – Text and Cases*. Bombay: Himalaya Publishing House.

Sawyer, R., and Pfeifer, J. (2005). Strategic Planning for First Responders: Lessons Learned from the NY Fire Department. In R. D. Howard, J. J. F. Forest, and J. C. Moore (eds), *Homeland Security and Terrorism: Readings and Interpretations* (pp. 246–258). New York: McGraw-Hill.

Shiv, B., and Fedorikhin, A. (1999). Heart and mind in conflict: The interplay of affect and cognition in consumer decision making. *Journal of Consumer Research*, *26*, 278–292.

Smith, M. D., and Chamberlin, C. J. (1992). Effect of adding cognitively demanding tasks on soccer skill performance. *Perceptual and Motor Skills, 75*, 955–961.

Staal, M. A. (2004). *Stress, Cognition, and Human Performance: A Literature Review and Conceptual Framework.* (NASA/TM-2004–212824). Moffett Field, CA: NASA Ames Research Center.

U.S. Department of Defense, Office of the Under Secretary of Defense for Acquisition and Technology, and Joint Chiefs of Staff. (2000). *Joint Warfighting Science and Technology Plan.* Deputy Under Secretary of Defense (Science and Technology. Washington, DC: Author.

U.S. Department of Defense. (1992). *Conduct of the Persian Gulf War: Final report to Congress,* (Public Law 102–25). Washington, DC: Author.

U.S. Department of the Army. (1992). *Fratricide: Reducing Self-Inflicted Losses.* Center for Army Lessons Learned (CALL) Newsletter No. 92(4). Fort Leavenworth, KS: Center for Army Lessons Learned, U.S. Army Combined Arms Command. Retrieved February 10, 2008, from http://www.globalsecurity.org/military/library/report/call/call_92–4_tblcon.htm.

U.S. Department of the Army. (1993). *Military Operations: U.S. Army Operations Concept for Combat Identification* (TRADOC Pam 525–58). Fort Monroe, VA: Training and Doctrine Command.

van Overschelde, J. P., and Healy, A. F. (2001). Learning of nondomain facts in high- and low-knowledge domains. *Journal of Experimental Psychology: Learning, Memory, and Cognition, 27*, 1160–1171.

Wickens, C. D., Stokes, A., Barnett, B., and Hyman, F. (1993). The effects of stress on pilot judgment in a MIDIS simulator. In O. Svenson, and A. J. Maule (eds), *Time Pressure and Stress in Human Judgment and Decision Making* (pp. 271–292). New York: Plenum Press.

Wilson, K. A., Salas, E., Priest, H. A., and Andrews, D. H. (2007). Errors in the heat of battle: Taking a closer look at shared cognition breakdowns through teamwork. *Human Factors*, *49*, 243–256.

Zajonc, R. B. (1965). Social facilitation. *Science, 149*, 269–274.

Chapter 12
Comparing Individual and Team Judgment Accuracy for Target Identification under Heavy Cognitive Demand

Verlin B. Hinsz
Dana M. Wallace
North Dakota State University

The proper and effective use of information is critical in various command and control situations. When information is unavailable or is used improperly, a variety of problems arise such as fratricide, when friendly forces are inaccurately identified as enemy combatants and are attacked and killed by their compatriots (e.g., BBC News, 2003, 2004; CBS News, 2006; Morgret, 2002). Fratricide has long been a concern of military and police forces (American War Library, 1996; Department of Defense, 1992; Garamone, 1999), and ways of eliminating attacks on friendly forces have attracted much attention. Many of these efforts have focused on technological advancements and devices that are used to identify friendly forces during combat (e.g., radio-frequency identification tags and GPS; Kime, 2003; PMM, 2003). In this chapter, we consider how the processing of available information and how command and control can be modified to supplement technology. In particular, we apply models of information processing in teams to show how teams can use and remember more information and thus make more accurate judgments than individuals (also see; Hinsz, 1990; Hinsz, Tindale, and Vollrath, 1997).

One purpose of command and control is to gather information that leads to decisions to act. Consequently, information processing and decision-making are two critical components of command and control. When flawed decisions are made, such as mistaking friends for foes (e.g., two U.S. Army helicopters downed by U.S. fighters in northern Iraq; Gordon, 1994; Snook, 2000), information processing is often to blame. Similarly, in target identification during combat, if the information provided is erroneous, a mistake may occur (e.g., the attack on the Chinese embassy in Belgrade; BBC, 1999). In the midst of combat, the accurate identification of a target as friendly or hostile is a vital issue for pursuit of the enemy as well as protection of friendly forces. The proper identification of a target involves cognitive processes associated with knowing, recognizing, and remembering details about the target. However, these cognitive processes can be overwhelmed when identification of a large number of targets is required in a dynamically changing and threatening battlespace.

Airborne Warning and Control Station (AWACS) operations are one arena in which target identification occurs routinely (Fahey, Rowe, Dunlap, and deBoom, 2000; Klinger, Thordsen, and Copeland, 1998). AWACS weapons directors are Air Force personnel who control and direct aircraft assets (e.g., fighters, refueling tankers) from airborne AWACS platforms (Fahey et al., 2000). AWACS operations involve both the information processing and decision-making aspects of command and control. In modern battle situations, AWACS weapons directors have to gather and monitor information about a large number of friendly and enemy targets, identify and discriminate between the types of targets, and then make decisions about the aircraft that will be directed to certain targets. The misidentification of a target can mean a fighter will not get refueled in a timely fashion, an enemy incursion will not be met with force, or ground troops will not receive appropriate air support. Also, a friendly force could be misidentified as an enemy target and the decision made to engage the target (i.e., friendly fire). Misidentifications of targets are more likely when massive amounts of information have to be digested by AWACS operators (Klinger et al., 1998). This overwhelming amount of information is also further complicated by the reality that targets may be moving quickly through the battlespace.

How can AWACS weapons directors cope with so much constantly changing information? What can operators do when there are too many targets to accurately identify and discriminate friend from foe? Given this overwhelming situation, how do AWACS weapons directors use available information to make decisions about which weapons will be released and to which targets they will be directed? When weapons directors face a large number of targets of unknown designation, they need to have techniques and technologies that help them deal with this cognitively taxing situation (Fahey et al., 2000). One way that AWACS weapons directors might effectively confront a cognitively demanding situation is to collaborate and cooperate as a team. By working as a team, the members can apply a division of labor (e.g., specific weapons directors direct their attention to particular sectors). In addition, working as a team allows the members to draw upon their larger, collective cognitive resources to deal with cognitively demanding situations. Therefore, teams might be effective when confronted with large amounts of information because they can draw on a larger pool of cognitive resources as well as use a division of labor.

Weapons directors collaborate and coordinate with other weapons directors so they work as a team. Weapons directors monitor the activities of aircraft that are their responsibility from screen displays at workstations. However, at particular times, weapons directors can lose their screen displays and have to perform their tasks 'blind.' Memory is therefore a critical cognitive process in the multiple tasks performed by weapons directors (Fahey, et al., 2000). For example, weapons directors may need to compensate for equipment glitches or environmental factors that require them to rely on their memory to be able to monitor, command, control, and communicate with designated aircraft. Failure to remember the location, attributes, or mission of an aircraft can be considered target misidentification. Members of a weapons director team may use other team members to aid them in remembering where aircraft are, what their missions are, where they are heading, possible conflicts that might exist in the flight corridor, and the existence of threatening conditions (e.g., fighter running out of fuel). In this sense, teams may act as a 'technology' to improve target identification. We investigated how collaboration in target identification may enhance weapons directors' performance under conditions of overwhelming information, thereby demonstrating how teams can be used to improve target identification and reduce potential misidentification errors.

There is a substantial literature demonstrating that teams deal more effectively with information than individuals (Hinsz et al., 1997). In the context of AWACS operations, a single weapons director might be responsible for processing the available information and making a decision to release weapons. Alternatively, a weapons director team could be assigned to collectively process the information and make decisions. There are a number of reasons why teams might be preferred to individuals in this situation (Hinsz, 1990). Teams are expected to be superior because they can pool information from different members (Hastie, 1986), teams are likely to correct errors that individual members may make (Laughlin, 1980), and teams can engage in better strategies for coming to a response (Hinsz, 2004; Laughlin, Bonner, and Miner, 2002). Consequently, in cognitive task performance, teams will remember more of the information (Hinsz, 1990), more exhaustively deliberate the information (Hinsz et al., 1997), and remain less biased in their judgments (Kerr, MacCoun, and Kramer, 1996). For these reasons, it is reasonable to believe that involving teams in the processing of information and making judgments will result in better target identification and discrimination than individuals.

A number of theoretical models predict that teams will outperform individuals on the cognitive tasks that confront teams such as weapons directors (e.g., ideal group model, Sorkin and Dai, 1994; social decision scheme theory, Hinsz, 1990). Using some assumptions based on social decision scheme theory, Figure 12.1 illustrates how team performance is expected to be superior to that of individuals. The x-axis shows the probability of an individual responding appropriately on the cognitive task. The y-axis shows the probability of a team responding appropriately under similar circumstances. (For illustrative purposes, a five-person team is assumed to perform a task for which it is difficult to tell if a response is correct.) Note that across a wide range, teams are predicted to outperform individuals. Therefore, there is a research and theoretical basis for believing that teams will make better target identification judgments.

The comparison of team and individual judgments about targets in the weapons directors' zones of operation also provides a means of examining differences between individual and team situation awareness (Endsley, 1995; Endsley and Jones, 2001). The situation awareness of weapons directors reflects their perceptions of the targets and the existing threats, an understanding of the relationships among various targets, and an ability to predict upcoming events that involve the targets. In the case of weapons directors, situation awareness is greatly influenced by the information processing of the individual or weapons director teams (Hinsz, 2001). By focusing on the processing of information and judgments of individuals and teams in a situation with heavy cognitive demands, it is possible to examine critical features of command and control. Moreover, such an examination can inform us how the proper identification of targets might be facilitated by the use of teams in command and control.

The general conclusion is that teams should be superior to individuals at identification and discrimination of targets, but how broadly does this conclusion apply? Many of the cases of misidentification of targets in combat occur in situations typified by overwhelming information in a rapidly changing environment that requires decisions made under intense time pressure. Are teams better at using information and likely to correct errors regarding target identification? Figure 12.1 indicates that when individuals are about equally likely as unlikely to make appropriate target identifications (approaching .50), teams would not perform much better. However, other models of team performance suggest that teams would still outperform individuals because of error correction and information pooling (e.g., Laughlin and Ellis, 1986). Whether teams and individuals differ in their ability to identify and discriminate targets and their attributes (e.g., speed, heading) under conditions of overwhelming information is a key question we investigated in the experiment described below.

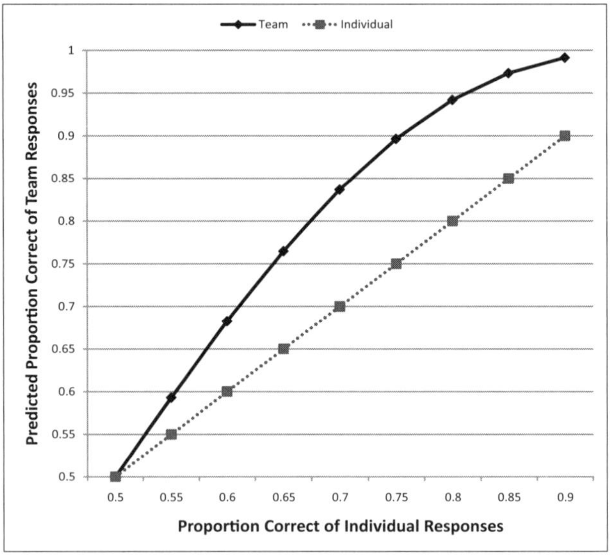

Figure 12.1 Predicted team proportion of correct responses given values for the proportion of correct individual responses

In our experiment, we asked participants to act as weapons directors in a simulated AWACS task environment in which their ability to remember attributes and make judgments about specific targets was tested. The weapons directors acted alone or as members of a team. To assess target identification, after they were presented with a situation with a large number of targets having varying attributes, we asked the participants to respond to a large set of statements about the targets and their attributes. The participants were asked whether the statement was true or false. As illustrated in Figure 12.1, we can use the probability of individuals and teams accurately answering the statement to test the prediction that teams make more accurate judgments than individuals. Moreover, the methodology of this experiment allows us to construct the *d'* index derived from signal detection theory (Banks, 1970; Green and Swets, 1966) to show how well the individuals and teams discriminate accurate from inaccurate information in their judgments about these targets. By using these measures we are able to examine how teams and individuals compare in their identification judgments about targets when they are presented with a cognitively demanding situation involving a number of targets.

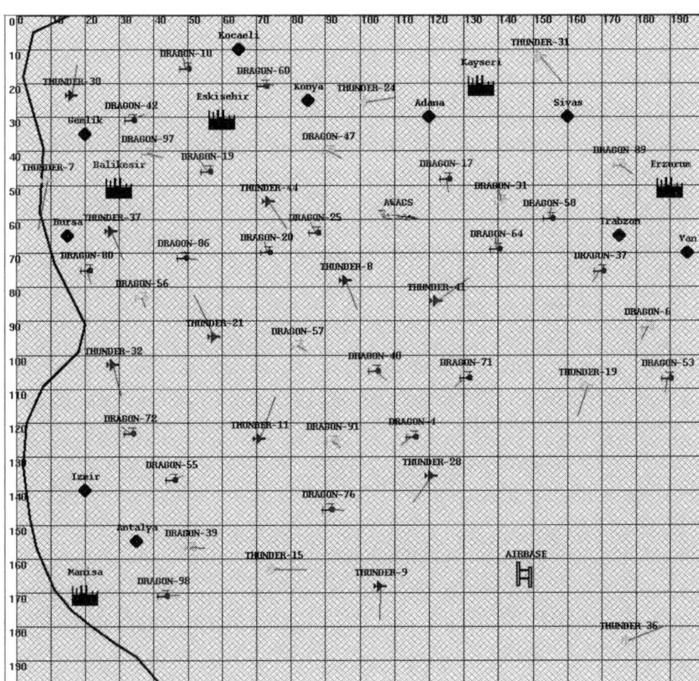

Figure 12.2 The layout of the grid for participants at the beginning of the scenario

Note: The larger screen display provides the controls for the weapons directors' activities as well as the grid reflecting the locations of the aircraft and defined locations on the map.

Method

Participants and Design

Participants were 116 students from North Dakota State University with about equal numbers of males and females. Participants either received extra credit for their lower level psychology classes or were paid for their participation. This experiment involved comparisons of weapons directors performing alone against teams of three or five members. There were 26 participants in the individual condition, 15 three-person teams (45 participants), and 9 five-person teams (45 participants). All participants performed the target identification test initially alone. Following that, the individuals performed a filler task, and then performed the target identification test again. Members of the teams performed the target identification test the second time as an interacting team.

Task Environment

An AWACS synthetic task environment that simulates features of weapons director tasks served as the research platform (Entin and Rubineau, 2002). The research scenario involved little contact with enemy forces and focused on a humanitarian relief effort in an allied nation after it suffered a devastating earthquake. A number of villages and cities were located in the region of responsibility, and humanitarian relief (e.g., food and medical supplies) needed to be delivered by air transports. This complex scenario required the efforts of a number of AWACS weapons directors so that it would be reasonable to compare performance of three and five person teams.

In general, the information displayed to the weapons directors fell into four categories. (1) Aircraft that were under the supervision of the weapons directors on the AWACS (e.g., fighters, helicopters). (2) Targets that were friendly, enemy, or of unknown disposition. (3) Background information about the lay of the land and other features of the place and space in the AWACS area of responsibility (e.g., national boundaries, villages). Much of this information was displayed on a grid with aircraft identified with icons and alpha-numerical labels (see Figure 12.2). In addition, colors were used to indicate the aircraft under a specific weapon director's supervision. (4) Operators could also open pop-up menus that provided specific information about the aircraft (i.e., speed, heading, mission, altitude, and coordinate location). Based on these seven attributes of the aircraft for a large number of targets (call sign, speed, heading, mission, altitude, location on grid map, and coordinate location), we generated a large number of unique target identification test items (i.e., 114) for testing the hypotheses and deriving measures.

Procedure

The training of participants in performing the tasks associated with the weapons directors had a number of components. The participants initially received a general description of the synthetic task environment and were told about the different features of the workstation, screen, and controls. After this, the specific features of weapons director's operations were described. This training was interactive allowing the participants to learn about different types of aircraft, their symbols, the functions the aircraft serves, and how a weapons director interacts with different types of aircraft.

After this initial training, the participants performed the weapons director tasks in a number of exercises and were given time to become familiar with their workstations. During these exercises, the participants were quizzed to make sure they understood the correct way to perform their functions. Problems, errors, and questions were addressed by the experimenter. Finally, a set of exercises were conducted to ensure that the participants achieved criterion levels of competence on the different tasks a weapons director performs in this synthetic task environment. Once the participant reached these criterion levels, participation in the actual experiment began.

The experimental session continued with a pre-briefing regarding the situation the weapons directors would confront. The participants were told that an allied nation had suffered a devastating earthquake and requested assistance from the United States. The U.S. was providing food and water, medical assistance, construction equipment, and search and rescue teams. A country neighboring the allied nation had shown hostility in the past, requiring patrol aircraft for that border. The aircraft under the AWACS's control pursued five different missions: 1) air-drop food and water, 2) combat air patrol, 3) transport construction equipment, 4) transport medical teams and supplies, and 5) transport search and rescue teams. The cover story suggested that another AWACS was transferring control to the participants' AWACS because it was shifting off mission. During this exchange period, no critical events occurred so the participants would gain situation awareness in their task environment as well as focus on the attributes of the aircraft upon which they would be tested. The transfer of aircraft from the departing AWACS to the weapons directors occurred at a methodical pace so that the weapons directors had time to interact and respond to each of the aircraft operating in the scenario.

We used the number of aircraft as a way of influencing the amount of information presented to produce a cognitively demanding situation for the weapons directors. Based on pilot work, we determined that 45 aircraft would create a cognitively demanding situation. These aircraft were present in the scenario at all times. In the five-person teams, nine aircraft were made the responsibility of each team member and were color coded with the responsible weapons director's color. For the three-person team, 15 aircraft were the assigned responsibility of each weapons director and were also similarly color-coded. Participants were told to monitor all of the aircraft on the map, but that they had primary responsibility for the subset of aircraft that had an icon with their color. Participants were also told that they were eligible to win a monetary bonus related to their ability to monitor and retain knowledge concerning the attributes of each aircraft during the simulation. Specifically, participants were told that they would complete examinations following the scenario and that the top 20 percent of performers on each of the examinations would receive a bonus of $10.

Once all of the aircraft were transferred to the weapons directors, the scenario continued relatively uneventfully for approximately 90 seconds. At this time, the participants' display screens went blank. The participants were told that their AWACS had suffered an equipment failure. Because the weapons directors still needed to control their aircraft, it was important to know the situation they faced when screen information was lost. The participants were asked for their knowledge of the situation just before their screens went blank (first target identification test). No discussion was allowed during this test period.

The target identification test involved 114 questions presented on the weapons directors' screens. Of these items, half were true items (e.g., The altitude of the aircraft at this location was 2200 feet.) and half were false (e.g., The aircraft at this location was THUNDER 27.). The participants were asked to indicate whether the statement presented was true or false.

After completing the first target identification test, the participants in the individual condition completed a filler task before completing the second target identification test. Participants in the three- and five-person team conditions were also asked to respond to the second target identification test, but as members of a weapons director team. The member of the team seated at the center was selected to enter the team responses for an item. The teams were told they could reach their team responses in any way they desired, but they were to make sure that their responses represented the collective opinion of all members of the team. After completing all the second target identification test items, the participants' questions were answered, they were debriefed about the study, thanked for their participation, information for distributing the incentives was gathered, and they were excused from the experimental session.

Results

The responses of the participants in this experiment were analyzed in a number of similar ways. The target identification responses were aggregated to produce indices of target identification (i.e., *d'*, proportion correct, errors of omission, errors of commission). These indices were calculated for both the first and second target identification tests. Mean values are presented in the two panels of Table 12.1. An important general pattern of the data is that both individuals and teams achieved very low levels of accuracy on the target identification tests. These rates of target identification were at or just above chance levels, which suggests that the cognitive load that the weapons directors' faced may have been very high.

It was predicted that the individuals and the participants who responded as a team would not differ on any of these indices on the first target identification test because random assignment to conditions would equate the conditions. Results were consistent with these expectations (see top panel of Table 12.1). There were no significant differences between conditions for the proportion of items answered correctly, values of *d'*, or rates of errors of commission and omission. The results also indicated that these indices did not differ between the first and second test for the individuals.

Table 12.1 Mean values and standard deviations of the measures for target identification tests 1 and 2

First Target Identification Test			
Dependent Variable	**Individual M (SD)**	**Three Coactors M (SD)**	**Five Coactors M (SD)**
Proportion Correct	0.51 (0.05)	0.52 (0.05)	0.50 (0.05)
d'	0.06 (0.26)	0.11 (0.26)	0.03 (0.33)
Errors of Commission	.54 (.10)	.57 (.12)	.57 (.15)
Errors of Omission	.43 (.11)	.39 (.12)	.42 (.12)
Second Recognition Test			
Dependent Variable	**Individual M (SD)**	**Three-Person Team M (SD)**	**Five-Person Team M (SD)**
Proportion Correct	0.51 (0.06)	0.55 (0.05)	0.51 (0.04)
d'	0.04 (0.30)	0.27 (0.27)	0.05 (0.22)
Errors of Commission	.55 (.12)	.55 (.13)	.51 (.09)
Errors of Omission	.42 (.12)	.35 (.11)	.46 (.11)

The team responses on the second target identification test provided values for the indices of team performance on the AWACS synthetic task environment (see bottom panel of Table 12.1). Results indicated that the three-person teams outperformed the comparable individuals on the second test in measures of target identification: proportion of items answered correctly, $F(2, 47) = 3.64$, $p < .05$, and values of *d'*, $F(2, 47) = 3.50$, $p < .05$. However, the five-person teams did not differ from the individuals or three-person teams on any of these measures.

Another way performance on the target identification test was examined was in terms of the errors observed. It was hypothesized that teams would have fewer errors than individuals due to the error correction process inherent in teams. There were no differences in the rate in which individuals and teams were likely to produce errors of commission (incorrectly saying that a false statement is true/all false statements). Only somewhat supporting the hypothesis, three-person teams were less likely to produce errors of omission (incorrectly saying a true statement is false/all true statements) than individuals and five-person teams, $F(2, 47) = 3.62$, $p < .05$. Given that three-person teams achieved higher levels of performance, this observed error correction probably contributed to their better target identification.

These results suggest that five-member teams had lower levels on the measures of target identification performance (e.g., proportion correct) than the three-member teams. Likewise, the five-person weapons director teams had more errors than three-person teams. These findings are inconsistent with predictions that the five-person teams would be more accurate than individuals. However, there was a relatively small sample of five-person teams for this analysis (nine). Moreover, the probability of individuals responding correctly approached chance levels (.50). Consequently, as seen in Figure 12.1, teams were not expected to differ from individuals when they were at chance levels of performance. The results presented in Table 12.1 demonstrate that the five-person teams did not improve on the initial individual levels of target identification performance. The results presented in Table 12.1 also indicate that three-person teams did improve their performance above that of the initial member responses, but not by much.

Discussion

This experiment investigated the accuracy of identifications of individuals and teams for targets displayed to weapons directors in an AWACS synthetic task environment. Perhaps the most striking aspect of this study was the low levels of accurate target identifications achieved on both the first and second target identification tests by the individuals and teams. This poor performance probably resulted from the overwhelming amount of information that the participants confronted. With 45 aircraft having seven attributes each, a total of 315 pieces of information were presented. This appears to have been a very challenging situation for participants when they responded to the target identification test. Participants apparently were unable to cope with the high demand of the situation. Performance was barely above chance levels, and many participants may have been guessing for a large number of the items. The analyses displayed in Figure 12.1 indicate that teams would not have better target identifications than individuals when they appeared to be guessing.

Research has consistently demonstrated teams to be superior to individuals in cognitive task performance (Hinsz et al., 1997). However, the observed results for this study differed in important ways from those associated with the comparison of team and individual judgment accuracy. One prediction held that collaboration in target identification by teams would enhance performance over that of single weapons directors. A small performance increment was observed for the three-person teams, but not the five-person teams. Consequently, there is no clear evidence that target identification of the teams in this study was superior to that of individuals.

An interesting implication of the differences in the patterns of predicted and observed results is that the nature of the target identification task itself influences the types of team processes that emerge. If accuracy on target identifications becomes sufficiently difficult, these processes might change such that the mechanisms that contribute to team superiority may be eliminated or obscured by the difficulty of the items. Clearly, the nature of the task along with features of teams has important influences on the ways teams perform cognitive tasks.

One feature of this experiment is that the weapons directors had responsibility for specific aircraft. Consequently, different weapons directors had specific knowledge or expertise with regard to the different aircraft on the display. This differential knowledge could lead teams to allow a weapons director with responsibility for controlling an aircraft to have greater influence on the team responses for information about that aircraft. This suggests an expertise-based perspective. If a weapons director with responsibility for an aircraft remembers a piece of information about that aircraft, then the team tends to accept that piece of information as correct (see Kirchler and Davis, 1986). Further research examining the role expertise has on effective team performance is warranted.

The current research investigated target identification of teams and individuals under highly cognitively demanding conditions that may more accurately mirror a typical battlespace. It is under these conditions that errors of judgment about targets often occur. These errors can be ones in which the friendly targets are misidentified, inaccurate information is held about it, a target goes unnoticed, or critical information about it is not considered. Under all types of errors, improper target identification can occur which can have devastating consequences. We found that when there was an overwhelming amount of information about the targets to consider, individuals and teams did not differ greatly in terms of errors, in spite of research suggesting that teams should correct errors and hence have better performance. Thus, under high cognitive demand, perhaps another approach or technology is necessary to reduce the kinds of errors that would contribute to friendly fire incidents. Moreover, teams placed under these cognitively demanding situations may benefit from additional training that increases expertise and situation awareness.

We argue that one must be cautious if additional technology is provided to personnel to improve target identification in combat. Before making these systems common issue to personnel, they must be trained to handle the increased information, and methods to improve the display of information will be required (Bates and Singer, 2003). This study suggests that a point can be reached at which both individuals and teams are unable to utilize additional information. If a division of labor within teams is properly implemented, it may help to lighten the cognitive load, but research would need to demonstrate that the division of labor actually enhances target identification.

In modern combat situations, a team of allied forces are directed at opposing forces. The command and control of these allied forces requires processing information and making decisions about appropriate actions to be taken. This requires sufficient accurate information to make appropriate decisions. If insufficient information or inaccurate information is used, errors in judgment during combat can occur. Moreover, if an overwhelming amount of information is provided, the human operators may not be able to respond appropriately, and mistakes can be made. Also, situation awareness can diminish. Of interest to this volume, when inappropriate or overwhelming information is provided, inaccurate identifications are made, and incidents of friendly fire may result. Not surprisingly, the most common causes of fratricide are human error (i.e., information processing) and misapplied technology (CSC, 1995). Thus, methods that promote error correction and appropriately apply technology may contribute to the reduction of fratricide. Research provides a strong theoretical and conceptual basis for considering the use of teams in processing information and making judgments which could have an additional contribution. Also, situation awareness might be enhanced by having teams respond to demanding situations. By having individuals come together to both train and act as teams, we may promote error correction and maximize the effectiveness of available technology.

Acknowledgments

This material is based on research sponsored by the Air Force Research Laboratory, under agreement number F49620–02–1-0234 with the first author. The U.S. Government is authorized to reproduce and distribute reprints for Governmental purposes notwithstanding any copyright notation thereon. The views and conclusions contained herein are those of the authors and should not be interpreted as necessarily representing the official policies or endorsements, either expressed or implied, of the Air Force Research Laboratory or the U.S. Government. We appreciate the comments of Renee Magnan, Kevin Betts, and Jared Ladbury on an earlier draft of this chapter.

References

American War Library. (1996). *The Michael Eugene Mullen American friendly Fire Notebook.* Retrieved online April 1, 2008, from: http://members.aol.com/veterans/.

Banks, W. P. (1970). Signal detection theory and human memory. *Psychological Bulletin, 74,* 81–99.

Bates, J., and Singer, J. (2003). *GPS Devices Proving Key to Avoiding Fratricide.* Retrieved online April 1, 2008, from: http://www.space.com/spacenews/archive03/gpsarch_092303.html.

BBC News. (1999). *Europe Embassy Strike 'a Mistake'.* Retrieved online April 1, 2008, from: http://new.bbc.co.uk/2/hi/europe/338557.stm.

BBC News. (2003). *'Friendly Fire' Hits Kurdish Convoy.* Retrieved online April 1, 2008, from: http://newsvote.bbc.co.uk.

BBC News. (2004). *Marine Killed in 'Friendly Fire'.* Retrieved online April 1, 2008, from: http://newsvote.bbc.co.uk.

CBS News (2006). *Friendly Fire that Killed Canadian was 'Freak Accident': Major*. Retrieved online April 1, 2008, from http://www.cbc.ca/story/world/national/2006/09/04/afghanfriendly.html.

CSC. (1995). *Fratricide: Avoiding The Silver Bullet.* Retrieved online May 21, 2004, from http://www.globalsecurity.org.

Department of Defense. (1992). *Conduct of the Persian Gulf War: Final Report to Congress* (Pubic Law 102–25). Washington, DC: Author.

Endsley, M. R. (1995). Toward a theory of situation awareness in dynamic systems. *Human Factors*, 37, 32–64.

Endsley, M. R., and Jones, W. M. (2001). A model of inter- and intrateam situation awareness: Implications for design, training and measurement. In M. McNeese, E. Salas, and M. Endsley (eds), *New Trends in Cooperative Activities: Understanding System Dynamics in Complex Environments*. Santa Monica, CA: Human Factors and Ergonomics Society.

Entin, E. E., and Rubineau, B. (2002). *AWACS Dynamic Distributed Decision Making Synthetic Task Environment*. Technical Report. Woburn, MA: Aptima, Inc.

Fahey, R. P., Rowe, A. L., Dunlap, K. L., and deBoom, D. O. (2000). *Synthetic Task Design (1): Preliminary Cognitive Task Analysis of Awacs Weapons Director Teams*. Report submitted to the Air Force Research Laboratory, Human Effectiveness Directorate, Brooks Air Force Base, TX.

Green, D. M., and Swets, J. A. (1966). *Signal Detection Theory and Psychophysics*. New York: Wiley.

Garamone, J. (1999). Fixes touted to combat friendly fire causalities. *American Forces Information Service.* Retrieved online April 1, 2008, from http://www.defenselink.mil/news/Feb1999/n02021999_9902027.html.

Gordon, M. (1994). U.S. jets over Iraq attack own helicopters in error; All 26 on board are killed. *The New York Times.* Retrieved online April 1, 2008, from http://query.nytimes.com/gst/fullpage.html?res=9D01E6DB1E3EF936A2575 7C0A962958260.

Hastie, R. (1986). Experimental evidence on group accuracy. In B. Grofman and G. Owen (eds) *Decision Research* (Vol. 2, pp. 129–157). Greenwich, CT: JAI Press.

Hinsz, V. B. (1990). Cognitive and consensus processes in group recognition memory performance. *Journal of Personality and Social Psychology, 59,* 705–718.

Hinsz, V. B. (2001). A groups-as-information-processors perspective for technological support of intellectual teamwork. In M. D. McNeese, E. Salas, and M. R. Endsley (eds), *New Trends in Collaborative Activities: Understanding System Dynamics in Complex Settings* (pp. 22–45). Santa Monica, CA: Human Factors and Ergonomics Society.

Hinsz, V. B. (2004). Metacognition and mental models in groups: An illustration with metamemory of group recognition memory. In E. Salas and S. M. Fiore (eds), *Team Cognition: Understanding the Factors that Drive Process and Performance* (pp. 33–58). Washington, DC: American Psychological Association.

Hinsz, V. B., Tindale, R. S., and Vollrath, D. A. (1997). The emerging conceptualization of groups as information processors. *Psychological Bulletin, 121,* 43–64.

Kerr, N. L., MacCoun, R., and Kramer, G. P. (1996). Bias in judgment: Comparing individuals and groups. *Psychological Review, 103,* 687–719

Kime, P. (2003). Marines focus on portable ID gear to reduce fratricide. *Sea Power, 46,* 23–24.

Kirchler, E., and Davis, J. H. (1986). The influence of member status differences and task type on group consensus and member position change. *Journal of Personality and Social Psychology, 51,* 83–91.

Klinger, D. W., Thordsen, M. L., and Copeland, R. R. (1998). *A Cognitive Task Analysis of Surveillance Personnel Onboard AWACS Aircraft.* Final Technical Report submitted to the Air Force Systems Research Laboratory, Brooks Air Force Base, TX.

Laughlin, P. R. (1980). Social combination processes in cooperative problem-solving groups on verbal intellective tasks. In M. Fishbein (ed.), *Progress in Social Psychology* (pp. 127–155). Hillsdale, NJ: Lawrence Erlbaum.

Laughlin, P. R., Bonner, B. L., and Miner, A. G. (2002). Groups perform better than the best individuals on Letters-to-Numbers problems. *Organizational Behavior and Human Decision Processes, 88,* 605–620.

Laughlin, P. R., and Ellis, A. L. (1986). Demonstrability and social combination processes on mathematical intellective tasks. *Journal of Experimental Social Psychology, 22,* 177–189.

Morgret, F. (2002). Friendly fire kills allies and support. *Proceedings of the United States Naval Institute, 128,* 86.

PMM. (2003). CIPS, CDS important too. *PS: Preventive Maintenance Monthly, 609,* 10.

Snook, S. A. (2000). *Friendly Fire: The Accidental Shootdown of U.S. Black Hawks Over Northern Iraq.* Princeton, NJ: Princeton University Press.

Sorkin, R. D., and Dai, H. (1994). Signal detection analysis of the ideal group. *Organizational Behavior and Human Decision Processes, 60,* 1–13.

Chapter 13
A Team Training Paradigm for Better Combat Identification

Wayne Shebilske
Wright State University

Georgiy Levchuk
Jared Freeman
Kevin Gildea
Aptima, Inc.

Combat Identification in the Air Force often involves more than an individual pilot or crew. For example, when a fighter pilot follows an Air Tasking Order (ATO) to find a specific target in a specific location, many others were involved in forming and implementing the ATO that specified the target. Similarly, when an Intelligence Surveillance and Reconnaissance (ISR) pilot encounters an object of interest that seems to be unexpected with respect to the ATO, the Air Force has standard procedures for coordinating a complex Offensive Team to assist in identification and response. As a result, combat identification entails, not only recognizing what an object of interest is, but also whether the object is an expected or unexpected threat or opportunity that demands a response. Expected threats and opportunities as well as planned responses to them are included in the ATO, which is updated every eight to twelve hours. Unexpected threats and opportunities that demand immediate action are called time-sensitive targets (TSTs). The time sensitivity and related risk of error can be appreciated when one considers that multiple TSTs often occur simultaneously and the following process is expected to be completed within 10 min. for each one. After ISR nominates an object of interest as a potential TST, the Dynamic Targeting Cell (DTC) differentiates TSTs from other unexpected objects, designates an object of interest as a TST, and submits plans to prosecute all designated TSTs. The plans include the order for attacking TSTs and strike packages for the attacks. A Senior Offensive Duty Officer (SODO) and the Chief Combat Officer (CCO) evaluate whether the proposed plans achieve the goals of prosecuting the TST efficiently and effectively without jeopardizing the ATO. Accomplishing these goals requires the DTC to identify enemy targets as well as friendly assets that have a potential contribution to a strike package for a TST greater than the potential cost of the asset not being available to execute the ATO. In this context, the risk of errors includes errors of omission, not attacking targets that should have been attacked, and errors of commission, attacking objects that should not have been attacked, such as friendly forces (see Greitzer and Andrews, Chapter 11; Barnett, Chapter 20, this volume).

The present experiment is part of a more general investigation aimed at reducing both kinds of errors. We investigated the human factors issues related to ISR and DTC teams who confront events that are challenging with respect to their past experiences. We observed, for example, human factors issues related to displays. On the one hand, computer scientists and engineers have succeeded in storing within computer systems all the information that the Air Force Office team needs to execute an ATO and to respond to TSTs. On the other hand, DTC teams often cannot find and use the information fast enough to prosecute TSTs in time. Our general investigation therefore includes efforts to develop better displays and performance support systems. Similarly, our present investigation of training experienced teams addresses only a subset of the human factors issues related to training in this context. For example, our general investigation also includes human factors issues related to training novices to become experts in these complex tasks. We will return to these issues in our discussion section where we will consider how our present analysis of human factors issues in experienced teams learning new experiences relates to a more general perspective.

Experienced Air Force Offensive teams train to address their current enemy escalating or changing tactics and to confront their current and/or new enemies in different contexts. Such training usually happens on the job. Training with realistic simulations could have at least three advantages over training on the job. First, training systems would be highly accessible to promote frequent practice. Second, practice scenarios in simulations could be systematically structured and focused on training objectives to promote deliberate practice. Third, assessments relevant to training objectives could be systematically delivered to train and maintain competencies.

A large body of literature on which human factors psychology is founded supports the use of computer-based training simulations and provides valuable training principles (e.g., Piorolli and Anderson, 1985; Schmidt and Bjork, 1992). Much of this literature, however, concerns basic training on tasks that are well defined (e.g., mathematics) and that are executed mainly by individuals. In contrast, Air Force Offensive Teams and other command and control operations involve human performance by teams in domains that are ill defined. Furthermore, the training under consideration is not intended for new operators, but advanced training for experienced operators learning something new. The historical focus of rigorous research on beginners learning well defined tasks made sense not only because it provided theoretical foundations for understanding learning and practical foundations for applying learning principles to training in operational settings, but also because tools were not available to enable comparable research on experts learning ill-defined advanced skills. Until recently, research on complex domains was limited to explorations of operational settings with little rigorous hypothesis testing.

Using modern simulation tools, we are attempting to extend to more complex domains the historically established path of using computer-based training systems for rigorous laboratory research aimed at building a foundation for supplementing on-the-job training with realistic simulation-based training systems. Specifically, we are using Aptima's Dynamic Distributed Decision-making (DDD) synthetic task environment to simulate and explore complex domains, precisely manipulate and measure variables that will guide modeling of human information processing during complex activities, and rigorously test grounded hypotheses about complex processes.

Our manipulations, measures, and hypotheses relate to our initial goal of developing and testing a paradigm to train experienced teams for combat identification in simulated ISR and DTC teams. In a previous experiment, Shebilske, Gildea, Freeman, and Levchuk (2007) manipulated a factorial combination of two independent variables: (a) Phase II vs. Phase III training and (b) Pretest vs. Posttest before and after each phase. In Phase I (this was not included in the hypotheses), simulated ISR and DTC teams performed 50 hours of background training. In Phase II, the same teams performed simulated missions for 49 hours. In Phase III, the same teams performed 18 more hours of simulated missions that were more difficult because the enemy increased the number of TSTs and the number of threats protecting each TST. The dependent variable was expert ratings of the DTC coordinating a strike package plan. Two experts made reliable independent ratings on a six-point scale. These ratings were enabled by a precise playback tool in Aptima's DDD 4.0. The tool integrated, time stamped, and played back all stimulus events and all responses in realistic simulations of ISR and DTC operations.

If college students can learn realistic ISR/DTC simulations, if they can be challenged by enemy escalations, and if they can learn to overcome this challenge, then the following hypotheses should be supported:

1. Posttest scores should be higher than the pretest scores in Phase II.
2. Pretest scores on Phase III should be lower than posttest scores in Phase II.
3. Posttest scores should be higher than the pretest scores in Phase III.

Shebilske et al. (2007) supported hypotheses 2 and 3, but not Hypothesis 1. They proposed that the pretest in Phase II was higher than expected because too much training had been given before the pretest. In addition to the quantitative results, a qualitative result was interesting. Scientists who had performed task analysis of DTC dynamics in operational settings, observed the simulated DTC near the end of Phase II. The scientists observed that the simulated DTC team (college students) performed DTC dynamics at about the level of an experienced operational DTC with a medium-high skill level.

The present experiment reduced the amount of background training and other training in an effort to support Hypothesis 1, and it retested Hypotheses 2 and 3 to determine whether they would be supported with reduced training.

Methods

Participants

The participants were 7 undergraduate college students (2 women and 5 men, mean age = 20 years), who were paid $7.25 per hour for 36 hours each. Their participation was part of their responsibilities as research assistants. Although the trainees were research assistants, they did not know the purpose of the experiment.

Materials

Office dividers separated seven work stations with four stations in one row and three in the other. Each station had an IBM compatible PC with a 17 in. monitor, a mouse, keyboard for inputs, and a headset linked with an Aardvark sound system audio net that enabled open and recorded communication within teams and isolation of sounds outside the team.

We employed Aptima's DDD synthetic task environment to simulate the ISR/DTC task because of its relevancy to modern Air Force operations, to combat identification, and to general online command and control operations. The synthetic task environment was based on a task analysis of operational DTC teams. The Air Force structures DTC operations as a kill chain, which includes multiple stages and multiple task objectives (TOs). Our three TOs were: TO1, detect and differentiate TSTs; TO2, prioritize the TSTs; and TO3, coordinate attack assets. They relate to complex teamwork occurring in three of the kill-chain stages: Find, Fix, and Target. Operations occurring during the Find stage are limited to the ISR who initiates operations related to TO1. The ISR role was filled by one of our trainees. DTC operations in the Fix phase include complex communications among the four DTC team roles, all of which were filled by our trainees: DTC Chief, Ground Track Coordinator (GTC), Attack Coordinator (AC), and Target Duty Officer (TDO). This teamwork relates to TO1 and TO2. DTC operations in the Target phase also include complex communications among the four DTC team members. This teamwork relates to TO3. In many operational settings and in the DDD simulation, the proposed order of attacking TSTs and the proposed strike package for each attack is approved or disapproved first by the SODO and then by the CCO. These two officers also control the execution of accepted proposals. In the present experiment, expert confederates played the roles of CCO and SODO proficiently so that variability related to approvals and to executions was minimal.

The confederates and the trainee who played the role of the DTC Chief sat in one row separated by a barrier from a row of the other 4 trainees. During operational missions and during the DDD simulated missions, these offensive team members usually communicate by text messaging and occasionally communicate by voice. The background training for the present trainees included standard Air Force brevity procedures for both communication modes. For example, the trainees were taught that the standard text or verbal acknowledgment to a message is 'copy.'

The DTC must operate in the face of uncertain information and within the constraints of the ATO, which we changed for each mission. Missions were usually separated by about 24 hours, but occasionally separated by 8, 12, or 0.5 hours in order to accommodate individual schedule demands. The ATO includes guidance on which objects of interests should not be prosecuted, what kinds of TSTs might be encountered, and how they should be prioritized relative to the ATO and relative to each other.

Design

The within-participant, independent variables were Phase (II vs. III) × Test Type (pretest vs. posttest). Phase I was 16 hours of background training. Phase II was 11 hours of performing missions. Phase III was 9 hours of performing more difficult missions. The planned comparisons were the pretest versus the posttest in Phase II, which reflected the change in performance during that phase; the posttest in Phase II versus the pretest in Phase III, which reflected the drop in performance during the initial exposure to the new experience; and the pretest versus the posttest in Phase III, which reflected the change in performance during that phase.

The dependant variable was the quality of the proposed strike package (TO3) for each TST, which was determined by expert ratings of the strike package. Specifically, two experts evaluated each strike package. They independently rated the strike package as correct or incorrect with high or low confidence (correct with high confidence = 4, correct with low confidence = 3, incorrect with low confidence = 2, incorrect with high confidence = 1, no plan before the TST disappeared = 0, and a plan for striking a target on the no strike list or on the ATO = -1). The expert ratings were the same for 96 percent of the ratings, and the few disagreements were resolved by conference so that one number was entered into the analysis for each TST. This measure is comprehensive in that the DTC had to accomplish TO1, TO2, and TO3 in order to get a high score.

The ratings evaluated the plan of the whole team, as opposed to evaluating each individual, making the experiment a single team design. This design is analogous to single-person designs for which Anderson (2001) discussed advantages and disadvantages. Anderson noted 'In practical affairs, the simple A-B design is sometimes all that is available' (p. 313). Single-person A-B designs are common in medicine and in behavior modification (e.g., Matyas and Greenwood, 1996). An advantage of these designs is analyzing individuals in depth. Similarly, an advantage of the present single-team design is that it enabled the controlled development of a highly experienced team for a task that demanded extended training. A potential disadvantage of A-B designs is that they can confound A and B with the material tested in A and B. We addressed this shortcoming by counterbalancing materials across the previous experiment (Shebilske et al., 2007) and the present experiment. Across the two experiments, therefore, support for the hypotheses cannot be attributed to differences in the materials.

Procedure

Training missions included planning and debrief sessions. During planning sessions, operational teams commit to plans, including the ATO and possible TSTs. During mission execution, the offensive team judiciously adapts plans when unexpected events occur, and the DTC plays the key role of proposing specific adaptations. During debriefs, the team extracts lessons from previous plans and adaptations. The team then incorporates these lessons during subsequent planning, missions, and debriefs. We simulated these operations in the present experiment, each session of which included planning (10 min), mission execution (40 min), and debriefing (10 min). To assist them with their work, participants used checklists analogous to those used by the Air Force (Elliott, Cardenas, and Schiflett, 1999).

During Phase I (16 hours), we taught background information using materials from Air Force publications. The background information included the characteristics of enemy weapons and friendly assets, communication brevity procedures, and communication dynamics among the team members. The communication dynamics came from an Aptima task analysis. It included why and when teammates communicate with one another and the protocol for who communicates what to whom. Phase I ended with sessions during which the trainees or the instructors could stop the action to address issues.

During Phase II (11 hours), the trainees performed 11 unassisted sessions. The debriefing included feedback on how well the team was accomplishing its task objectives. Each mission was unique, but they had consistent patterns of differences and similarities, which were created by simulating the same enemy forces attacking with a consistent strategy. The mission difficulty was affected by the number of TSTs per mission and the number of threats that protected the TSTs. Phase II started with warm up trials, which had three to seven TSTs and ten threats. Two pre-test missions had eight and nine TSTs and ten threats. These threats were four fighters, one long rang SAM, and five short range SAMS. Each of two post-test missions in Phase II had eight and nine TSTs and 37 threats to the strike package. These threats were 24 fighters, three long range SAMS and ten short range SAMS. These enemy threats required the strike package to include enough assets to take out not only the TST but also the threats to the strike package. Difficulty increased gradually.

During Phase III (9 hours), the trainees performed 9 more unassisted missions, during which they trained for new experiences. The main change at the beginning of Phase III was a strategic variation. For example, during the original training, the DTC developed the strategy of using refueling tankers to support their proposed attack packages. This strategy was helpful because the proposed attack packages utilized some assets that were low on fuel due to their execution of operations related to the ATO. A challenging change in enemy strategy was to destroy the refueling tankers, which forced attack assets to return to base to refuel. This requirement forced the DTC to choose different attack package patterns from those to which they had become accustomed. The enemy used its new strategy consistently between the pretest and posttest in Phase III, so that the trainees had an opportunity to practice defending against the new strategy. The number of TSTs (eight and nine) and Threats (37) for the posttest in Phase II was similar to the number of TSTs (nine and ten) and Threats (33 and 35) for pretest missions in Phase III. In contrast, each of the posttest missions for Phase III had 13 TSTs and 45 Threats.

Results

Figure 13.1 shows that TO3 accuracy increased between the pre and posttests in Phase II (support for Hypothesis 1), fell between Phase II and III (support for Hypothesis 2), and then increased slightly between the pre and post tests in Phase III (support for Hypothesis 3). As had been done in the previous experiment by Shebilske et al. (2007), we used SPSS to conduct conservative t-tests that do not assume equal variance. The mean accuracy for TO3 in Phase II increased from 1.5 on the Pretest to 3.2 on the Posttest, as predicted ($t(35) = 3.13, p < .01$). Between the posttest in Phase II and the Control pretest in Phase III, the decrease in TO3 accuracy from 3.2 to 1.7 was significant ($t(34) = 2.84, p < .01$). In Phase III, TO3 accuracy rose from 1.7 on the pretest to 2.2 on the posttest, but the change was not significant ($t(41) = .86, p > .05$).

Discussion

The pattern of results in the present experiment and across the present experiment and the previous experiment (Shebilske et al., 2007) is consistent with Hypotheses 1 (improved performance during Phase II, in which the enemy used a consistent strategy), Hypotheses 2 (drop in performance when the simulated enemy changed strategies in Phase III), and Hypotheses 3 (improved performance while training to do a new strategy in Phase III). Statistical tests supported Hypotheses 1 and 2 in the present experiment and Hypotheses 2 and 3 in the previous experiment. Hypothesis 2, which was supported in the both experiments, is important because it is necessary for developing a paradigm to train experienced teams for combat identification in simulated Air Force ISR and DTC teams. That is, teams that are experienced in defending against one enemy strategy will have to learn to defend against new strategies. Although performance dropped significantly when enemy strategies changed in both experiments, the improvement in performance after the strategy change was significant in Experiment 1, but not in Experiment 2. A possible reason for the difference is that, in Experiment 2, the training before the strategic change was substantially less than it had been in Experiment 1. Future

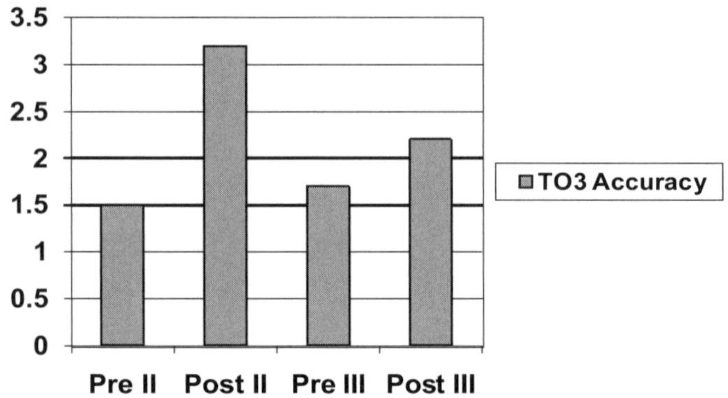

Figure 13.1 Mean accuracy ratings for training objective 3 (TO3) for two sets of pre-tests and post-tests: Phase II training (II) and phase III training (III)

experiments will have to investigate learning to defend against a strategic change as a function of the amount of prior training.

Future experiments will also investigate the type of prior training. The present experiment also provided pilot data for exploring the training of beginners in this complex domain. Comparisons of our previous and present experiment suggest that the ideal amount of background training using the specific procedures that we developed for teaching beginners what to do and how to do it is closer to 16 hours than 50 hours. Future experiments will systematically investigate the procedures that we are developing for training at both beginner and advanced levels.

In the present experiment, the selection of training scenarios was chosen by the trainer. The specific sequence, which gradually increased TSTs and then gradually increased threats, was a hierarchical-part-task training protocol (Fredericksen and White, 1989) because a DTC team must detect TSTs before it can design strike packages that take into account threats. Although this sequence is efficient, the present system lacked the instructional intelligence to improve training through adaptability (cf. Bell and Kozlowski, 2002; Freeman et al., 2006). We are currently doing research that will expand this predetermined sequence into an adaptive sequence that will change as a function of the trainees' performance.

Future experiments will also systematically compare training on traditional displays with training on new displays that we are developing in other research. The goals of these future experiments will be to facilitate transfer of training from beginner to advanced levels and from traditional displays to new displays.

Future directions will also include replicating this experiment with actual DTC teams. We will collaborate with practitioners to investigate whether simulating a specific new experience that is anticipated will help actual DTC teams adapt more quickly to the real change. We will also investigate whether learning-to-learn effects for actual DTC teams simulating many new experiences make the DTC teams generally more adaptive to new experiences in operational settings. Ness, Tepe, and Ritzer (2004) reviewed analogous specific and general training benefits to land, air, and naval warfare applications. The present experiment suggests that desktop simulation of new experiences may extend these advantages to combat identification in Air Force ISR/DTC teams.

We will address these future directions from a general perspective of reducing disastrous errors (see Greitzer and Andrews, Chapter 11, this volume) not only by improving the ability of the Air Force Offensive Team to respond to unexpected events, but also by minimizing unexpected events through better planning and better coordination among those who make plans, those who implement plans, and those who respond to unexpected events relative to plans. In related projects, we are researching Sensor Aided Vigilance Environments (SAVE; e.g., Levchuk, Djuana, and Pattipati, 2008; Levchuk, Kleinman, Ruan and Pattipati, 2003; Levchuk, Levchuk, Luo, Pattipati, and Kleinman, 2002) with the goal of developing performance support tools to assist planners. Specifically, Georgiy Levchuk and his colleagues are using a Partially Observable Markov Decision Process (POMDP) to model enemy organizations and missions from sensor data gathered at three levels, satellites, reconnaissance planes, and ground sensors. These models reason about, forecast, and assess events leading to terrorist attacks. Gildea and Shebilske are testing the effectiveness of these models in assisting planners. The importance of working concurrently on improving plans and on improving responses to unplanned events is appreciated best in the context of understanding that neither is likely to be perfect in the fog and friction of war (e.g., Sumida, 2000).

This more general perspective has the advantage of providing a framework for addressing the coordination among those who make plans, those who implement plans, and those who respond to unexpected events during implementation. Such coordination entails reciprocal information exchange. That is, planners both give information to those who implement missions and to those who respond to unexpected events and receive information from them for subsequent plans. To investigate and train this team of teams, our proposed expansion of the present simulations to experienced actual DTC teams will include experienced actual teams planning ATOs, and experienced actual troops executing the simulations. We will investigate whether simulating a specific new experience will help expert team of teams adapt more quickly to the real change and whether learning-to-learn effects for expert team of teams simulating many new experiences make the team of teams generally more adaptive to new experiences in operational settings. The investigation will also be aimed at understanding better how experts in this team of teams operate, and the training will also be aimed at conveying this better understanding to beginners in this team of teams. In pursuing this last goal, we will expand from individuals to the team of teams approach that Hall and Aungst (Chapter 10, this volume) used to convey to beginners what experts had learned about combat identification.

This more general team-of-teams' perspective, in the context of this volume, also has the advantage of suggesting interesting links to other research on combat identification beyond the links that we already discussed, such as to errors and to transfer of strategies from experts to novices. For example, Neyedli, Wang, Jamieson, and Hollands (Chapter 16, this volume) reviewed important insights about variables that promote optimal reliance on automation in combat identification systems as opposed to over or under reliance. From the team-of-teams' perspective, we might ask whether or not these variables would promote optimal reliance among teams. The analogy that raises this question is to regard the team of teams and all their tools as the system and to explore whether each team in this system is analogous to an automated component for the other teams. Accordingly, one can ask whether the variables that predict trust in automated components also predict trust among the team of teams.

Other interesting links are to situation awareness. For example, vigilance is an important variable in operational settings because many hours of low activity can come between bursts of high activity. In the present synthetic task environment, vigilance demands were reduced by filling all missions with high activity. One might ask, therefore, whether the research by Shingledecker and his colleagues (Chapter 3, this volume) could be used to add meaningful vigilance demands to the present synthetic task environment or at least to specify consequences and boundary conditions for including or not including vigilance demands. Similarly, one might ask whether the research by Bolstad, Endsley, and Cuevas (Chapter 9, this volume) on shared situation awareness could guide the promotion of team coordination in the present team-of–teams' context.

Another set of interesting links in the team-of–teams' perspective compares the operator communication dynamics reviewed in the present chapter with those in a more general context. For example, operators of unmanned aerial vehicles and other ISR operators typically communicate with the DTC as discussed in the present chapter (see Figure 13.2), but fighter pilots rarely do so according to the pilots who participated in the panel sessions for the workshop on which the present chapter is based. Expanding the present synthetic task environment to the proposed team of teams would provide a framework for exploring conditions under which it might be beneficial for fighter pilots who encounter an unexpected object of interest to draw on resources available in the DTC and for testing hypotheses about relevant procedures, such as, communicating directly with the DTC versus communicating indirectly through an ISR who is already experienced in obtaining the information within minutes. Similarly, one could explore whether, how, and in what conditions it might be beneficial to open potential reciprocal communication paths among those who plan missions, those who implement the plans, and those who respond to events that are unexpected with respect to the plans. Our concurrent but separate research on the DTC and SAVE has laid a foundation for the proposed integrative research for a team-of–teams' perspective.

In summary, a more general team-of–teams' perspective in the context of this volume has the advantage of suggesting interesting links to other research on combat identification:

- reducing errors of omission and commission;
- transferring strategies from experts to novices;
- predicting trust of automation and among the team of teams;
- evaluating vigilance demands in the team of teams;
- promoting shared situation awareness and coordination among the team of teams; and
- opening potential reciprocal communication paths among the team of teams.

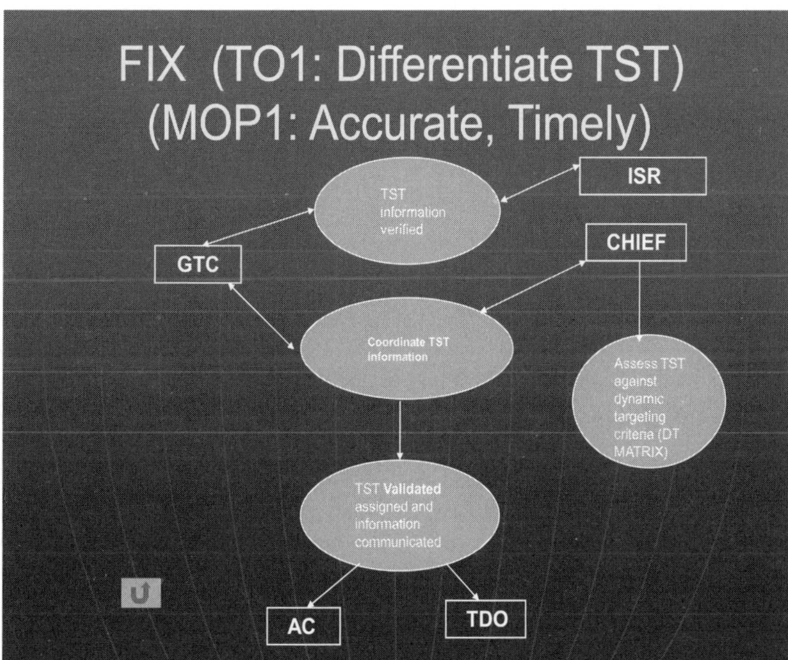

Figure 13.2 Communication dynamics for ISR and DTC

References

Anderson, N.H. (2001). *Empirical Direction in Design and Analysis.* Mahwah, NJ: Lawrence Erlbaum.

Bell, B. S., and Kozlowski, S. W. J. (2002). Adaptive guidance: Enhancing self-regulation, knowledge and performance in technology-based training. *Personnel Psychology, 55,* 267–306.

Elliott, L. R., Cardenas, R., and Schiflett, S. G. (1999). *Measurement of AWACS Team Performance in Distributed Mission Scenarios.* Retrieved August 28, 2003, from http://www.dodccrp.org/1999CCRTS/pdf_files/track_3/013ellio.pdf.

Eliot, C., and Woolf, B. P. (1995). An adaptive student centered curriculum for an intelligent training system. *User Modeling and User-Adapted Instruction, 5,* 67–86.

Fredericksen, J. and White, B. (1989). An approach to training based upon principled task decomposition. *Acta Psychologica, 71*, 89–146.

Freeman, J., MacMillan, J., Haimson, C., Weil, S., Stacy, W., and Diedrich, F. (2006). From gaming to training. *Proceedings of the Society for Applied Learning Technologies*, Orlando, FL.

Levchuk, G. M., Kleinman, D. E., Ruan, S., and Pattipati, K. R. (2003). Congruence of human organizations and missions: Theory versus data. *Proceedings of the 8th International Command and Control Research and Technology Symposium,* Washington, DC.

Levchuk, G. M., Lea, D., and Pattipati, K. R. (2008). Recognition of coordinated adversarial behaviors from multi-source information. *SPIE, 6943*, 305–316.

Levchuk, G. M. Levchuk, Y., Luo, J., Pattipati, K. and Kleinman, D. (2002). Normative Design of organizations – Part I: Mission planning. *IEEE Transactions on Systems, Man, and Cybernetics – Part A: Systems and Humans, 32*, 346–359.

Matyas, T. A., and Greenwood, K. M. (1996). Serial dependency in single-case times series. In R. D. Franklin, D. B. Allison, and B. S. Gorman (eds), *Design and Analysis of Single-case Research* (pp. 215–243). Mahwah, NJ: Lawrence Erlbaum.

Ness, J. W., Tepe, V., and Ritzer, D. R., (2004). *The Science and Simulation of Human Performance*. Amsterdam: Elsevier.

Pirolli, P. L., and Anderson, J. R. (1985). The acquisition of skill in the domain of programming recursion. *Canadian Journal of Psychology, 39,* 240–272.

Schmidt, R. A. and Bjork, R. A. (1992). New conceptualization of practice: Common principles in three paradigms suggest new concepts for training. *Psychological Science, 3,* 207–217.

Shebilske, W., Gildea, K., Freeman, J., and Levchuk, G. (2007) Training experienced teams for new experiences. *Proceedings of the Human Factors and Ergonomics Society 51st Annual Meeting*. Baltimore, MA: Human Factors and Ergonomics Society.

Sumida, J. (2000). History and theory: The Clausewitzian ideal and its implications. *Journal of the Royal United Services Institute of Australia, 21,* 75–90.

Chapter 14

Analysis of the Tasks Conducted by Forward Air Controllers and Pilots during Simulated Close Air Support Missions: Supporting the Development of the INCIDER Model[1]

Beejal Mistry
Defence Science and Technology Laboratory

Gareth Croft
QinetiQ

David Dean
Julie Gadsden
Gareth Conway
Katherine Cornes
Defence Science and Technology Laboratory

Overview

In this chapter we provide a brief overview of a Hierarchical Task Analysis (HTA) that was carried out to determine the tasks carried out by Forward Air Controllers (FACs) and Pilots during simulated Close Air Support (CAS) missions. The HTA was completed by a military judgment panel to inform the future development of the Integrative Combat Identification Entity Relationship (INCIDER) model. The INCIDER model integrates physical representations of sensors and identification friend or foe (IFF) systems with human cognitive and behavioral characteristics to provide a simplified representation of detection and classification processes set within an operational context. At present, the model represents the decision-making process of a single decision maker. The HTA described within this report is being used to develop specific air domain processes for the individual model, and is intended to provide a basis for extended representations that will enable more complex pairings and groupings of individuals involved in the same decision-making process.

Introduction

One critical aspect of warfare is the ability to correctly identify entities that are encountered in the battlespace. Successful combat identification allows individuals to prosecute, avoid, or track hostile agents so that they will not pose a threat to friendly or neutral entities. In contrast, poor combat identification can result in fratricide (a unit fatally engaging a friendly unit in error) or a missed opportunity (failing to engage an enemy when it is possible to do so). Both fratricide and missed opportunities can result in loss of life, loss of equipment, reduced morale, political repercussions, and extended mission prosecution time all of which can impact significantly on operational effectiveness.

Here we will consider combat identification during simulated CAS missions. CAS missions are missions involving air action against hostile forces that are in close proximity to friendly forces. Because of the close proximity of friendly forces during CAS missions, accurate combat identification is particularly important. For this reason, much of the research on combat identification within the UK Ministry of Defence has centered on CAS missions (see Gadsden, Krause, Dixson, and Lewis, 2008, for a review of the principal research studies).

Although a number of technological approaches are being developed to help improve combat identification, the UK military is also concerned with understanding the human decision-making process which underpins the identification of battlespace entities. A number of different methods have been used to examine the human-related factors pertinent to the decision-making process when an individual is attempting to identify a battlefield entity.

One method of examining human-related factors involves modeling. Models are helpful tools for examining the effect of new equipment, developing training, and assessing the effects of new procedures. In addition, models are also helpful in investigating the root causes of battlefield incidents. One model that has recently been developed within The Defence Science and Technology Laboratory (Dstl) since 2004 is the Integrative Combat Identification Entity Relationship (INCIDER) model.

The INCIDER Model

The INCIDER model integrates physical representations of sensors and identification friend or foe and Situational Awareness systems[2] with human cognitive and behavioral characteristics to provide a simplified representation of detection and

2 Systems which provide information that can be used to develop individual and shared situational awareness – often in the form of a tactical picture.

classification processes set within an operational context. The INCIDER model is made up of a conceptual framework and an encounter model. The Conceptual framework ('INCIDER Relationships' in Figure 14.1) is a collection of interrelated factors (including human factors) that could have an impact upon combat identification depending on the situation and broader environmental context. The Conceptual Model was used to develop the Encounter Model which attempts to represent combat identification encounters within specific operational contexts or scenarios. The approach taken to calibrate and validate these models involved the use of synthetic environment experimentation and observations during live exercises (specifically the Coalition Combat ID Advanced Concept Technology Demonstrator (ACTD) exercises URGENT QUEST and BOLD QUEST).

The conceptual framework captures the decision-making process that is carried out by personnel on the battlefield when they encounter a battlespace entity. The model outputs the time to identify the target as a friend or foe and the accuracy of the identification process (i.e., whether an entity was correctly identified as a friend or foe). A number of Dstl reports provide the background and history of the model (Dean, Handley, and Vincent, 2006; Dean, Vincent, Mistry, Hynd, and Syms, 2005; Dean, Vincent, and Smith, 2007; Smith, Allen, and Lovett, 2006) and Dean, Vincent, Mistry, Spaans, and Petiet (2008) have provided an open-source description of the work.

The model contains representations of a number of human factors, such as fatigue, stress, experience, expectation and personality. The precise effect(s) that these factors have on decision-making is not well understood at this time, therefore these factors function within the model as simple moderating factors. In brief, for each run of the model, after detecting the entity, the decision maker undertakes an iterative process to gain more information about the entity. At the end of this decision-making process, either a confidence threshold is reached and the identity of the entity is decided upon or alternatively, the decision maker runs out of time and the entity remains unidentified. The absolute level of the confidence threshold that has to be exceeded before an entity can be identified is dependent on which human factors are included within the model. In addition, the model assumes that the decision maker will enter the encounter with a preconception (prior belief) about the identity of the target. This expectation will be based on previous history and the pre-mission briefing. Confirmatory bias is also represented in the model. Confirmation biases occur when individuals seek to retrieve information based on their preconceptions of the context (e.g., Darley and Gross, 1983). Effectively, individuals attend to information which supports their preconception and ignore contradictory information. These preconceptions may be the result of information gleaned through mission briefs.

A key feature of the development of INCIDER has been the use of an integrated analysis and experimentation programme to simultaneously develop model concepts and analysis tools whilst undertaking experiments to validate the model. This process is represented in Figure 14.1. A significant part of this validation programme has involved examining how certain human factors influence the decision-making process. For example, Smith, Allen, and Lovett (2007) used the NEO-Five Factor Inventory (Costa and McCrae, 1985) to examine whether the levels of particular personality characteristics were correlated with performance measures generated through experimentation (e.g., whether the target was correctly identified, the time to engage the enemy, the distance from the enemy at which it was engaged, the amount of information the decision-maker requested) using operational scenarios represented within a synthetic environment. The number of participants that were recruited for this study was small and so the reliability of these results cannot be assured. However, the results did demonstrate that two of the personality variables correlated significantly with performance measures. Specifically, the level of trait neuroticism was positively correlated with the number of fratricide incidents that were experienced and the level of trait extraversion was negatively correlated with the length of time the decision maker took before engaging an entity. The information gleaned from this research has been represented within the model to increase its validity.

Presently the INCIDER model represents a single decision maker; however, CAS missions are conducted by a team, typically two individuals – the Forward Air Controller (FAC) who is responsible for selecting the target and conveying key targeting information, and the pilot who is responsible for delivering the desired effect. Therefore, to represent this process, extensions to the model are required beyond a single decision maker. HTA was selected as a method to investigate the interactions between pilot and FAC that needed to be represented within the INCDIER model.

It is first necessary to understand the tasks that FACs and pilots do as individuals and the tasks that FACs and pilots have to work together to achieve. Once the tasks that FACs and pilots have to work together to achieve have been identified, it will then be possible to examine how human and technological issues impact firstly

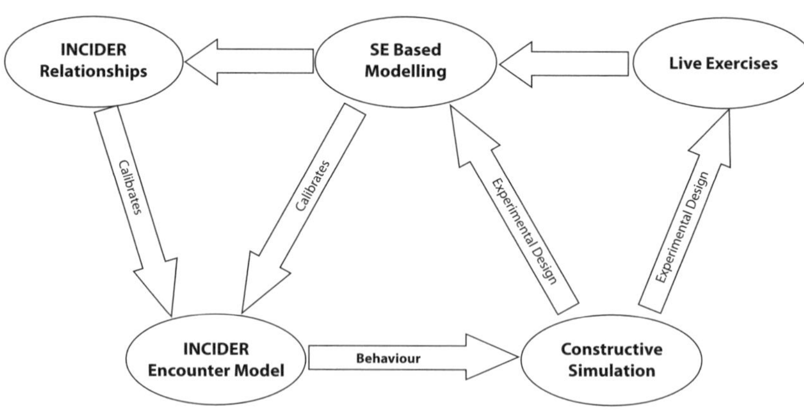

Figure 14.1 The INCIDER integrated analysis and experimentation programme

upon their ability to work together and secondly their individual decision-making processes. The ability to separate individual and joint tasks affords the opportunity to include another comparison improving the model in terms of representation and validity.

Procedure and Layout

Four military Subject Matter Experts (SMEs) were recruited for this study. One participant was a Royal Air Force (RAF) pilot, two were FACs (one was from the RAF [and also a former pilot] and one was from Territorial Army Royal Artillery), and one was a Royal Artillery Mortar Fire Controller.

Three scenarios were developed for a Synthetic Environment. The Synthetic Environment used was the Forward Air Controller Synthetic Environment (FACSE), which is based on the commercial game 'Lock-On™'.[3] The first scenario was used as baseline training for the participants and therefore only the data from Scenarios 2 and 3 were analyzed to generate the HTA. Scenarios 2 and 3 were each run twice with different pairs of FAC and pilot. Each scenario began when the FAC was already in position on the battlefield and an appropriate distance away from the target. The FAC and pilot were required to detect, identify, and prosecute a hostile target within the synthetic environment. Each trial ran until the 'kill chain' had been completed.

The experiment used two workstations: one for the pilot and the other for the FAC as detailed within Figure 14.2. Both the pilot and FAC workstations were accommodated within the same laboratory. Open microphone communications were installed to allow communication between the FAC and pilot, and all verbal communications were recorded for further offline analysis.

Key to Layout Interfaces:

1. Pilot workstation;
2. Delayed Blue Force Situational Awareness pilot screen;
3. FAC map/Blue Force SA screen;
4. FAC workstation;
5. Observer workstation: SA control;
6. Server workstation: voice communications recording.

As indicated within Figure 14.2, an observer was assigned to both the FAC and the Pilot to observe all actions and record online communications in real time. The observers used independent time-recording to log actions and verbal communications to the nearest minute throughout the course of the scenario. This

3 Lock-On was developed by the Fighter Collection which is an organization based at Duxford Airfield in the UK (www.fighter-collection.com). FACSE is a training aid developed in conjunction with HVR Consulting (www.hvr-csl.co.uk).

level of granularity was sufficient to provide an accurate record of what each of the operators did within the scenario and crucially to provide an outline structure from which to conduct a more detailed interview with the FAC and pilot after each scenario. A military adviser also observed the proceedings, specifically noting inconsistencies with standard procedures and any factors which affected the general realism of the Synthetic Environment (i.e., the ecological validity).

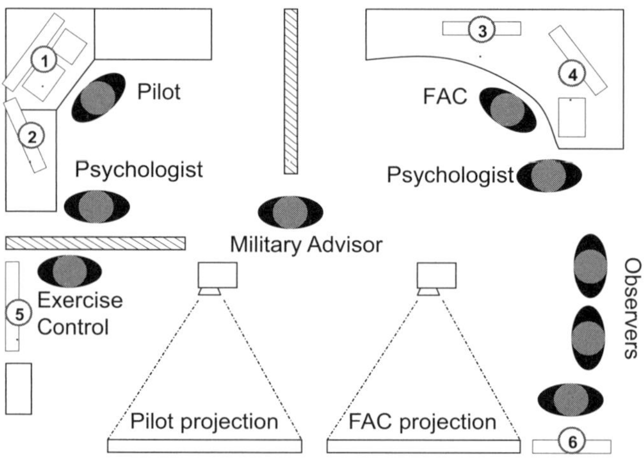

Figure 14.2 Laboratory layout

Data Analysis

After each session the observer notes were used to guide the generation of open-ended questions that were put to the FAC and pilot in order to glean a clearer understanding of the tasks and events undertaken during the scenario including why certain things occurred at specific time points. This semi-structured open-interview technique, as also used by Watson and Wright (2008), appears to provide an appropriate means to elicit the relevant details from the FAC and Pilot concerning the scenario (Stanton, Salmon, Walker, Barber, and Jenkins, 2005).

A draft HTA was generated using the behavioral observations gathered by the observers, audio transcripts of the communications during the scenario, and the data captured during the post-scenario interviews. Drafts of the HTA were further verified and clarified where relevant by the FACs and Pilots during follow-up one-on-one interviews, in order to produce the final HTA.

Results

Figure 14.3 summarizes the top level tasks for both FAC and pilot in parallel. Essentially, the tasks are comprised of actions completed by the FAC and pilot and the communications between them. The component boxes are numbered to reflect the general order of the tasks.

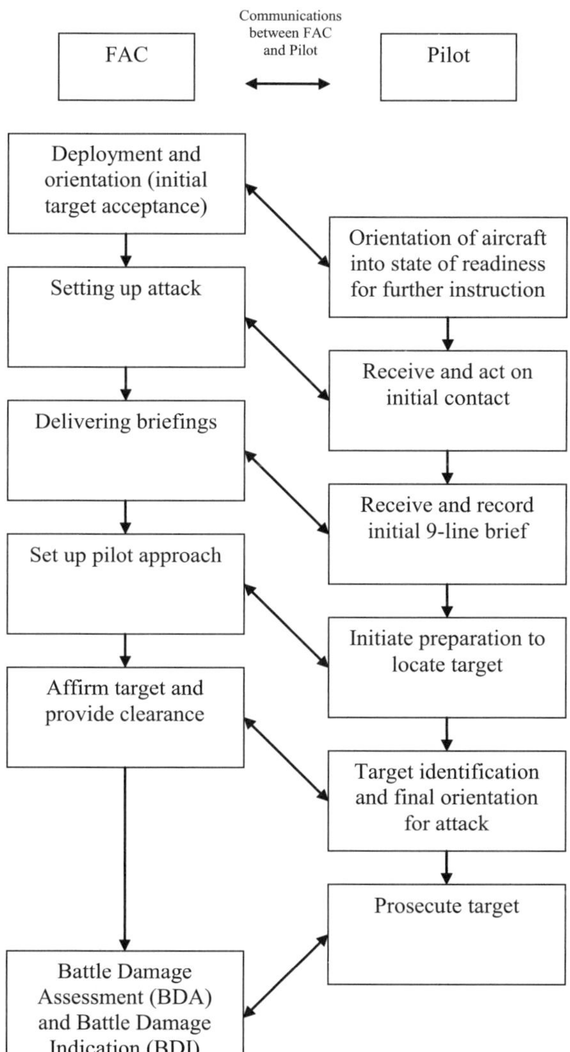

Figure 14.3 Summary task diagram for FAC-pilot tasks during a generic scenario with the synthetic environment

FAC Task Summary

At a high level, the tasks that the FAC had to carry out were to set up the attack, deliver briefings to the pilot, set up the pilot's approach, affirm the target and provide the pilot clearance, and to provide Battle Damage Assessment. More specifically, once the target was located and the pilot was contacted verbally (FAC task 1.0), the FAC proceeded to collect the necessary information to prepare the pilot for further instructions and compile the 9-line brief (FAC task 2.0). This task involved checking the location of friendly forces, generating the co-ordinates of the target, deciding how best to mark the target (i.e., whether it was best to use a laser pointer, IR pointer, or smoke), checking for enemy air defences, and selecting the best approach for the pilot to make sure that friendly forces were avoided and the pilot could visually acquire the target as easily as possible. Once the 9-line brief was compiled it was communicated to the pilot (FAC task 3.0). This involved speaking to the pilot, listening to the pilot reading back the 9-line and affirming or clarifying depending on whether the pilot had correctly understood the 9-line brief. Once the FAC was confident that the pilot had understood the 9-line brief he guided the pilot to an appropriate location to commence the final talk-on to the target (FAC task 4.0). The final talk-on involved a discussion between the FAC and pilot where the FAC describes the location of the target using descriptions that the pilot would understand based on visual cues available to the pilot, (e.g. the target is situated north of the woods adjacent to a fire). For this final talk-on to be successful it was essential that the FAC be able to take the pilot's perspective so that only descriptors of the target location that can be seen from the air are used. The FAC then affirmed the target with the pilot to ensure requisite consistency and provided clearance (FAC task 5.0). Finally, following the attack prosecution by the pilot, the FAC provided a battle damage assessment and further instructions to the pilot (FAC task 6.0), as necessary.

Pilot Task Summary

At a high level, the tasks that the pilot had to carry out were to orient the aircraft so that it was in a position to receive communications from the FAC, receive information from the FAC, receive and record the 9-line brief, locate the target, orient the plan into a position to be able to perform the final attack and perform the attack. The pilot initially maneuvered the aircraft within a pre-set holding area and awaited instructions from the FAC (Pilot task 1.0). The pilot then received communication from the FAC and travelled to the appropriate location (Pilot task 2.0). Next, the pilot received and recorded the 9-line brief (Pilot task 3.0), and oriented the aircraft in co-ordination with the FAC around an appropriate anchor point (agreed visible location on the ground, Pilot task 4.0). Units of measure, which are commonly understood map-referenced landmarks which are used to provide distance information (e.g., an airfield runway) or other localising features were exchanged between the FAC and pilot to ensure that the pilot understood exactly where the FAC wanted effects to be delivered (Pilot task 5.0). The FAC not only describes the position of the target to the pilot, the pilot also describes the position of the target to the FAC in order to double check that both members of the team understand exactly where the effects need to be delivered. Sometimes there are slight differences in the landmarks referenced, contingent on the two different viewpoints (air to ground and ground to ground for the pilot and FAC respectively). Finally, once the pilot has located the target and received attack clearance, the target is prosecuted (Pilot task 6.0).

Team Working Requirements

From this summary of the task analysis it is clear that the FAC and pilot have to work closely together to ensure that they both understand what the target is and where it is located. To do this successfully, the FAC must be able to take account of the pilot's perspective so that a common set of descriptors can be identified and used which 'make sense' to both the pilot and the FAC. The outcome of the HTA has highlighted the tasks that FACs and pilots have to work together on to achieve success. Future experimental designs should seek to investigate further how human-related variables affect the ability of the FAC and pilot to work together and how this ability affects the decision-making process.

Conclusions

The research presented here represents the first step towards the inclusion of team representations within the INCIDER model. The military judgment panel has allowed us to capture the key elements of the interaction between a FAC and a pilot. The information gleaned through the HTA can be used to refine the steps that the INCIDER model goes through, in particular the sequence in which the user accesses various information sources, and how the pilot and FAC interact. In the future, further HTAs will be conducted to help generate a multi-agent version of the INCIDER model. In particular these HTAs will focus on how groups make decisions. Future experimentation is required in order to understand how human variables such as fatigue, stress, experience, expectation, and personality might influence how the FAC and pilot work together. Once the effect of these variables have been understood future work can consider these factors influence the decision-making process in combat identification tasks and can be incorporated within the model to improve its validity.

Acknowledgements

The work was sponsored by Director Equipment Capability Command, Control and Information and Infrastructure (DEC CCII) Directorate of Analysis, Experimentation and Simulation (DAES).

References

Costa, P. T., Jr., and McCrae, R. R. (1985). *The NEO Personality Inventory Manual*. Odessa, FL: Pyschological Assessment Resources.

Darley, J. M., and Gross, P. H. (1983). A hypothesis-confirming bias in labeling effects. *Journal of Personality and Social Psychology, 44*, 20–33.

Dean, D., Handley, A., and Vincent, A. (2006). *Progress Made on the INCIDER Model, FY 2005/2006* (Dstl internal report WP23845). Salisbury, Wiltshire: Defence Science and Technology Laboratory.

Dean, D., Vincent, A., Mistry, B., Hynd, K. and Syms, P. R. (2005). *The Integrative Combat Identification Entity Relationship (INCIDER) Conceptual Model* (Dstl internal report TR14377). Salisbury, Wiltshire: Defence Science and Technology Laboratory.

Dean, D., Vincent, A., Mistry, B., Spaans, M., and Petiet, P. (2008). Representing a combat ID analysis tool within an agent based constructive simulation. *The International C2 Journal, 2,* 1938–6044.

Dean, D., Vincent, A., and Smith, S. J. (2007). *Progress on the INCIDER Integrated Analysis and Experimentation Programme* (Dstl internal report CR23927). Salisbury, Wiltshire: Defence Science and Technology Laboratory.

Gadsden, J., Krause, D., Dixson, M., and Lewis, L. (2008, May). *Human Factors in Combat ID – An International Research Perspective.* Paper presented at The Human Factors Issues in Combat Identification Workshop, Gold Canyon, AZ.

Smith, S. J., Allen, J., and Lovett, R. (2006). *Pilot Study to Evaluate Human Factors Data from a Synthetic Environment Experiment to Validate the INCIDER Model* (Dstl internal report TR23780). Salisbury, Wiltshire: Defence Science and Technology Laboratory.

Smith, S. J., Allen, J., and Lovett, R. (2007) *Evaluation of Human Factors Data from a Synthetic Environment Experiment to Validate the INCIDER Model* (Dstl internal report TR23782). Salisbury, Wiltshire: Defence Science and Technology Laboratory.

Stanton, N. A., Salmon, P. M., Walker, G. H., Baber, C., and Jenkins, D. P. (2005). *Human Factors Methods. A Practical Guide for Engineering and Design.* Aldershot, Hampshire: Ashgate.

Watson, S. K., and Wright, M. D. (2008). *Identification of Combat ID Failures and Across DLODs Mitigating Actions* (QINETIQ internal report 08/00736). London: QINETIQ.

Chapter 15
Team Cognition During a Simulated Close Air Support Exercise: Results from a New Behavioral Rating Instrument[1]

Jerzy Jarmasz
Defence Research and Development Canada

Richard Zobarich
Lora Bruyn-Martin
Tab Lamoureux
HumanSystems, Inc.

Introduction

The nature of recent wars, such as the one in Afghanistan, has caused Coalition forces to depart from or even completely rewrite doctrine in a number of areas. Such is the case with Close Air Support (CAS). United States Joint doctrine defines CAS as 'air action by fixed- and rotary-wing aircraft against hostile targets that are in close proximity to friendly forces and that require detailed integration of each air mission with the fire and movement of those forces' (United States Department of Defense [DOD], 2005, p. I-1). North Atlantic Treaty Organization (NATO) doctrine also defines CAS in very similar terms (NATO, 2005). Integration between air assets and the supported ground forces is typically performed by an air controller attached to ground forces, called a Joint Terminal Attack Controller (JTAC) in US doctrine and often called a Forward Air Controller (FAC) in other NATO countries.

The US and NATO CAS doctrine publications, which cover CAS doctrine for the majority of forces involved in the Afghanistan conflict (as well as a similar conflict in Iraq), provide guidance for CAS mostly with respect to cold-war-type scenarios, where the purpose of CAS is primarily to deliver kinetic effects (i.e., firepower) onto well-defined, mechanized enemy targets in an environment requiring the establishment of air superiority, defeat of enemy air defenses, and close integration of munitions delivery by air assets with other joint fires (e.g., land or naval artillery; see also Barber et al., 1991, for a human factors analysis of CAS supporting this view). However, as Haun (2006) argued, the realities of current conflicts have forced a shift in CAS away from the doctrinal focus on delivering kinetic support to ground troops in a non-permissive air environment to a focus on delivering non-kinetic support (e.g., intelligence, surveillance and reconnaissance, shows of force, convoy escort) in an environment where air superiority and the absence of air defenses can be taken for granted. More importantly for our purposes, perhaps, Haun pointed out that in this new operational environment, CAS 'targets' are more often than not individuals (dismounted or in civilian vehicles) or small structures, rather than formations of mechanized infantry or large military headquarters. These targets are exceedingly difficult to distinguish from people or buildings belonging to the population that coalition troops are supposed to be supporting. Haun called this type of CAS 'low intensity conflict CAS' or 'LIC CAS' to distinguish it from doctrinal CAS. Haun's comments highlight the crucial role that the ability to effectively and quickly tell friend from foe from bystander, in order to prevent fratricide and civilian casualties ('collateral damage'), plays in contemporary CAS operations. This ability is what is commonly known as combat identification or 'combat ID.'

Fratricide is certainly not ignored in the doctrine manuals; the term itself appears many times in the manuals, and the DOD and NATO manuals devote whole sections to the issue. However, fratricide is a growing concern, with Wilson, Salas, Priest and Andrews (2007) reporting that close to 20 percent of casualties in recent conflicts have been estimated to be the result of friendly fire. The Afghanistan conflict alone has produced some very high-profile fratricide incidents involving Canadian soldiers (Department of National Defense of Canada [DND], 2006) and American soldiers (Wilson et al., 2007) among others. Furthermore, the issue of civilian casualties stemming from CAS missions is a common theme in news coverage of the Afghanistan and Iraq wars, as a cursory survey of even mainstream news media demonstrates.

Thus, effective combat ID is a key factor, perhaps even *the* key factor, in successful CAS in low-intensity environments, whether the mission involves dropping ordnance, performing surveillance and reconnaissance, or providing overwatch for a convoy. Furthermore, as task analyses of CAS (e.g., Zobarich, Lamoureux, and Bruyn-Martin, 2007) show, combat ID and maintaining situation awareness (SA) in general, are highly distributed, collective tasks where a number of players must communicate effectively to construct a joint awareness of friend, foe, and bystander. This is all the more important in coalition missions (US DOD, 2005). Thus, as Wilson et al. (2007) point out, ensuring effective team cognition (e.g., team communications, team situation awareness) is imperative for effective combat ID and for minimizing the risk of fratricide and civilian casualties in CAS in current operational environments.

Improving Team Cognition: Rating Scales for Distributed Simulation Exercises

Distributed Mission Operations (DMO, also called Distributed Mission Training, and 'UK Mission training through Distributed Simulation' in the United Kingdom) is a simulation-based training technique with recognized potential as a method for collective training for geographically dispersed, joint and/or multinational operations. DMO could thus be a useful tool for improving CAS performance in general and combat ID performance in CAS in particular. In DMO, trainees located in disparate geographic locations interact in a common virtual space to train team skills, using simulators linked up via high-speed networks. The US Air Force (USAF) has been researching DMO for a number of years and has developed a number of tools for assessing individual and team performance in DMO settings (Schreiber and Bennett, 2006). Because the trainees are not co-located in physical space, and are often not co-located in virtual space either (e.g., pilots flying different aircraft), assessment of team performance can present special challenges.

The Canadian Forces (CF) Air Warfare Centre is currently standing up its Distributed Mission Operations Centre (DMOC) and developing its DMO expertise. As part of this effort, we were tasked with developing measures of team effectiveness in the context of Exercise Northern Goshawk, a simulated coalition distributed air operations exercise that was facilitated under the auspices of the Technical Cooperation Program (TTCP) Coalition Mission Training Research (CMTR) Project Arrangement. The exercise was designed as a CAS, Time Sensitive Targeting (TST), and Troops in Contact (TIC) training event involving participants and researchers at simulation sites in Canada, the US, and the UK August 6–10, 2007. We focused on the Canadian component of the exercise, which simulated a CAS mission involving a Canadian FAC and coalition (US and UK) pilots. The FAC was supporting a CF Commander (not a member of the primary training audience) who was commanding a small convoy transiting through sparse terrain populated with a few simple structures (small dwellings and a religious building). The scenarios were meant to represent situations typical of the low intensity conflict situations, including TIC events that the CF currently encounters in operations in Afghanistan.

In this chapter, we discuss our efforts to capture team cognition processes during the Canadian portion of Exercise Northern Goshawk with a Behaviorally Anchored Rating Scale (BARS) we designed for this purpose. We start by summarizing the development process for the BARS, which we discuss in more detail in Jarmasz, Zobarich, Bruyn-Martin, and Lamoureux (2008) and Zobarich et al. (2007). We then discuss the results we obtained with the scale, which were applied for the first time at Exercise Northern Goshawk. Finally we discuss challenges and issues that were encountered during this first trial of the scale, many of which are also treated in more detail in Jarmasz et al. (2008) and Zobarich et al. (2007).

Development of the Rating Instrument

General Considerations

A BARS was chosen for assessing team cognition largely because it is an unobtrusive method that does not require self-reports by the trainees (time constraints did not allow for a self-report type of assessment during this exercise). Furthermore, BARSs have been shown to be effective for assessing team effectiveness in a number of settings (e.g., military command-and-control, see Murphy, Grynovicki, and Kysor, 2003; medical emergency department training, see Morey et al., 2002). However, because CAS involves the coordination of different but interrelated tasks by a distributed team, we first needed to identify the aspects of CAS missions that involved team interaction, especially between the FAC, the pilot, and the supported commander. This was accomplished by performing a Hierarchical Task Analysis (HTA) of CAS missions. Having identified the phases of CAS where team behaviors would be most evident, we set about developing anchors for the BARS based on the behavioral markers of team cognition breakdown proposed by Wilson et al. (2007). This was one way of ensuring the scale focused on the desired construct, namely team cognition.

Task Analyses

The HTA method was chosen because of its origins in systems theory (Annett and Cunningham, 2000; Shepherd, 1998), thus making it suitable for capturing system-level (such as team) behavior. The HTA for the overall CAS mission was conducted and validated based on extensive interviews with Subject Matter Experts (SMEs) knowledgeable in CF procedures in CAS (active FACs and FAC instructors from the CF as well as a CF-18 pilot with CAS experience). The team HTA identified the different members of the broad CAS team (including the Pilots, the FAC, the Fire Support Coordination Centre, the Air Support Operations Centre, the Forward Observation Officer, the Signals Officer, and the Convoy Commander), but only the FAC and Pilot branches were developed in detail. Each team member's task breakdown included a number of tasks that fed into, received input from, or required an explicit appreciation of the tasks from other team members. Those tasks were explicitly represented in both team members' branches. In practice, we used the tasks in the FAC branch which fed into or were received from the Pilot task to form the 'skeleton' of the Pilot's branch. This allowed us to identify the critical team coordination points that occur generically across CAS missions. We could then develop sets of behaviorally-anchored items to assess the quality of team cognition at each of these coordination points. The full details of the HTA are given in Zobarich et al. (2007).

Development of BARS Items

To keep matters relatively simple, we developed the individual items from the perspective of the FAC's tasks that involved other team members (thus capturing team cognition from the perspective of the person coordinating the teamwork). The development of the items proceeded along two main lines: (1) the identification of suitable measurement points in the tasks of the FAC, and (2) the development of behavioral anchors for these points.

The first line of work focused on those tasks that contributed to the task of another team member, received information from the actions of another team member, or required a significant understanding of the perspective or activities of another team member. The identification of measurement points indicates where measurement should take place, not what should be measured. Typically, the measurement point selected was not the lowest level of decomposition of the HTA. The lowest level of decomposition was often used to inform the scale anchor behavioral descriptions.

The second line started with the team cognition construct (Wilson et al., 2007), which is presented as having three main dimensions: communication, coordination and cooperation. Wilson et al. (2007) further break these categories down into generic team behaviors, and proposed a number of behavioral markers (worded as questions) for probing the quality of each behavior. These generic behaviors constituted good candidates for the behaviors that needed to be assessed (for example, which items needed to be developed) at the measurement points identified in the HTA. Since a rating is essentially an answer to the question, we reworded the generic behaviors in each of the team cognition categories as questions that were relevant to CAS missions based on our HTA. These items, shown in Table 15.1, could then serve as a basis for developing specific items in the BARS.

Table 15.1 Dimensions of team cognition as adapted to the BARS

Communication	How effective was information exchange? Was information exchange economical? Did closed looped communication go as expected?
Coordination	How well were team members' knowledge requirements managed? How well did team members monitor each other's performance? How effective was back-up behaviour? How adaptable were team members to the changing demands of the situation?
Cooperation	To what extent were team members working toward the same ends? How effective were FAC/others as a team?

Thus, a total of ten items, capturing three main dimensions of team cognition, could potentially be formulated at each measurement point identified in the HTA. In practice, not all were found to be applicable to each measurement point. Table 15.2 presents the measurement points and the BARS items selected for development.

Table 15.2 BARS items selection table

Measurement Point	Communication			Coordination				Cooperation		
	1	2	3	4	5	6	7	8	9	10
Determine air assets	✓	✓		✓				✓	✓	
Understand blue situation[1]	✓	✓	✓	✓	✓		✓	✓	✓	
Understand red situation[2]	✓	✓	✓	✓	✓		✓	✓	✓	
Understand white situation[3]	✓	✓	✓	✓	✓		✓	✓	✓	
Understand brown situation[4]	✓	✓	✓	✓	✓		✓	✓	✓	
Understand time	✓	✓	✓	✓	✓	✓	✓			
Maintain personal safety	✓				✓	✓				✓
Transmit immediate CAS request	✓	✓	✓	✓	✓	✓	✓	✓		✓
Receive pilot's scheduled check-in	✓	✓	✓	✓	✓		✓			
Deconflict target area and airspace	✓	✓	✓	✓	✓	✓	✓	✓	✓	✓
Transmit CAS brief	✓	✓	✓	✓		✓	✓	✓	✓	
Communicate remarks	✓	✓	✓	✓		✓	✓	✓	✓	
Communicate options with pilot	✓	✓	✓	✓			✓			
Designate target	✓	✓	✓	✓	✓	✓	✓	✓	✓	✓
Coordinate with FOO	✓	✓	✓	✓			✓	✓	✓	✓
Transmit talk-on	✓	✓	✓	✓	✓	✓	✓	✓	✓	✓
Perform BDA	✓			✓						
Abort CAS mission	✓	✓		✓	✓					✓

[1] Blue situation = friendly forces (e.g., location, movement, intent).

[2] Red situation = situation and intent of adversary.

[3] White situation = civilian situation, including individuals and landmarks (e.g., culturally significant buildings).

[4] Brown situation = weather and terrain factors.

Having identified appropriate measurement points, it was then necessary to develop the anchors for each point. We settled upon a 5-point scale as giving sufficient sensitivity to changes in perceived performance while not overwhelming raters with choices. To construct the anchors we started with the behavioral markers, originally phrased as questions, that Wilson et al. (2007) developed for each sub-dimension of Table 15.1 (e.g., Did team members seek information from all available resources? Did team members pass information within a timely manner before being asked?). Using these, reformulated as statements and combined with information from the lower level of description in the HTA (if there was one), we decided upon what would reflect 'perfect' performance (a '5' on the scale) and what would reflect very poor performance (a '1' on the scale), and then developed complementary anchors for the intermediate scale points. Even though the measurement points in Table 15.2 represented different tasks, the behavioral anchors generally followed similar patterns depending on the generic team behavior they were based on. A complete listing of all the behavioral anchors we developed is provided in Zobarich et al. (2007). Due to space limitations, here we present sample items from each team cognition category for a particular task (FAC transmits the CAS brief and communicates remarks to the pilot) in Table 15.3.

Table 15.3 Sample BARS items for each team cognition category from the 'Transmit CAS brief and communicate remarks' task

Communication: How effective was information exchange?	
5	FAC passed complete and accurate brief (following theatre standard) and provided all key remarks info (e.g., weapons effects, attack geometry, ACA measures, number of attempts, level of risk for blue and white forces, danger close initials).
4	FAC passed all items of brief and most key remarks, but provided some remarks only when prompted (e.g., danger close initials).
3	FAC passed all items of brief and remarks available to him, but had to communicate with others to obtain missing info requested by the pilot (e.g., MAXORD, ACA measures).
2	FAC omitted important brief and remarks items that were available to him.
1	FAC failed to provide sufficient brief and remarks for pilot to complete mission.
Coordination: How well were team members' knowledge requirements managed?	
5	FAC/pilot implicitly coordinated in an effective manner (e.g., did not require special coordination or discussion beyond standard turn-taking), and displayed a common understanding of CAS brief SOPs and of the brief/remarks.
4	FAC/pilot coordinated explicitly and effectively (e.g., discussed coordinates formats), and achieved a common understanding of the CAS brief SOPs and brief/remarks with little effort.
3	FAC/pilot made explicit attempts at coordinating knowledge (e.g., asked each other questions), and at great effort (e.g., much time spent in discussion) achieved common understanding of the SOPs and brief/remarks.
2	FAC/pilot made explicit attempts at coordinating knowledge, and achieved an incomplete common understanding of the SOPs and briefs/remarks.
1	FAC/pilot failed to display a common understanding of the SOPs and brief/remarks.

Table 15.3 *Concluded*

Cooperation: To what extent were FACs/Pilots working towards the same ends?	
5	FAC/Pilot collaborated to ensure CAS brief SOPs (mandatory CAS brief, or that collective SA did not require brief) were adhered to and that all required info was passed and understood for the attack.
4	FAC/Pilot collaborated to understand all relevant CAS brief and remarks info but both had slightly different priorities on brief SOPs, which were easily resolved or accepted.
3	FAC/Pilot collaborated to understand all mandatory CAS brief info, but did not cooperate fully on understanding remarks or had a significant disagreement about brief SOPs; pilot has most but not all info required for talk-on and attack.
2	FAC/Pilot collaborated poorly to achieve joint understanding of brief and remarks, and disagreed significantly about brief SOPs; pilot had only a fraction of the required information for the talk-on and attack.
1	FAC/Pilot could not agree on brief SOPs and did not collaborate to ensure pilot received and understood brief and remarks; pilot did not have any useable information to proceed with talk-on and attack.

Protocol for Applying the BARS

Prior to piloting the BARS during Exercise Northern Goshawk, we organized the items within the BARS according to five generic phases of the CAS missions that we expected to see during the exercise based on our knowledge of the daily scenarios and the CAS HTA: (1) pre-check in, (2) pilot check-in, (3) generic non-kinetic CAS support tasks covering reconnaissance, shows of force, convoy escort and so on, (4) CAS target prosecution tasks, and (5) generic post-attack tasks (e.g., battle damage assessment, re-tasking, egress). The tasks groupings that served to organize the BARS items are shown in Table 15.4 (see Zobarich et al., 2007 for a complete list).

Two of the authors (JJ and RZ) were assigned to perform the ratings. After a dry run applying the BARS on the first day of the exercise (see exercise schedule below), we determined some basic ground rules for applying the scale. To the extent possible, ratings would be made every time a task was performed (for example, every time the FAC talked a pilot onto a target) rather than giving an overall rating to the task for the mission. Also, during the target prosecution phase of a mission, where CAS team members were likely to perform many tasks in parallel and the number of possible simultaneous ratings was expected to overwhelm the abilities of a single rater, one rater (JJ) was designated to rate primarily the CAS brief and target-talk on tasks, and rate the other tasks as resources allowed, while the second rater (RZ) was designated to rate primarily the Deconflict Target Area and Designate Target tasks. In cases where a number of tasks were being performed simultaneously by the team (for example, designating target while deconflicting air space), each rater focused on a designated subset of the task categories from

Table 15.4. Minor changes were also made to the BARS after the dry run on the first day (the wording of items that assessed the team cognition same dimension in different tasks was harmonized, and a few items were either added or dropped from specific tasks depending on their perceived utility).

Table 15.4 CAS tasks to be rated during Northern Goshawk, grouped into generic CAS mission phases

CAS mission phase	Tasks
1. Pre-check in	Understand situation updates Transmit immediate CAS request (as needed)
2. Pilot check-in	Receive pilot's check-in (includes situation brief)
3. Non-kinetic CAS support	Communicate options with Pilot Situation updates
4. Target prosecution	Transmit CAS brief and communicate remarks Transmit talk-on Deconflict Target Area Designate Target Abort Mission (as needed)
5. Post-attack	Perform Battle Damage Assessment Communicate options with Pilot Situation updates

Ratings and Observations during Northern Goshawk

Collecting the Ratings

Ratings were collected by the raters on each day of the exercise for the duration of the Canadian portion of the missions (approximately 2 hours per day for 4 days). As discussed below, Day 1 was used to perform a dry run of the ratings and to finalize the scale. The raters were able to directly observe the FAC and the Convoy Commander while listening to the radio communications between the FAC and the air radio net (pilots and ASOC).

A number of challenges were encountered in applying the BARS. These challenges are discussed in detail in Jarmasz et al. (2008). In general terms, the raters found it challenging to perform all the ratings prescribed by the protocol in real-time. This was partly due to the quick pace of events and could be remedied by reviewing the recordings of the exercise. The raters found that their ability to follow events and thus perform more of the ratings increased each day of the exercise, despite the scenarios becoming more complicated each day.

Due to these challenges and to the fact that neither rater produced a complete set of ratings on any day, we cannot perform inferential statistics on the ratings that were collected. Thus, the inter-rater reliability of the scale and any day-to-day changes in team cognition ratings cannot be statistically assessed at this time. Instead, we describe possible team cognition patterns by presenting summary statistics for each day of the exercise (except Day 1), as well as individual ratings for selected incidents in the exercise that had combat ID implications.

Summary Statistics

For each day of the exercise where ratings were collected with the BARS (as described for Days 2, 3, and 4), we computed average ratings for each team cognition category in each major task of the CAS missions identified in Table 15.4, by combining ratings from both raters across all items in the major team cognition categories (communication, coordination, cooperation) for all separate instances of a given task (e.g., all ratings for the communication category from all talk-ons in a given day were combined into one average). Since the sample size for each average rating is different (ranging from $n = 27$ for one item applied to multiple instances of the same task by both raters, to $n = 1$ in a few cases) we do not attempt statistical comparisons of the rating averages. Given the sheer number of values that were obtained even with this procedure, we report (Table 15.5) only the averages for tasks in Phases 3 (non-kinetic CAS support) and 4 (Target prosecution) of the CAS missions, which are the ones which directly led up to attacks on targets in CAS missions, and therefore are likely the ones most relevant to combat ID and fratricide reduction.

A visual inspection of Table 15.5 suggests that the rating averages were on the whole relatively high with most ratings being higher than 4.0 and none lower than 3.83. A calculation of the ‘grand average’ of all the ratings for each day (including those not included in Table 15.5) reinforces this impression, with the lowest grand average being 4.22 on Day 3. There are a few noteworthy ‘large-scale’ patterns in evidence in the ratings. The presence of ratings for the Abort mission tasks on Days 2 and 4 indicate aborted attacks on both of those days. One of the aborted attacks was due to a technical problem with one of the simulated instrument displays the FAC was using. The second aborted mission will be discussed in more detail below. Also, almost all of the rating averages in the cooperation category achieved the highest possible value of five, suggesting high degrees of cooperation between participants throughout the exercise. Also note that all of the sub-4 ratings for the Phase 3 and 4 tasks occur on Days 2 and 3, mainly for communications ratings and to a lesser extent for coordination ratings, which may be an indication that team communications and coordination during these phases of the CAS missions improved over the duration of the exercise. We note that no fratricide or collateral damage events were observed during the exercise.

Table 15.5 Summary statistics of ratings for Phase 3 and 4 tasks in Exercise Northern Goshawk

Task	Team cognition category	Day 2	Day 3	Day 4
3.1 Communicate options with pilot	Communication	3.83 (1.17)[1]	3.67 (0.98)	4.44 (0.73)
	Coordination	3.50 (0.84)	4.00 (0.00)	4.33 (0.52)
3.2 Understand situation updates	Communication	3.89 (0.93)	4.33 (1.03)	4.13 (0.52)
	Coordination	4.00 (0.00)	3.17 (0.41)	4.17 (0.39)
	Cooperation	4.67 (0.58)	4.00 (1.15)	4.50 (0.58)
4.1 Transmit CAS brief and communicate remarks	Communication	4.22 (1.20)	4.33 (1.03)	4.50 (0.84)
	Coordination	4.50 (0.76)	4.75 (0.50)	4.50 (0.55)
	Cooperation	5.00 (0.00)	5.00 (0.00)	5.00 (0.00)
4.2 Transmit talk-on	Communication	4.22 (0.97)	3.83 (1.17)	4.29 (0.69)
	Coordination	3.83 (0.83)	4.13 (0.83)	4.08 (0.90)
	Cooperation	5.00 (0.00)	5.00 (0.00)	5.00 (0.00)
4.3 Deconflict target area	Communication	4.33 (1.00)	4.00 (1.00)	4.33 (1.12)
	Coordination	4.13 (0.75)	4.00 (0.00)	4.33 (0.52)
	Cooperation	5.00 (0.00)	4.00 (0.00)	4.00 (0.00)
4.4 Designate target	Communication	4.14 (1.07)	4.33 (0.58)	4.50 (0.84)
	Coordination	NR[2]	4.50 (0.58)	4.43 (0.55)
	Cooperation	NR	NR	NR
4.5 Abort mission	Communication	5.00 (0.00)	NR	5.00 (0.00)
	Coordination	4.67 (0.58)	NR	4.75 (0.50)
	Cooperation	5.00 (NA)[3]	NR	5.00 (0.00)

[1] Values in parentheses represent standard deviations.

[2] NR = no rating was made for this measure during the event.

[3] NA = standard deviation cannot be computed because mean was based on a single sample.

Ratings for Specific Events

The summary statistics above hide a number of events during the exercise that had combat ID implications. We provide three examples here. The first event involved ordnance dropped on a target by a pilot without clearance from the FAC, in clear violation of CAS procedure (NATO, 2005; US DOD, 2005). At the end of the talk-on phase, at a point when the FAC was already satisfied that the pilot had visually acquired the correct target, the pilot stated his intention to drop ordnance on target with terminology that was unknown to the FAC. The FAC requested the pilot repeat his transmission a number of times, but due to the low quality of the simulated radio channel (see discussion below), the FAC could not make out what the pilot said. The pilot then proceeded to drop the ordnance, which destroyed the target and surprised the FAC. At the time, the rater who was responsible for rating the talk-on tasks (as per the protocol above) rated one of the three communications items for the task as a 3, and two of the four coordination items as a 3 and a 2, respectively, (see Table 15.6).

Thus ratings for the interaction between the FAC and the pilot on this attack suggest poor performance on some aspects of team communication and team coordination. Subsequent discussion of the incident by the FAC and the pilot, as well as discussion at the mission debrief, established that the pilot had interpreted the FAC's confirmation that the correct target had been identified as clearance to attack, whereas the FAC was unfamiliar with the terminology the pilot used to signal his intention to expend ordnance. Poor transmission quality on the simulated radio channel also made it difficult for the players to understand each other, even though on the whole they seemed to be trying to cooperate. Thus, the incident seemed to involve a breakdown in communications (non-standard terminology, need for frequent repetition, and failure to ensure complete mutual understanding) as well as coordination (failure to ensure everyone fully agreed on procedures before the attack). The ratings for this incident made during the event (Table 15.6) are consistent with the interpretation of events in the debrief, suggesting the BARS is able to track such breakdowns. Fortunately the correct target was hit and no collateral damage was incurred, but the situation could have had serious consequences if, for instance, this had been a real attack and the FAC had been concerned that the pilot's angle of attack or chosen ordnance might have effects on nearby friendly forces or civilians.

Table 15.6 Ratings for talk-on task of Incident 1

Team cognition category	Items	Rating
Communication	How effective was information exchange?	5
	Was communication economical?	3
	Did closed looped communication go as expected?	4
Coordination	How well were team members' knowledge requirements managed?	3
	How well did team members monitor each other's performance?	2
	How effective was back-up behavior?	5
	How adaptable were team members to the changing demands of the situation?	4
Cooperation	To what extent were team members working toward the same ends?	5
	How effective were FAC/others as a team?	5

The second event we focus on involves the second aborted attack of the exercise. In this incident, the FAC had been prosecuting a cluster of targets (small buildings and vehicles) with multiple aircraft simultaneously. This was a complicated attack with the FAC and the pilots spending a lot of time questioning each other to ensure each pilot had positive visual identification of the correct target. The FAC seemed to lose track of which pilot he had just cleared for an attack (he subsequently explained in an informal debrief that this occurred due to having too many pilots talking to him on the same channel). Rather than risk the wrong target being hit, the FAC aborted one attack, re-assessed the situation with the pilots, resumed the attack, and prosecuted the remaining targets successfully, with no fratricide or collateral damage. Thus in this incident, team cognition was heavily taxed by the complexity of the attack, yet the willingness of the players to try to cooperate and coordinate seems to have contributed to avoiding an inappropriate drop. The ratings for the talk-on and abort tasks of this attack, shown in Table 15.7, appear to be consistent with this interpretation: the communication and coordination categories for the talk-on contain a number of '3' ratings, whereas the cooperation categories, as well as the ratings for the Abort task contain mostly '5' ratings (the one '4' rating for the Abort task reflects the fact that the FAC and pilots explicitly coordinated by re-discussing the abort codes during the task).

The third incident involved a situation where a technical problem with some simulators (a so-called 'sim-ism') impaired some of the participants' situation awareness and ability to perform combat ID. The scenario involved a target area that included a number of static objects (buildings) and mobile objects (vehicles). The pilots were reporting a failure to see some of the expected objects at the coordinates indicated by the FAC, generating some confusion. After requesting that the pilots describe the general features of the target area, the FAC was satisfied that they were in the right location and could see at least some of the target area objects, and suspected (correctly) that a 'sim-ism' was preventing them from seeing some of the targets. Given the absence of nearby friendly forces or civilians, the FAC was able to talk the pilots onto the right targets by marking the target area and by referring the pilots to the targets they were able to see. The FAC was able to gauge their situation awareness enough to realize there was a problem, and take corrective measures to ensure that all team members then supported each other in successfully prosecuting an attack on a target. As a result, the ratings for the talk-on for this attack, shown in Table 15.8, are generally high, with only one item, 'Was communication economical', obtaining a rating below 3 due to extra discussion needed and some ongoing transmission problems.

Table 15.7 Ratings for talk-on and abort tasks for Incident 2

Task	Team cognition category	Item	Rating
Transmit talk-on	Communication	How effective was information exchange?	5
		Was communication economical?	3
		Did closed looped communication go as expected?	5
	Coordination	How well were team members' knowledge requirements managed?	3
		How well did team members monitor each other's performance?	3
		How effective was back-up behavior?	5
		How adaptable were team members to the changing demands of the situation?	3
	Cooperation	To what extent were team members working toward the same ends?	5
		How effective were FAC/others as a team?	5
Abort mission	Communication	How well were team members' knowledge requirements managed?	4
	Coordination	How well did team members monitor each other's performance?	5
	Cooperation	To what extent did team members display mutual trust?	5

Table 15.8 Ratings for talk-on task for Incident 3

Team cognition category	Item	Rating
Communication	How effective was information exchange?	5
	Was communication economical?	3
	Did closed looped communication go as expected?	5
Coordination	How well were team members' knowledge requirements managed?	4
	How well did team members monitor each other's performance?	4
	How effective was back-up behavior?	5
Cooperation	How adaptable were team members to the changing demands of the situation?	4
	To what extent were team members working toward the same ends?	5
	How effective were FAC/others as a team?	5

The three incidents we previously described all represented challenging situations for team cognition, with the first two being more challenging than the last one. Accordingly, the ratings for the last incident were generally higher than for the first two. More generally, many of the attacks and non-kinetic CAS support tasks in the exercise came off with few or no problems, and the ratings for those events were generally higher than for the three presented above (almost all '4' and '5' ratings). Thus, it seems that the BARS is able to capture differences in team cognition processes that seem to correspond to actual differences in team performance in CAS, some of which, as we discussed, could have significant implications for combat ID effectiveness and the prevention of fratricide and civilian casualties.

Discussion and Conclusions

Congruence of the Ratings with Exercise Events

On the whole, the ratings produced by the BARS we developed to rate team-cognition breakdowns seems to have produced results consistent with the subjective observations of observers. The lower average ratings that were obtained for the communications categories discussed above were reflected in comments collected from local observers of the event who seemed to agree that the simulated radio transmissions were at times of poor quality (noisy or otherwise distorted) and that the high number of players on the same channel also adversely affected the ability to understand transmissions at times. The high ratings for almost all items assessing cooperation and many of the ones assessing coordination were also echoed by the comments of many observers to the effect that participants seemed quite willing to cooperate and 'play' together. The seeming improvement for the ratings for the non-kinetic and target prosecution tasks over the last three days was also consistent with the comments of the local participants that the participants seemed to be developing a 'rapport' and cooperating better.

The examination of the ratings for a few selected events also seems to suggest that the BARS seems to be congruent with events in the exercise at a 'micro' level, in that the ratings seemed to capture different degrees and different types of team cognition impairment for different events in the exercise. Thus, there is reason to think that the scale we developed is capable of tracking differences in the quality of team processes at both a 'macro' (averaged ratings) and 'micro' (ratings for specific events) levels, at least under certain circumstances.

Challenges in Applying the BARS

Despite the promising results discussed so far, we have already noted that due to the way the ratings were collected, we were not able to perform any inferential statistics on our ratings. We were able to find parallels between the ratings we collected and the subjective impressions of the observers and participants who were involved in the Canadian portion of the exercise, but we lacked objective measures with which to compare the ratings. Thus, we are not in a position to state with much certainty that our scale tracks real changes in the quality of team cognition during a team task or that whatever changes are tracked correspond to objective differences in combat ID performance or the ability to avoid fratricide or unwanted casualties.

These are but two of a number of challenges and issues that were raised in this initial attempt to apply the BARS to a distributed CAS exercise. These challenges and potential solutions are discussed in detail in Jarmasz et al. (2008). We summarize them below.

Convergent and discriminant validity Some aspects of the scenarios affected the degree to which the properties of the BARS (namely discriminant validity) could be assessed. Many of the participants had extensive CAS experience (the Canadian FAC was a qualified CAS instructor) and all seemed willing to cooperate with each other for the sake of the exercise. Also, friendly forces and civilian entities were deliberately kept at some distance from CAS targets by exercise organizers, due mainly to technical issues affecting some of the simulators. Thus, a performance-ceiling limited opportunities to verify the scale's ability to discriminate between good and bad team performance. This could be remedied by using the BARS on CAS exercises involving wide ranges of trainee experience and task difficulty (especially regarding combat ID issues).

Problems with the BARS itself One problem was the fact that performing the ratings in real-time during the exercise was difficult for two reasons: the sheer number of items to be applied (especially in cases involving multiple simultaneous tasks) and the fact that applying them required familiarity with CAS exercises. One of the raters was observing a CAS event for the first time during Exercise Northern Goshawk, and often found it challenging to make required ratings during the exercise, especially in the early days of the event. Providing prospective raters with prior experience with CAS events (e.g. via an audio recording of a previous CAS exercise) could ensure raters are more comfortable applying the ratings in real time. The BARS could also be made easier to apply by reducing the number of items, or by applying it to after-action reviews or audio recordings of CAS exercises so raters can apply it at their leisure after first observing the event in real time.

The exercise also revealed some deficiencies in specific items (some of these were addressed after the dry run of the BARS on Day 1), which could be addressed by refining them by reviewing data from other exercises and more consultation with SMEs. We avoided the critical incident (Flanagan, 1954) and retranslation of expectations methodologies (Smith and Kendall, 1963), which base behavioral ratings on the opinion of SMEs rather than a theoretical construct. However, applying these to the existing scale could help improve the diagnostic value of some of its items, as well as determine whether some of the items or even some of the tasks used in the current form of the scale have less diagnostic value than others. These could then be eliminated, shortening the scale and making it easier to use.

Inter-rater reliability Ultimately, the construct validity and the inter-rater reliability of the BARS will have to be assessed and improved by having a number of raters apply it to a number of CAS missions or exercises. Fortunately, Exercise Northern Goshawk was part of a broader international research effort under The Technical Cooperation Program, and other performance measures and digital logs from a number of similar DMO exercises have been collected by our research partners (see Schreiber and Bennett, 2006). Thus, there is an extant body of data which could potentially be used for improving and refining the BARS.

Conclusion

The BARS we developed to assess the quality of team cognition in CAS exercises appears to have captured some team performance patterns which seem to be consistent with subjective appraisals of the participants' collective performance in the exercise. Thus, we feel that once work required to refine the scale and improve its inter-rater reliability has been performed, the scale has the potential to be very useful for capturing team cognitive processes that underlie effective combat ID and team performance in CAS. This is particularly important with respect to avoiding friendly fire incidents, which are a very real and serious issue in CAS as some recent events in Afghanistan and Iraq have shown. Further, we feel it could play a useful role in supporting the development and assessment of distributed simulation exercises for CAS, and by extension CAS mission safety and effectiveness, especially for low-intensity conflicts, for Canada and allied nations. Finally, the general methodology we present (start with an HTA of a team task, generate behavioral anchors for key interaction points using Wilson et al.'s [2007] behavioral markers, and validate and refine with SMEs) could be a valuable methodology for assessing team performance and training in other distributed team tasks, in particular those where failure of team cognition could lead to friendly fire or other lethal consequences.

Acknowledgements

The authors thank the members of 1 Royal Canadian Horse Artillery, Canadian Forces Base Shilo, especially Lieutenant-Colonel Williams (then Commanding Officer), Warrant Officer Cyr (FAC Supervisor), Major Allen and Sergeant Everett, as well as Captain Williams of the Artillery School at Combat Training Centre, Gagetown, for their support and expertise in the development of the BARS and execution of the exercise.

References

Annett, J., and Cunningham, D. (2000). Analysing command team skills. In J. M. Schraagen, S. F. Chipman, and V. L. Shalin (eds). *Cognitive Task Analysis*. Mahwah, NJ: Lawrence Erlbaum.

Barber, A., Brown, D., Chandler, E., Davis, D., Dye, C., Moyer, J., et al. (1991). *Human-centered Analysis of the Future Close Air Support/Battlefield Air Interdiction (CAS/BAI) Mission* (DTIC Technical Report HSD-TR-1991–0021). Cameron Station, Alexandria, VA: Human Systems Division.

Department of National Defence (2006). *Board of Inquiry Minutes of Proceedings, A-10 Friendly Fire Incident 4 September 2006 Panjwayi District, Afghanistan*. Retrieved from http://www.forces.gc.ca/site/focus/opmedusa/A10_BOI_Report_e.pdf.

Flanagan, J. C. (1954). The critical incident technique. *Psychological Bulletin, 51,* 327–357.

Haun, P. M. (2006). The nature of close air support in low intensity conflict. *Air and Space Power, 20*, 107–110.

Jarmasz, J., Zobarich, R., Lamoureux, T., and Bruyn Martin, L. (2008, September). Avoiding friendly fire: Constructing behaviorally-anchored rating scales to assess team cognition in distributed mission training for close air support. *Proceedings of the Human Factors and Ergonomics Society 52nd Annual Meeting,* New York City, NY.

Morey, J. C., Simon, R., Jay, G. D., Wears, R. L., Salisbury, M., Dukes, K. A., and Berns, S. D. (2002). Error reduction and performance improvement in the emergency department through formal teamwork training: evaluation results of the MedTeams Project. *Health Services Research, 37,* 1553–1581.

Murphy, J., Grynovicki, J. O, and Kysor, K. P. (2003, June). Case study of a prototype set of Behaviorally Anchored Rating Scales (BARS) for C2 assessment. *Proceedings of the 8th International Command and Control Research and Technology Symposium,* Washington, DC.

North Atlantic Treaty Organization (2005). *Tactics, Techniques and Procedures for Close Air Support Operations* (ATP-3.3.2.1(A)). Brussels, Belgium.

Schreiber, B. T., and Bennett Jr, W. (2006). *Distributed Mission Operations Within-simulator Training Effectiveness Baseline Study: Summary Report* (AFRL-HE-AZ-TR-2006–0015-Vol I). Mesa, AZ: Air Force Research Laboratory.

Shepherd, A. (1998). HTA as a framework for task analysis. *Ergonomics, 41,* 1537–1552.

Smith, P. C., and Kendall, L. M. (1963). Retranslation of expectations: an approach to the construction of unambiguous anchors for rating scales. *Journal of Applied Psychology, 47,* 149–155.

United States Department of Defense (2005). *Joint Tactics, Techniques and Procedures for Close Air Support* (JP 3–09.3). Washington, DC: Joint Chiefs of Staff.

Wilson, K. A., Salas, E., Priest, H. A., and Andrews, D. (2007). Errors in the heat of battle: taking a closer look at shared cognition breakdowns through teamwork. *Human Factors, 49*, 243–256.

Zobarich, R., Bruyn-Martin, L., and Lamoureux, T. (2007). *Forward Air Controller: Task Analysis and Development of Team Training Measures for Close Air Support* (DRDC Toronto CR 2007–156). Toronto: Defence Research and Development Canada.

SECTION 5
Automation

Successful combat identification does not result from a set of operating procedures or a piece of equipment, but rather from the combination of various elements that contribute to the identification process (Defense Science Board 1996). Necessary elements of successful identification have been discussed in earlier chapters, including basic cognitive processes and visual discrimination, situation awareness, and teamwork. Now the focus turns toward the technology utilized for combat identification.

With increasing frequency, the US military relies on automation to identify targets as friends, foes, or neutrals (Hawley, Mares, and Marcon, Chapter 19). Automation constitutes the use of control systems, such as a computer, to perform functions previously performed by humans (Parasuraman and Riley, 1997). Although combat identification systems may reduce the need for human intervention, human participation remains an essential element in any combat identification process. Moreover, the use of automation may not improve target identification performance (Rice, Clayton, and McCarley, Chapter 17, this volume). Automation may fail to enhance performance either due to a failure of the technology or human error. Operators may grow complacent with the use of the automation and engage in misuse—overreliance on the system (Parasuraman and Riley, 1997). Conversely, operators may disuse automation by failing to rely on an identification system when use of the system would have enhanced target identification performance. And finally, the automated system may be designed, implemented, or managed inappropriately leading to what is referred to as abuse. This section focuses on the design, implementation, and use of automation intended to improve combat identification decisions.

In an effort to enhance the appropriate use of automation and reduce incidence of fratricide, Heather F. Neyedli, Lu Wang, Greg A. Jamieson, and Justin G. Hollands (Chapter 16) investigate how the design of a combat identification system affects reliance on the system. In the first experiment, the researchers examine the importance of presenting the operational reliability of the combat identification system. The second experiment compares different methods of presenting the reliability information. In Chapter 17, Stephen Rice, Krisstal Clayton, and Jason McCarley examine how a biased and imperfectly reliable aid affects human-automation performance. Using a visual search task, they observe how a bias toward either false alarms or misses affects operator compliance and reliance with the combat identification system. Seeking to determine the appropriate use of an automated combat identification aid, Mary T. Dzindolet, Linda G. Pierce, and Hall P. Beck (Chapter 18) employ the Framework of Automation Use Model to organize their review of studies on how operators use automation. John K. Hawley, Anna L. Mares, and Jessica L. Marcon (Chapter 19) discuss the combat identification problem with a focus on the human factors issues of target identification decision aids. As the use of automation often has unintended consequences during active combat, John Barnett (Chapter 20) discusses factors to consider in the design and application of automation to help ensure a reduction in fratricide and collateral damage. Jean W. Pharaon (Chapter 21) evaluates the primary US Army friendly fire inhibitor system, the Force XXI Battle Command Brigade-and-Below (FBCB2), and recommends system enhancements to improve the fratricide prevention effectiveness of the FBCB2. He conducted a literature review to obtain background information on existing combat identification systems that could supplement the FBCB2 and a user survey to assess the FBCB2 system.

References

Defense Science Board. (1996). *Report of the Defense Science Board Task Force on Combat Identification.* Office of the Under Secretary of Defense for Acquisition and Technology. Retrieved February 23, 2009, from http://www.acq.osd.mil/dsb/reports/combatidentification.pdf.

Parasuraman, R., and Riley, V. (1997). Humans and Automation: Use, Misuse, Disuse, and Abuse. *Human Factors, 39*, 230–253.

Chapter 16
Evaluating Reliance on Combat Identification Systems: The Role of Reliability Feedback[1]

Heather F. Neyedli
Lu Wang
Greg A. Jamieson
University of Toronto

Justin G. Hollands
Defence Research and Development Canada

Introduction

A variety of technical solutions have been developed to improve soldiers' combat identification (CID) performance. For instance, individual CID systems (Lowe, 2007), which operate similarly to identification friend or foe (IFF) systems, offer some support for this difficult task. In these systems, each soldier is equipped with a gun-mounted interrogator and a helmet-mounted transponder. The interrogator emits a laser inquiry, to which the transponder replies using an encrypted radio signal. If the signal is returned, the interrogator indicates (usually through a small light) that the target is friendly. The key drawback of this automated system is that it cannot positively identify a target without a working transponder (Sherman, 2000; Sherman, 2002; SIMLAS, 2006; Zari et al., 1997). If the interrogator receives no transponder signal, the target could be hostile, neutral, or friendly.

Humans are prone to misuse and disuse imperfect automation (Parasuraman, Molloy, and Singh, 1993), and CID systems are no exception (Briggs and Goldberg, 1995). Misuse occurs when individuals rely on automation inappropriately, usually by over-relying on an imperfect aid. Disuse occurs when participants under-use or reject the capabilities of automation (Parasuraman and Riley, 1997). Informing individuals about the reliability of an aid can assist them in relying on an aid more appropriately (Lee and See, 2004); although, several studies in the CID domain have shown that this is not always the case (Dzindolet et al., 2000, 2001a, 2001b;

Karsh et al., 1995). Moreover, little research has explored the possible means of providing this reliability information.

The current research explores factors that affect reliance on CID systems, with the end goal of helping soldiers to use the systems more effectively and reduce fratricide incidents. To achieve this, two experiments were performed. Experiment 1 examined whether informing participants about the reliability of the aid affected participants' reliance on the aid. Experiment 2 examined how different reliability display formats affected participants' reliance on the aid.

Experiment 1

The goal of the first study was to examine how knowledge of aid reliability affects reliance on automation. While the general human-automation interaction research indicates that providing information about automation performance generally improves users' reliance strategy (Lee and See, 2004), some studies suggest that participants tend to rely inappropriately on CID aids even when they are informed of the aid's reliability (Dzindolet et al., 2000, 2001a, 2001b; Karsh et al., 1995). For example, Dzindolet et al. (2001a) performed a study where the reliability of the feedback from a CID system was 60, 75, or 90 percent. The overall performance was similar among the three aided groups and an unaided group. Although participants were informed of the aid reliability, they were more likely to misuse (i.e., follow the incorrect feedback) than disuse (i.e., reject the correct feedback) the aid regardless of its reliability. The researchers concluded that the aid reliability instruction alone did not encourage appropriate reliance on the aid. However, we question whether this conclusion may have been premature. A participant's decision to rely on the aid more often than ignoring it may be appropriate, especially in the case where the aid outperforms manual control. We anticipate that an alternative approach to the data analysis would reveal more insight into reliance behavior.

We propose a new method of measuring reliance on automation based on Signal Detection Theory (SDT; Macmillan and Creelman, 1991; Wickens and Hollands, 2000). In SDT, two performance indicators—sensitivity (d') and response bias (β)—characterize the participants' performance. Sensitivity refers to the keenness of the detector with d' measuring the separation of the signal and noise distributions. The response bias measure β indicates the level of the evidence variable that will produce a signal present decision. β can be influenced by the probability of a signal occurring or the value or cost of an outcome (e.g., a correct rejection). When a soldier receives aid feedback, expectations of the probability that the target is friendly or hostile should change. This change, according to SDT, should influence the setting of response bias but not sensitivity. If this premise is true, then reliance on the aid can be measured based on the change in response bias. Therefore, we call our method the *response bias* approach.

In our first experiment, we revisited the question of reliance on imperfect CID aids. The objectives of the experiment were (1) to examine the effectiveness of using aid reliability information to support appropriate reliance on a simulated, rifle mounted, CID system, and (2) to test the feasibility of using response bias as an indication of participants' reliance on the CID aid.

Experimental Design

The protocol used a 3 (aid reliability: no aid, 67 percent, 80 percent) × 2 (instruction of aid reliability: informed, uninformed) mixed design. Aid reliability was manipulated as a within-subjects factor. For example, a reliability of 67 percent indicates that when the aid responds with 'unknown' feedback to a target, 67 percent of the time an unknown target is an enemy. In the no-aid condition, the participants conducted the CID task through visual identification alone. The instruction of aid reliability, informed or uninformed, was a between-subjects factor.

The experiment was comprised of three mission blocks, one at each level of aid reliability. The order of conditions was counterbalanced. Each block consisted of 120 trials, with one target appearing in each trial. For each block, half of the targets were friendly and half were hostile.

CID Simulation

The IMMERSIVE (Instrumented Military Modeling Engine for Research using Simulation and Virtual Environments) synthetic task environment served as the test bed. Developed by Defence Research and Development Canada-Valcartier, IMMERSIVE uses the modules of a commercial first-person shooter game–Unreal Tournament 2004. Experimenters can create scenarios by setting terrains, combat activities, and force characteristics. Friendly and hostile forces are distinguished by differences in uniforms, weapons, actions, and feedback from the CID system.

Data Collection

Participants Twenty-six University of Toronto students with normal visual acuity were recruited. Complete data were collected from 24 participants and used for analysis. To take part in this experiment, each participant was required to pass a visual acuity test (measured with a Snellen eye chart), and an ocular dominance test (measured using the Porta Test; Roth, Lora, and Riley 2002). Each participant was paid $30 CAD for participating, and a bonus $10 CAD was given to the top performer.

Tasks and procedures The experiment took approximately 2.5 hours to complete. The participants were instructed to imagine themselves in a battlefield. Their primary task was to identify targets in the scene and shoot the enemies. Their score was determined by the accuracy and speed of their engagement decisions. The participants were advised that they would have an aid to assist them in two of the three blocks. When the aid identified a friendly soldier, it would respond with 'friend' feedback – a blue light. Otherwise it would respond with 'unknown' feedback – a red light. The blue light would never appear when a target was hostile. However, the participants were told that the 'unknown' feedback was set to be less than 100 percent reliable to mimic system failures. It was possible that a red light could be shown when a target was actually friendly. Before each mission block, the participants received a set of instructions indicating whether the aid was on or off, and for the informed group, the probability of the red light feedback being incorrect.

Data Analysis

The participants' reliance on the aid was analyzed using the response bias method. SDT parameters *d'* and β were calculated using the values contained in Table 16.1. According to SDT, response bias should vary with the expectation of the target probability, whereas sensitivity should remain constant (Macmillan and Creelman, 1991; Wickens and Hollands, 2000).

The optimal β values were calculated so they could be compared to the actual values. For the 67 percent reliability condition: $\beta_{optimal} = \frac{P(Friend \mid Unknown)}{P(Terrorist \mid Unknown)} \frac{33\%}{67\%} = 0.50$; for the 80 percent reliability condition: $\beta_{optimal} = \frac{P(Friend \mid Unknown)}{P(Terrorist \mid Unknown)} \frac{20\%}{80\%} = 0.25$.

Table 16.1 The outcome matrix in the condition that the aid gave 'unknown' feedback

		States of the world	
		P(Enemy|Unknown)	P(Friend|Unknown)
Participant Response	Shoot	Hit (H) P(H|Unknown) Value = V(H)	False Alarm (FA) P(FA|Unknown) Cost = C(FA)
	Not Shoot	Miss (M) P(M|Unknown) Cost = C(M)	Correct Rejection (CR) P(CR|Unknown) Value = V(CR)

Results

To increase the normality of the probability data for the FA rate and miss rate, an arcsine transformation was applied: Transformed Probability Data = 2 * arcsine [Probability Data]$^{1/2}$ (Dzindolet et al., 2001a; Howell, 1992; Winer, Brown, and Michels, 1991). A 3 (aid reliability: no aid, 67 percent, 80 percent) × 2 (group: uninformed, informed) mixed ANOVA was used to examine the transformed FA rate, transformed miss rate, response time, sensitivity, and response bias. Response time was measured from when the target appeared until the first shot was taken in trials that participants shot at the target.

Order effects The data were analyzed to examine whether the order in which the blocks were completed affected any of the dependent variables. Order was significant only for response time $F(2,44) = 4.02$, $p < 0.05$, with participants responding more quickly on blocks occurring later in the experiment.

False alarm (friendly fire) There was a significant main effect of aid reliability on false alarms, $F(1.506$[2]$,44) = 10.75$, $p = .001$, $r = .44$. As seen in Figure 16.1, participants made fewer false alarm errors when they had the CID aid, $F(1,22) = 9.86$, $p = .005$, $r = .56$, and fewer errors in the 80 percent reliability condition compared to the 67 percent reliability condition, $F(1,22) = 13.95$, $p = .001$, $r = .62$.

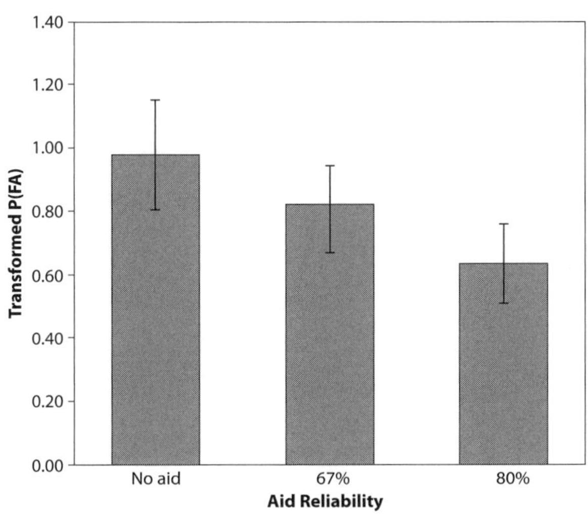

Figure 16.1 The effect of aid reliability on transformed FA rate

2 When the assumption of sphericity was violated, the degrees of freedom were corrected using the conservative Greenhouse-Geisser value.

Miss (miss hostile target) There were no significant main effects of aid reliability or group on miss rate, and no significant interactions.

Response time There were no significant main effects of aid reliability or group on response time, and no significant interactions.

Sensitivity, d' There were no significant main effects of aid reliability or group on sensitivity, and no significant interactions, which is consistent with the hypothesis that sensitivity would not vary with the aid reliability or instruction group.

Response bias, β We conducted a natural logarithm transformation on the response bias (β) scores due to a violation of the normality assumption. These transformed values were negative; lower values of $ln\beta$ indicate a more liberal bias. The 3 (aid reliability: no aid, 67 percent, 80 percent) × 2 (group: uninformed, informed) ANOVA on participants' $ln\beta$ revealed a main effect of aid reliability, $F(2,44) = 5.44$, $p = .01$ $r = .33$ (see Figure 16.2). Contrasts revealed that response bias in the no aid condition was more conservative than in the 67 percent reliability condition, $F(1,22) = 4.42$, $p = .05$, $r = .41$, and in the 80 percent reliability condition, $F(1,22) = 13.05$, $p < .01$, $r = .61$; that is, participants were more liberal in making their engagement decision when they received aid feedback. The main effect of group was also significant, $F(1,22) = 7.55$, $p = .01$, $r = .51$. The informed group was more liberal than the uninformed group. The interaction was not significant.

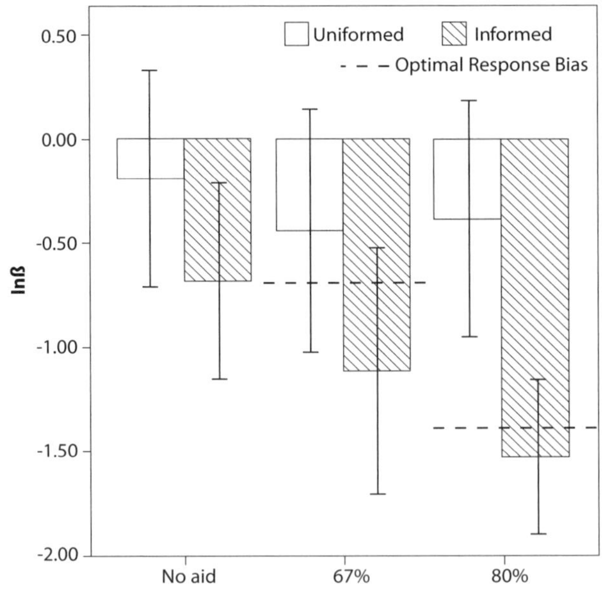

Figure 16.2 Response bias in each experimental condition

Appropriateness of reliance As shown in Figure 16.2, in the 67 percent reliability condition, whether informed of the failure rate of 'unknown' feedback or not, participants' response bias did not significantly deviate from the optimal value ($ln\beta_{optimal}$ = -.693): for the uninformed group, $t(11) = .96$, $p = .35$, $r = .28$; for the informed group, $t(11) = -1.57$, $p = .15$, $r = .43$. However, in the 80 percent reliability condition, while the informed group's response bias did not differ significantly from the optimal value ($ln\beta_{optimal}$ = -1.386), $t(11) = -.80$, $p = .443$, $r = .23$, while the response bias of the uninformed group was less than optimal, $t(11) = 3.91$, $p < .01$, $r = .76$. This result indicates that participants who were not informed of the failure rate of the 'unknown' feedback did not rely on the feedback enough in the 80 percent reliability condition.

Discussion

Experiment 1 found CID speed and accuracy were not significantly improved by imperfect CID systems, which is in partial agreement with previous findings (Dzindolet et al., 2000, 2001a, 2001b; Karsh et al., 1995). The present experiment found no significant differences in the speed of engagement decision or the number of missed hostile targets among all test conditions. However, in contrast to previous studies, the CID aid in this study contributed to a significant reduction in the number of friendly fire engagements. This improvement was found at two reliability levels, and increased with the aid's reliability.

The improved CID performance with the CID aid might be attributed to at least three factors. The first is the more effective training of the automation reliability. For example, Dzindolet et al. (2001b) found that instructions about aid reliability alone were not sufficient for participants to rely on feedback appropriately. In their study, participants were able to view the stimulus a second time if they were unsure whether the target was present. Even for fully reliable feedback (similar to the 'friend' feedback in the present study), participants requested a second viewing of the stimulus, indicating that they might have been doubtful about the reliability information or had not understood the instruction correctly. In the present experiment, in addition to the verbal instructions, participants were also given a short paper test to make sure they correctly understood the different reliabilities of the two types of feedback. This training method may have led to a better reliance strategy–participants almost invariably followed the perfectly reliable 'friend' feedback. The second factor is the cost of reliance which can increase the chance of automation disuse. In Karsh et al.'s (1995) study, participants were informed that the reliability of the CID aid was as high as 90 percent, yet they only activated the CID aid about 15 percent of the time. This might be because there was a 0.75-second delay in the aid feedback, increasing the cost of reliance. In contrast, there was no delay of feedback in this study and therefore the participants may have been more willing to use the aid. A third factor is workload. In Dzindolet et al.'s study, the participants were required to respond to audio stimuli in addition to performing the detection task. Therefore, it is likely that the workload was higher in their study. Some research suggests that misuse of automation is more likely to happen when the participants are responsible for tasks in addition to the automated task (Parasuraman et al., 1993).

Experiment 2

Providing participants with information regarding the aid's reliability was shown to improve reliance in Experiment 1. This information was provided through instruction and not through a human-machine interface (HMI); therefore, it is prudent to consider how the means of communicating reliability information might affect the appropriateness of reliance.

Reliability or uncertainty information is often probabilistic, and can be conveyed through numeric (e.g., 80 percent certain) or linguistic (e.g., unlikely, very likely) indicators with little to no difference in judgment performance (Budescu, Weinberg, and Wallsten, 1988). Numeric values can also be displayed in an analogue (e.g., a pie chart) or digital form. Finger and Bisantz (2002) proposed using an icon that degraded (becoming more pixilated) as the reliability decreased. In experiments using the degraded icons, no difference in target identification performance was observed between seemingly more 'precise' methods (e.g., digital-numeric) and the degraded icon (Finger and Bisantz, 2002; Bisantz, Marsiglio, and Munch, 2005).

The prospect of presenting reliability feedback along with the identification feedback raises another issue. Should these two sources of feedback be displayed separately or use an integrated multi-element display? The proximity compatibility principle (PCP; Wickens and Carswell, 1995) suggests that integrating the two pieces of information may assist the user in using them simultaneously in a time-pressured environment. However, PCP also cautions that integrating information can inhibit the focused attention required to determine a precise value. Therefore, the participant may not be able to glean the reliability information as accurately when using an integrated display. Our second experiment sought to tease out these potential advantages and disadvantages for the aided CID task.

Four prototypes were developed to display the inquiry feedback and the reliability of the feedback (see Figure 16.3). The prototypes differed in the method used to display reliability information as well as whether the feedback reliability was integrated or separated from the inquiry feedback. The first method to display reliability information, dubbed the mesh display, used a degrading stimulus. The second method, the pie chart display, used an analogue graphical representation.

The objectives of Experiment 2 were to determine: (1) which method of displaying feedback reliability information affords the best performance and most appropriate reliance on the CID system; and (2) whether integrating or separating feedback reliability and feedback identification information affords the best performance and most appropriate reliance on the CID system.

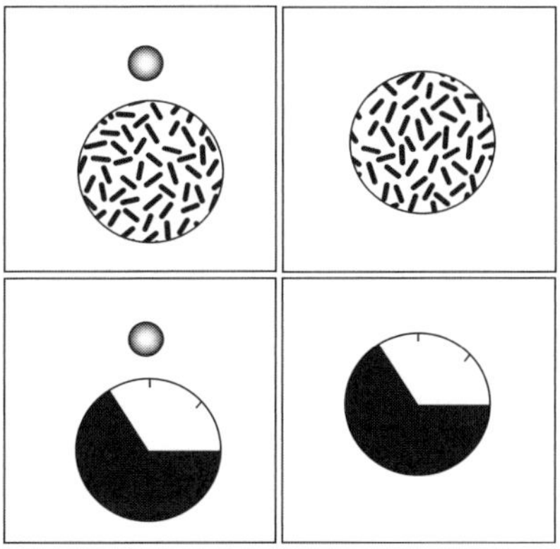

Figure 16.3 Mesh display: Separated and integrated and pie display versus separated and integrated

Experimental Design

A 2 (display type: pie, mesh) × 2 (display proximity: integrated, separated) × 5 (reliability level: 50 percent, 60 percent, 70 percent, 80 percent, 90 percent) mixed design was used. Display proximity and reliability level were within subjects factors, while display type was a between subjects factor. In other words, each participant experienced one of the display types in both an integrated and separated format at five different reliability levels. Each between-subjects group consisted of 14 participants.

The participants completed 840 trials, separated into eight blocks of 105 trials each. In four blocks the participants were presented with a separated HMI and in the other four blocks with an integrated HMI (the order was counterbalanced between participants). For each block, the targets in half of the trials were friendly and in the other half were hostile. The reliability levels varied randomly within block.

Data Collection

Participants Thirty University of Toronto students with normal acuity vision were recruited. Complete data were collected from 28 participants and used in the analysis. Participants were paid $40 CAD for their participation with a bonus of $10 CAD for 'good' performance.

Task and procedures The procedure was similar to that in Experiment 1 with one difference; in Experiment 2 participants were required to kill (as opposed to shoot at) a target if they considered it hostile to receive a score. This change was made to mimic the time pressure of an actual engagement. Given a fixed time, the participants had a better chance of killing a target if they could make a decision earlier. The trial was counted as a miss if the participants shot at but did not kill the target.

Data analysis As in Experiment 1, probability data was transformed using an arcsine function. A 2 (display proximity: integrated, separated) × 2 (display type: pie, mesh) × 5 (reliability level: 50 percent, 60 percent, 70 percent, 80 percent, 90 percent) mixed ANOVA was conducted on the transformed FA rate, transformed miss rate, time to kill a target, and *d*'. The time to kill the target was measured from when the target first appeared to when the participant successfully killed the target.

Results

False alarm (friendly fire) The main effects of display proximity and display type on false alarm rate were not significant. However, the main effect of reliability level was significant, $F(2.20, 5.67) = 21.3$, $p < 0.001$, $r = 0.41$, such that the FA rate increased with the reliability level. That is, when the display indicated an increased probability of an enemy soldier with unknown feedback, participants killed more friendlies. No interaction was significant.

Miss (miss hostile targets) There was a main effect of display type on miss rate, $F(1, 26) = 4.31$, $p <0.05$, $r = 0.38$ (Mesh: M = 0.12, SD = 0.14; Pie: M = 0.15, SD = 0.11). Participants who were shown the pie display missed more hostile targets than participants who were shown the mesh display. The main effect of reliability level was also significant, $F(1.54, 40.2) = 24.4$, $p < 0.001$, $r = 0.44$, with miss rate decreasing as reliability level increased (Figure 16.4). The main effect of display proximity was not significant, $F < 1$, nor was any interaction.

Time to kill target The main effect of reliability level had a significant effect on kill time, $F(4, 104) = 10.9$, $p < 0.001$, $r = 0.31$, with the kill time decreasing as the reliability level increased. The main effects of display proximity and display type were not significant, nor was any interaction.

Sensitivity d' The effect of display type on sensitivity was significant $F(1,22$[3]$) = 6.62$, $p < 0.05$, $r = 0.48$. The participants using the mesh display were better able to distinguish the enemy soldiers from friendly soldiers (see Figure 16.5). None of the other main effects (display proximity or reliability level) or interactions were significant.

3 If a participant had zero hits or FAs for a reliability level, the SDT measures could not be calculated for that level. At high reliability levels, corrections (such as the one suggested by Macmillan and Creelman [2004]) produced unstable estimates of sensitivity and the decision criterion and therefore were not used.

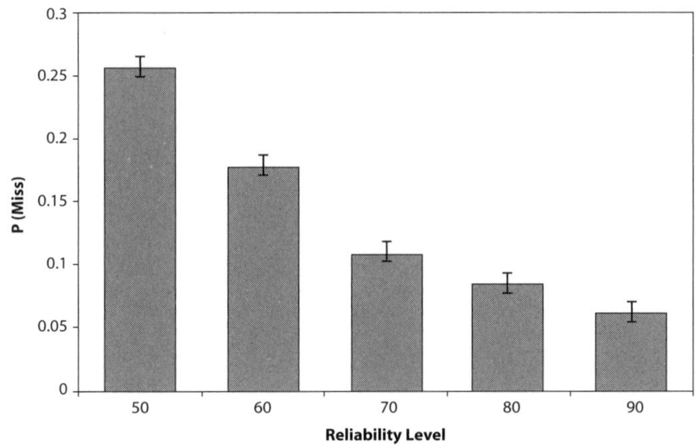

Figure 16.4 The main effect reliability level on miss rate

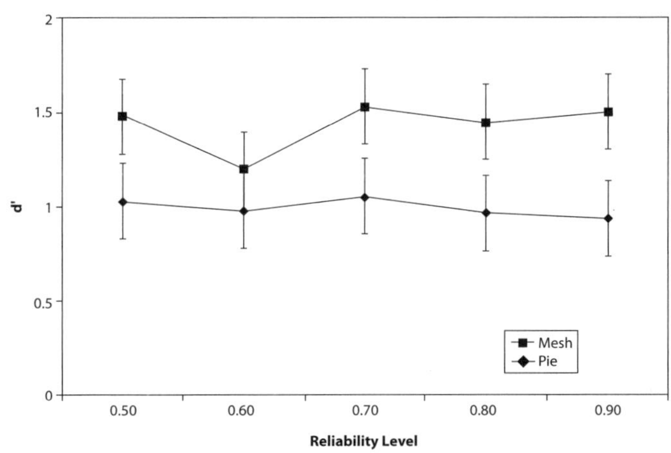

Figure 16.5 Main effect of display type on sensitivity, *d'*

Reliance To analyze the appropriateness of the participant's reliance on the aid given different displays, we fit a function relating the reliability level to the $\beta_{optimal}$ to the participants' β_{actual} for each combination of display proximity and display type. $\beta_{optimal}$ is equal to a cost function multiplied by the probability of noise divided by the probability of a signal occurring:

$$\beta_{\text{optimal}} = (V(CR) + C(M))/(V(H) + C(FA)) \times P\,(\text{friend}|\text{unknown feedback})/\,P\,(\text{enemy}|\text{unknown feedback})$$

In the present experiment when the participants received unknown feedback they killed 90 percent of the targets they shot at. Therefore, even if the participants correctly identified an enemy, only 90 percent of the time were they credited for the correct identification (as per instructions), which changed the value of a hit. Therefore: $V = \frac{V(CR)+C(M)}{V(H)+C(FA)} = \frac{1+0}{1\times 0.9+0} = 1.11$. The probability that the target is an enemy given unknown feedback is equivalent to the reliability level (RL), and the probability that the target is a friend given unknown feedback is therefore *1 – RL.* The equation for $\beta_{optimal}$ can therefore be rearranged to: $\beta_{optimal} = V\left(\frac{1-RL}{RL}\right) = V\left(\frac{1}{RL}-1\right)$. This non-linear inverse equation for $\beta_{optimal}$ was fit to the β_{actual} values for each combination of display type and display proximity. The parameter *V* for each combination was compared to the optimal *V* of 1.11 and R^2 was calculated. Because this was a constrained model, it was possible for R^2 to be negative if a horizontal line was a better fit to the data (see Table 16.2).

Both integrated displays had positive R^2 values and a *V* parameter close to the optimal value of 1.11. Both separated displays had negative R^2 values, indicating that a horizontal line was a better fit to β_{actual} than the optimal equation. Therefore, participants relied on the separated displays less optimally (see Figure 16.6).

Table 16.2 *V* and R^2 values for each display combination when the $\beta_{optimal}$ equation was fitted to β_{actual}

Display	V	R^2
Integrated-Pie	1.30	0.24
Integrated-Mesh	1.20	0.19
Separated-Pie	1.30	-0.04
Separated-Mesh	0.85	-0.27

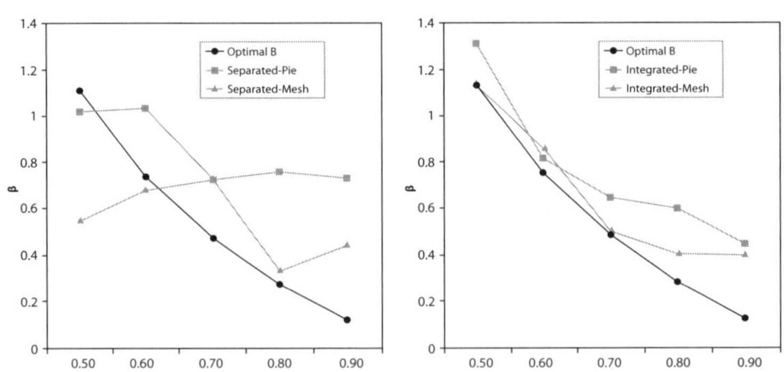

Figure 16.6 Comparison of $\beta_{optimal}$ to β_{actual} for separated and integrated displays

Discussion

Participants adjusted their reliance on the aid when the reliability information was displayed. However, the form of the display affected how this information was used to adjust reliance and identify targets.

Effect of Display Type

Unexpectedly, participants using the mesh display (whether integrated or separated) showed greater sensitivity in discriminating enemy soldiers from friendly soldiers as displaying reliability information was expected to affect only the response criterion.

The participants judged the presence of a target by examining both the display and the soldier in the environment. Because the targets remained identical, if we assume these form a cohesive signal, a change in the salience of the displays offers one possible source for the difference in sensitivity; that is, for a more salient display the participants would demonstrate greater sensitivity.

A more nuanced explanation considers the display and the target as separate channels. It is possible that, instead of these two information sources creating a cohesive signal, they actually compete for the participants' attention. Following this reasoning, a salient or distracting display could draw the participants' attention away from the target, which contains the veridical identity, thereby decreasing sensitivity. Consistent with this hypothesis, the group using the pie display not only had lower sensitivity than the mesh display group but also had a higher miss rate with no decrease in FA rate.

The mesh display also contained an emergent feature of the graphical element dimming and degrading as the reliability level decreased, which may have assisted the participants in determining the reliability information, perhaps without even looking directly at the display. In addition, multiple studies have found that individuals perform more quickly and accurately with mesh-like displays as compared with pie displays and well as numeric or linguistic indicators (Finger and Bisantz, 2002; Bisantz et al., 2005; Feldman-Stewart, Brundage, and Zotov, 2007). Because kill time did not differ between conditions, this left more time for participants to examine the target. Increasing stimulus exposure duration has been shown to increase sensitivity (Wickens and Hollands, 2000). The pie group may have spent more time sampling the display information, leaving them less time to sample the target.

Effect of Display Proximity

Participants were particularly insensitive to changes in reliability level when they were shown the separated-mesh display–failing to shift their decision criterion with the reliability level. Instead, participants maintained a decision criterion that was slightly more liberal than optimal for a reliability level of 70 percent, regardless of the displayed reliability level. It is worth noting that the mean reliability level across trials was 70 percent. It may be that separating the information was so detrimental that the participants ignored the feedback reliability information altogether. It is clear that the participants still used the aid for the inquiry feedback even while ignoring the reliability information because they held fire during the 'blue light' friend trials and activated the aid on most if not all trials. As discussed above, the participants have to consult two sources of information, both the aid and the stimulus in the time pressured scenario. Separating the information may have effectively created another information channel that the participants sometimes chose to disregard while under time pressure to kill the target. When the feedback reliability information was integrated with the feedback itself, it is possible the participant could more easily access the information while determining the results of the inquiry feedback.

Conclusion

Using a simulated CID environment and a new reliability assessment method, we showed that providing participants with a CID aid reduced the number of friendly fire engagements. Providing 'unknown' feedback reliability information led participants to more appropriately rely on the automated decision aid, whether this information was provided through verbal instruction or displayed in an interface. Furthermore, the form of the interface influenced the effectiveness of the information conveyed by affecting both participants' sensitivity in detecting the targets and reliance on the aid.

References

Bisantz, A. M., Marsiglio, S. S., and Munch, J. (2005). Displaying Uncertainty: Investigating the Effects of Display Format and Specificity. *Human Factors, 47,* 777–796.

Briggs, R. W., and Goldberg, J. H. (1995). Battlefield recognition of armored vehicles. *Human Factors, 37,* 596–610.

Budescu, D. V., Weinberg, S., and Wallsten, T. S. (1988). Decisions based on numerically and verbally expressed uncertainties. *Journal of Experimental Psychology: Human Perception and Performance, 14,* 281–294.

Dzindolet, M. T., Pierce, L. G., Beck, H. P., Dawe, L. A., and Anderson, B. W. (2000). Misuse of an automated decision making system. *Proceedings of Human Interaction with Complex Systems* (pp. 81–85). Urbana-Champaign, IL:IEEE.

Dzindolet, M. T., Pierce, L. G., Beck, H. P., Dawe, L. A., and Anderson, B. W. (2001a). Predicting misuse and disuse of combat identification systems. *Military Psychology, 13,* 147–164.

Dzindolet, M. T., Pierce, L., Pomranky, R., Peterson, S., and Beck, H. (2001b). Automation reliance on a combat identification system. *Proceedings of the 45th Annual Meeting of the Human Factors and Ergonomics Society* (pp. 532–536). Minneapolis/St. Paul, MN: Human Factors and Ergonomics Society.

Feldman-Stewart, D., Brundage, M.D. and Zotov, V. (2007) Further insight into the perception of quantitative information: Judgments of gist in treatment decisions. *Medical Decision Making, 27,* 34–43.

Finger, R. and Bisantz, A.M., (2002) Utilizing graphical formats to convey uncertainty in a decision making task. *Theoretical Issues in Ergonomics Science, 3,* 1–25.

Howell, D. C. (1992). *Statistical Methods for Psychology* (3rd ed.). Belmont, CA: Duxbury Press.

Karsh, R., Walrath, J. D., Swoboda, J. C., and Pillalamarri, K. (1995). *Effect of Battlefield Combat Identification System Information on Target Identification Time and Errors in a Simulated Tank Engagement Task* (Technical report ARL-TR-854). Aberdeen Proving Ground, MD: Army Research Lab.

Lee, J. D., and See, K. A. (2004). Trust in automation: Designing for appropriate reliance. *Human Factors, 46,* 50–80.

Lowe, C. (2007). Cutting through the fog of war. Retrieved Aug 20th, 2007, from http://www.defensetech.org/archives/003496.html.

Macmillan, N. A., and Creelman, C. D. (1991). *Detection Theory: A User's Guide.* New York: Cambridge University Press.

Parasuraman, R., Molloy, R., and Singh, I. (1993). Performance consequences of automation-induced 'complacency'. *International Journal of Aviation Psychology, 3,* 1–23.

Parasuraman, R., and Riley, V. (1997). Humans and automation: Use, misuse, disuse, abuse. *Human Factors, 39,* 230–253.

Roth, H. L., Lora, A. N., and Heilman, K. M. (2002). Effects of monocular viewing and eye dominance on spatial attention. *Brain, 125,* 2023–2035.

Sherman, K. (2000). Combat identification system for the dismounted soldier. *Proceedings of SPIE 2000: Digitization of the Battlespace V and Battlefield Biomedical Technologies II.* Orlando, FL: the International Society for Optical Engineering.

Sherman, K. B. (2002). Combat ID coming for individual soldiers. *Journal of Electronic Defense, 25,* 34–35.

SIMLAS: Manworn combat training and ID system. (2006). RUAG Electronics. Retrieved October 23, 2006, from http://www.ruag.com/ruag/binary?media=99587andopen=true.

Wickens, C. D., and Carswell, C. M. (1995). The proximity compatibility principle: Its psychological foundation and relevance to display design. *Human Factors, 373,* 473–494.

Wickens, C. D., and Hollands, J. G. (2000). *Engineering Psychology and Human Performance* (3rd edn). Upper Saddle River, NJ: Prentice Hall.

Winer, B. J., Brown, D. R., and Michels, K. M. (eds), (1991). *Statistical Principles in Experimental Design* (3rd edn). New York: McGraw-Hill.

Zari, M. C., Zwilling, A. F., Fikes, J. W., Hess, D. A., Ward, R. N., Anderson, C. S., and Chiang, D. J., (1997). Personnel identification system utilizing low probability-of-intercept techniques: prototype development and testing. *Proceedings of IEEE 1997 on Security Technology* (pp. 224–230). Canberra, Australia: IEEE.

Chapter 17

The Effects of Automation Bias on Operator Compliance and Reliance

Stephen Rice
New Mexico State University

Krisstal Clayton
Western Kentucky University

Jason McCarley
University of Illinois

Introduction

Automation has been described as the use of machines to augment or replace human activity (Wickens and Hollands, 2000). Over the past few decades, automation has become more pervasive in our society, to the point where it is virtually impossible to get through a day without the benefit of automated aids. Automation is used in both the home (alarm clock) and the office (palm pilot), with the goal of reducing human workload and allowing people to focus their cognitive resources on other tasks.

Unfortunately, the spread of automation through our everyday lives has outpaced the development of a theoretical understanding of human-automation interactions (Young, 1969). Although some progress has been made in developing these theories of automation (e.g., Bainbridge, 1982; Lee and Moray, 1994; Parasuraman and Riley, 1997), much remains to be done if we are to fully understand the ramifications of the spread of automation. More particularly, it would be desirable to explicate precisely how the spread of automation can both benefit and harm overall human-automation performance. Although automation can improve human performance, past research indicates that automated aids can also lull human operators into complacency or lead them into error (e.g., Dixon and Wickens, 2006; Parasuraman and Riley, 1997; Wickens and Dixon, 2007). It is not safe to assume that automation will improve performance (especially since most automation enables the possibility of human error), or that humans should blindly adopt the assistance of each new aid that is offered to them. The design of automation that truly benefits human users requires solid research that keeps pace with emerging technologies.

Two topics of recent concern have been the influence of automation on the control of perceptual and cognitive resources (Wickens and Holland, 2000), and the potential for automation to relieve these resources so that the operator can reallocate them to another task. Presumably, if automation can take full responsibility for a given task, then the human operator can ignore that task and focus his or her attention on another, thereby allowing the two tasks to be performed in parallel (Dixon, Wickens, and Chang, 2005). One area where this could be of great benefit is in visual search, a task common to the combat arena and specifically to unmanned aerial vehicle (UAV) environments, where the payload operator must search photographic images for enemy targets (e.g., Yeh and Wickens, 2001; Maltz and Shinar, 2003). Visual search involves both perceptual and cognitive resources and can be highly demanding, particularly in combat identification tasks, where targets may be very small and/or camouflaged. The visual search paradigm is also useful in that it offers an opportunity to generalize research findings to other paradigms that involve human-automation interaction.

Automation has been described as having four stages, including information synthesis, diagnosis, selection, and execution (Parasuraman, Sheridan, and Wickens, 2000). For the current purposes, we limit our interest to the consideration of diagnostic automation, frequently referred to as Stage 2 automation (Dixon, Wickens, and McCarley, 2007). Diagnostic automation provides the human operator with an assessment of the state of an environment; it does not actually decide what to do, but instead offers a diagnosis to inform the operator's choice of action. For example, in combat identification, a diagnostic aid might alert the operator when it determines that an enemy target is present within a particular scene, leaving it to the operator to choose the appropriate response. Thus, after receiving the diagnosis, the operator can choose to accept it immediately (e.g., if under time pressure), attempt to confirm it by checking the raw data, or simply ignore it. Clearly, this decision is made more difficult when the automation is known to be prone to errors. Intuitively, the more errors the automation makes, the less likely the human operator will trust it (Dixon and Wickens, 2006; Wickens and Dixon, 2007), to the point where an abundance of errors might cause the human operator to ignore the aid altogether (Breznitz, 1983; Sorkin and Woods, 1985).

Note that within the framework of signal detection theory (MacMillan and Creelman, 2005), the errors committed by a diagnostic system can take either of two forms: false alarms (FA) or misses. An automation FA occurs when the aid mistakenly diagnoses the presence of a target when none are actually present, whereas a miss occurs when the aid concludes that a target is absent when it is actually present. Although it is tempting to assume that these two types of errors cause equivalent harm, this is not the case. Data suggest not only that automation FAs cause more overall harm than misses (Bliss, 2003), but also that the two forms of error in fact have qualitatively different effects on operator trust and dependence (Dixon and Wickens, 2006; Maltz and Shinar, 2003; Meyer, 2001, 2004; Wickens and Dixon, 2007).

Meyer (2001, 2004) proposed that automation FAs degrade compliance, which he defined as the operator's willingness to act on a target-present diagnosis using the aid. In contrast, automation misses affect what Meyer called reliance, which he defined as the operator's tendency to trust the automation when it is silent, or otherwise indicates that there is no target. In its strongest form, this model essentially holds that FAs affect one underlying cognitive dimension (i.e., $\text{Trust}_{\text{Alert}}$) that determines the operator's responses to all automation hits and FAs, whereas misses affect a different cognitive dimension (i.e., $\text{Trust}_{\text{Non-alert}}$) that determines the operator's responses to automation correct rejections (CRs) and misses. This model is presented in Figure 17.1 (see Dunn and Kirsner, 1988 for a summary of the single-process and two-process models).

The model in Figure 17.1 predicts that compliance and reliance are independent of each other, and that the effects of automations misses and FAs are fully selective; FAs do not affect reliance, and misses do not affect compliance. Data appear to disconfirm this strong form of the compliance/reliance distinction, thereby indicating that automation FAs reduce operator reliance as well as compliance (Dixon and Wickens, 2006; Dixon, Wickens, and McCarley, 2007; Wickens, Dixon, Goh, and Hammer, 2005). Figure 17.2 presents this model. Furthermore, some research has shown that automation misses can also compromise both reliance and compliance (Rice and McCarley, 2008), which would be represented by Figure 17.3.

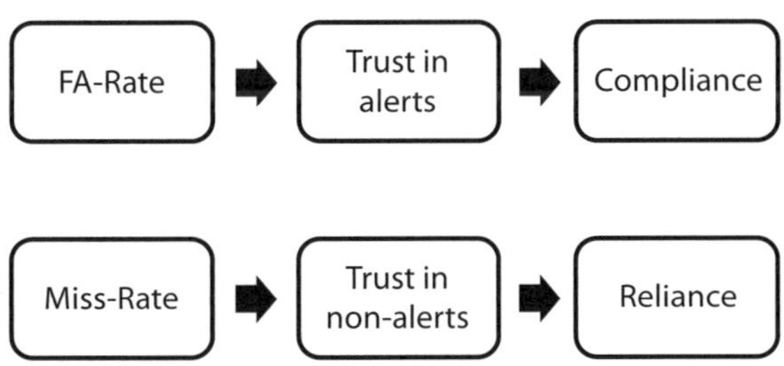

Figure 17.1 A selective two-process model

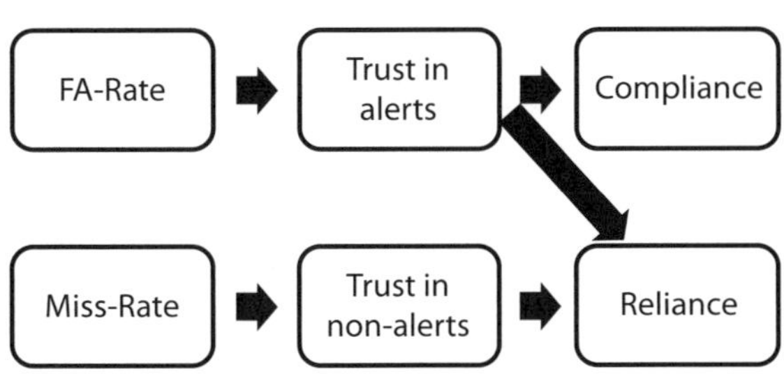

Figure 17.2 Mandler's two-process model

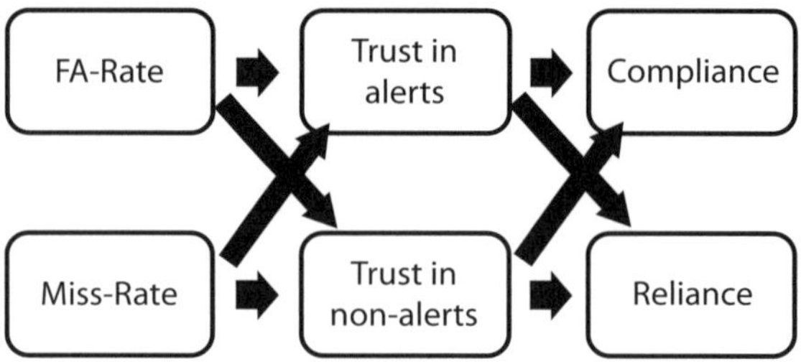

Figure 17.3 Non-selective two-process model

It remains an open question, though, whether compliance and reliance reflect independent psychological processes. One might assume from the non-selectivity in the Dixon and colleagues studies that FAs and misses affect a single cognitive process, as represented by Figure 17.4 which in turns regulates operator compliance and reliance. However, it is alternatively possible that $\text{Trust}_{\text{Alert}}$ and $\text{Trust}_{\text{Non-alert}}$ indeed exist as independent psychological dimensions, but that each of these dimensions is affected by both forms of automation error (see Figure 17.3).

If the influence of automation misses and FAs on $\text{Trust}_{\text{Alert}}$ and $\text{Trust}_{\text{Non-alert}}$ were simply weighted differently, the result would be a pattern like that described in Figure 17.3, with FAs degrading compliance more than reliance, and with misses degrading reliance more than compliance. It is the intention of the current study to address these questions with the use of state trace analysis.

In a state trace analysis, one dependent variable is plotted as a function of another. In the current study, for instance, a measure of compliance was plotted as a function of a measure of reliance. Of interest is the form of the relationship between the dependent variables if both are mediated by a single, common underlying mental value. Under the very weak assumption of a monotonic relationship between the underlying value and an observable variable, therefore, any manipulation that increases the underlying value would also increase the value of both dependent variables. Consequently, the relationship between the dependent measures will

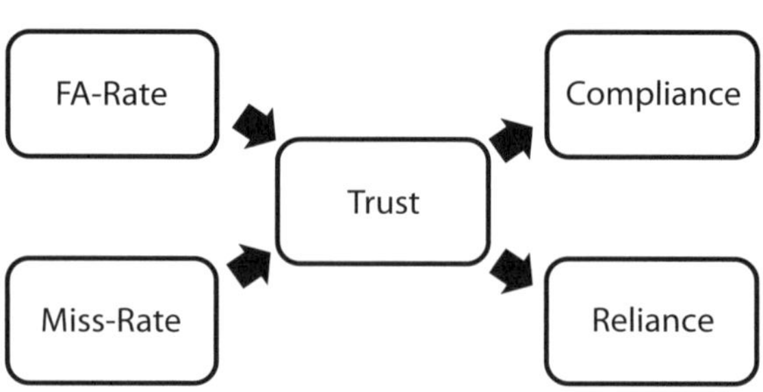

Figure 17.4 Single-process model

itself be monotonic. A non-monotonic relationship, sometimes called a reversed association (Dunn and Kirsner, 1988), disconfirms a model based on a single underlying dimension, demonstrating instead that at least two underlying mental values are necessary to account for variation in the dependent measures (Bamber, 1979; Loftus, Oberg, and Dillon, 2004; see the appendix of Harley, Dillon, and Loftus, 2004 for a brief introduction to state trace analysis).

Current Study

Participants in the current study were required to search aerial images of Baghdad for the presence of a designated target item (an enemy tank). As they performed the task they were assisted by a diagnostic aid that provided an assessment on each trial as to whether or not a target was present. The aid was imperfectly reliable, however, and the operator was free to agree or disagree with the aid as was felt appropriate. The bias and the reliability of the aid were both manipulated, such that the aid either could be FA-prone or miss-prone, with reliability levels of 95 percent, 90 percent, 85 percent, 80 percent, 75 percent, 70 percent, 65 percent, 60 percent, 55 percent, or 50 percent. Operators were informed of the aid's reliability and response bias (miss-prone or FA-prone) before beginning the search task. Furthermore, they were given feedback about their accuracy and RT after each trial. We predicted that (a) more reliable automation would result in higher agreement rates and quicker RTs; (b) less reliable automation would result in lower agreement rates and longer RTs; (c) and state trace analysis would reveal a non-monotonic function, which would indicate that a multiple-dimensional cognitive model is needed to explain the data.

Method

Participants

Participants were 400 New Mexico State University undergraduates (241 females) who received course credit. The mean age was 20.3. Participants were screened for normal or corrected-to-normal visual acuity and color vision.

Apparatus and Stimuli

Stimuli were displayed on a Dell computer with a 20" monitor using 1024×768 resolution and a refresh rate of 60 Hz. A set of 100 aerial photographs of Baghdad was created using GoogleEarth. These 100 unaltered images served as target-absent stimuli. Target-present stimuli were created by digitally inserting an image of a tank into each target-absent photograph which was approximately 2 degrees by 2 degrees in visual angle. Thus, there were 100 target-present and 100 target-absent images.

Procedure and Design

Participants began by signing a consent form and reading on-screen instructions. The instructions included a picture of the tank and information about the bias and reliability of the automation. Instructions asked the participants to be as accurate as possible in conducting their search task without wasting time. Once participants were comfortable with the instructions, they pressed a key to begin the experiment.

Each trial began with a fixation display, whereby participants were instructed to look at the fixation cross for 1000 ms. This display was then replaced with a display providing the automation's recommendation for that trial. This display read either, 'The automation has detected a tank!' or 'The automation has determined that there is no tank!' After 1500 ms, this display was replaced with a stimulus image, which remained until the participant made a response. Responses were made by either pressing the J key for target-present or the F key for target-absent. Following this, a feedback display appeared reporting the participant's accuracy and RT for that trial, along with a measure of cumulative accuracy.

As they performed the task, some participants were aided by a diagnostic aid which provided recommendations before each trial. The aid had a reliability level of 95 percent, 90 percent, 85 percent, 80 percent, 75 percent, 70 percent, 65 percent, 60 percent, 55 percent, or 50 percent and was either FA-prone or miss-prone. Participants were informed of the reliability of their automation before the experiment began. Thus, there were a total of 20 conditions. Twenty participants were randomly assigned to each condition in a between-subjects design.

Results

Sensitivity

The signal detection measure of sensitivity, *d'*, was used to measure participants' ability to discriminate target-present from target-absent images. Data are presented in Figure 17.3. A two-way ANOVA with Bias and Reliability as factors indicated that performance increased as the reliability of the automated aid increased, $F(8, 342) = 8.81$, $p < .001$, but showed no reliable effect of Bias, $F < 1.0$, nor a reliable interaction of Bias and Reliability, $F < 1.0$, $p > .05$.

Bias

The signal detection measure *C*—where $C = \frac{z(HR) + z(FAR)}{2}$, and *z(HR)* and *z(FAR)* are the z-scores for hit rates and false alarm rates, respectively—was used to analyze participants' response bias. A two-way ANOVA with Bias and Reliability indicated that participants in the FA-prone conditions (M = -0.05), had a more liberal bias than they did in the miss-prone conditions (M = 1.11), $F(1, 342)$ =

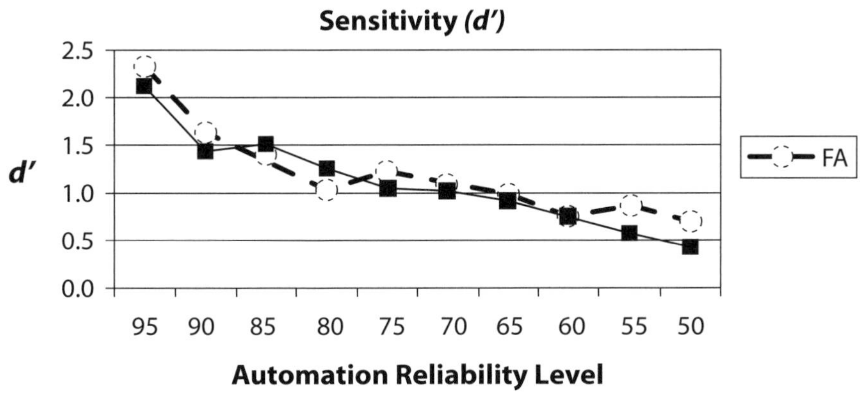

Figure 17.5 ***d'* as a function of bias and reliability**

20.00, $p < .001$; that is, they were more likely to say that there was a target present, independent of their actual performance sensitivity. There was no significant main effect of Reliability on response bias, $F(8, 342) = 1.84$, $p > .05$, nor was there an interaction between Bias and Reliability, $F(8, 342) < 1.0$, $p > .05$.

Agreement Rates and Response Times (RTs)

Agreement rates and RTs were used as a measure of trust in the automation. We assumed that when participants trust the automation, they will quickly agree with the aid. Thus, the following analyses investigated how often participants agreed with the aid and how quickly they did so.

Agreement Rates (Compliance)

Compliance agreement rates refers to how often participants agreed with the aid when it indicated that a target was present. A two-way ANOVA with Bias and Reliability as factors performed on data from the imperfect automation conditions revealed a main effect of Bias, $F(1, 342) = 32.10$, $p < .001$, and a main effect of Reliability, $F(8, 342) = 4.32$, $p < .001$, with no significant interaction, $F(2, 66) = 1.35$, $p > .05$. These effects indicate that participants in the FA-prone conditions were less likely than those in the miss-prone conditions to agree with the automation when it judged that a target was present. Participants in both the FA-prone conditions and miss-prone conditions were less likely to comply with the automation when the reliability of the aid was low than when it was high.

Agreement Rates (Reliance)

Reliance agreement rates refers to how often participants agreed with the aid when it indicated that a target was absent. A two-way ANOVA with Bias and Reliability as factors performed on data from the imperfect automation conditions revealed a main effect of Bias, $F(1, 342) = 33.89, p < .001$, but no main effect of Reliability, $F(8, 342) = 1.51, p > .05$, and no interaction, $F(2, 66) = 1.91, p > .05$. These results indicate that participants in the miss-prone conditions were less likely than those in the FA-prone conditions to agree with the automation when it judged that a target was absent.

Response Times (Compliance)

Compliance response times refers to how quickly participants agreed with the aid when it recommended that a target was present. A two-way ANOVA with Bias and Reliability as factors in the imperfect automation conditions revealed a main effect of Bias, $F(1, 342) = 30.58, p < .001$, no main effect of Reliability, $F(8, 342) < 1.0$, $p > .05$, and no significant interaction between Bias and Reliability, $F(2, 66) = 1.21, p > .05$. These data indicate that participants in the FA-prone conditions were slower to agree with the automation when it determined that a target was present, relative to participants in the miss-prone conditions.

Response Times (Reliance)

Reliance response times refers to how quickly participants agreed with the aid when it recommended that a target was absent. A two-way ANOVA with Bias and Reliability as factors in the imperfect automation conditions revealed a main effect of Bias, $F(1, 342) = 30.85, p < .001$, with no main effect of Reliability, $F(8, 342) < 1.0, p > .05$, and no significant interaction between Bias and Reliability, $F(2, 66) < 1.0, p > .05$. These data indicate that participants in the miss-prone conditions were slower to agree with the automation when it determined that a target was present, relative to participants in the FA-prone conditions.

State Trace Analyses

The pattern of non-selective effects in the behavioral data (above) indicate that FA-prone automation disrupted operator compliance more than reliance, while miss-prone automation disrupted operator reliance more than compliance. This pattern of effects would be difficult to account for with a single-process model in which automation dependence is regulated by a unitary underlying construct such as a general level of trust. The data thus appear to support a multiple-process model in which different measures of trust regulate compliance and reliance behavior. State trace analyses were conducted to confirm this interpretation. For these analyses, agreement rates and RTs were used, as seen in Figure 17.6.

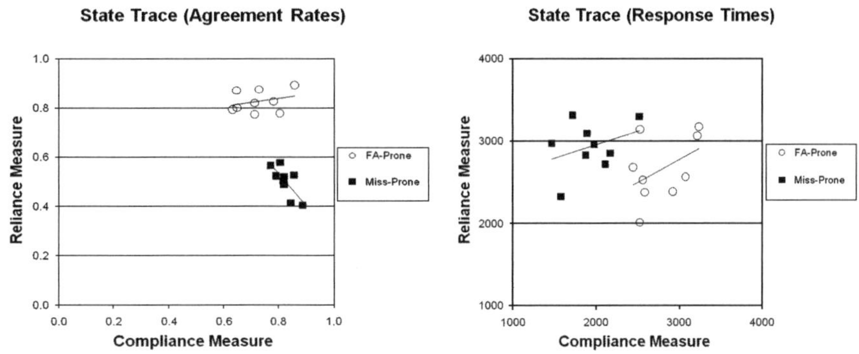

Figure 17.6 State Trace Analyses on a) Agreement Rates (%); and b) RTs (sec). Solid squares indicate miss-prone conditions, while clear circles indicate FA-prone conditions. Trend-lines are included

The figures clearly reveal a non-monotonic relationship between the dependent variables. This provides strong evidence against a single-process account of automation dependence. FA-prone and miss-prone automation appear to affect at least two different cognitive processes, which then in turn differentially affect operator behavior.

Discussion

Benefits of Automation

As expected, highly reliable automation increased human performance in comparison to automation that was less reliable. Although past research has found that FA-prone automation causes a decrease in human-automation performance in comparison to miss-prone automation (e.g., Maltz and Shinar, 2003; Dixon, Wickens, and McCarley, 2007), no such asymmetry was demonstrated in the current data. Instead, *d'* and RTs were similar across the two bias manipulations.

Compliance and Reliance

Using operator agreement rates and RTs as an index of the effects of error-prone automation on operator compliance and reliance demonstrated that automation reliability did indeed have strong effects on operator behavior. The results revealed that the effects of FA- and miss-prone automation on operator compliance were asymmetrical but not fully independent. FA-prone automation severely compromised operator compliance but also produced a weak decrease in reliance, while miss-prone automation strongly degraded reliance but simultaneously reduced compliance as well. Therefore, these results argue against the strongest form of a two-process model of automation dependence (Maltz and Shinar, 2003; Meyer, 2001, 2004).

Regardless of these results, it is important to mention that non-selective effects of automation FAs and misses do not necessarily disconfirm the compliance/reliance distinction. They can be easily reconciled, rather, with a two-process model like that of Figure 17.3, in which two distinct forms of trust influence dependence and reliance and behavior. State-trace analyses conducted over measures of reliance and compliance produced results consistent with this possibility, demonstrating a non-monotonic relationship between measures. This non-monotonic relationship precludes a model, like that of Figure 17.4, which holds that both types of automation errors affect a single cognitive process that in turn affects operator compliance and reliance behaviors. As such, the data in total indicate that the multi-trust model which fits the current data most closely is the generalized multiple-process model illustrated in Figure 17.3. In the current task, as described in the Introduction, the model would hold that automation errors affect two separate cognitive processes. We refer to these two separate cognitive processes as $\text{Trust}_{\text{Alert}}$ and $\text{Trust}_{\text{Non-alert}}$ These two processes differentially influence compliance (i.e., the operator's response to an alert from the aid) and reliance (i.e., the operator's response to an 'all clear' judgment from the aid). The results provide strong support for Meyer's (2001, 2004) compliance/reliance distinction.

Conclusions

In summary, data lead to a number of conclusions. First, highly reliable automation produces better human-automation performance than does less reliable automation (Wickens and Dixon, 2007). Second, the effects of automation misses and FAs and misses on compliance and reliance are not fully selective. Although FA-prone automation strongly reduces operator compliance, it also has weak effects on operator reliance. Automation misses, conversely, strongly compromise reliance, but also reduce compliance. Third, despite the non-selective effects of automation misses and FAs on operator dependence, the two types of automation errors affect at least two different cognitive processes, as revealed by the state trace analysis.

The practical implications of these findings are that designers must be aware of the differential effects of FA-prone and miss-prone automation on human dependence and behavior. Although intuition may suggest that FA-prone and miss-prone automation have equal effects on operator dependence and performance, there is in fact an important distinction to be made between the two types of automation errors and how they affect operator dependence and behavior.

Moreover, system designers may well face a tradeoff in designing an automated system for optimal human performance. Consider, for example, a case in which the automation designer wishes to minimize the possibility that a human operator will miss a target. The designer's understandable inclination may be to establish a liberal response bias for the automated aid, ensuring that the automation maintains a high hit rate. This hit rate, however, will come at the cost of a high FA rate. An unfortunate consequence of the aid's frequent FAs, in turn, will be a decrease in the operator's willingness to comply with the aid's alerts. The designer will thus have created a situation in which the aid is likely to detect a target but the operator, ironically, is unlikely to act on the aid's alerts. In establishing the response criterion for an automated diagnostic aid, therefore, the designer's goal should not be simply to optimize behavior of the aid itself, but to elicit an optimal pattern of dependence from the aid's user.

Acknowledgements

The authors wish to thank Amy Wells, Gayle Hunt, and Jackie Chavez for their help in collecting data. This research was funded by an Air Force grant (Index #111915). Any opinions, findings, and conclusions or recommendations expressed in this paper are those of the authors.

References

Bainbridge, L. (1982). Ironies of automation. G. Johansen and J. E. Rijnsdorp (eds), *Analysis, Design and Evaluation of Man-Machine Systems* (pp. 151–157). London: Pergamon Press.

Bamber, D. (1979). State trace analysis: A method of testing simple theories of causation. *Journal of Mathematical Psychology, 19*, 137–181.

Bliss, J. (2003). An investigation of alarm related accidents and incidents in aviation. *International Journal of Aviation Psychology, 13,* 249–268.

Breznitz, S. (1983). *Cry-wolf: The Psychology of False Alarms*. Hillsdale, NJ: Lawrence Erlbaum.

Dixon, S., and Wickens, C. (2006). Automation reliability in unmanned aerial vehicle control: A reliance-compliance model of automation dependence in high workload. *Human Factors, 48*, 474–486.

Dixon, S. R., Wickens, C. D., and Chang, D. (2005). Mission control of multiple unmanned aerial vehicles: A workload analysis. *Human Factors, 47*, 479–487.

Dixon, S. R., Wickens, C. D., and McCarley, J. S. (2007). On the independence of compliance and reliance: Are automation false alarms worse the misses? *Human Factors, 49*, 564–572.

Dunn, J.C., and Kirsner, K. (1988). Discovering functionally independent mental processes: The principle of reversed association. *Psychological Review, 95*, 91–101.

Harley, E. M., Dillon, A. M., and Loftus, G. R. (2004). Why is it difficult to see in the fog? How stimulus contrast affects visual perception and visual memory. *Psychonomic Bulletin and Review, 11*, 197–231.

Lee, J. D., and Moray, N. (1994). Trust, self-confidence, and operator's adaptation to automation. *International Journal of Human-Computer Studies, 40*, 153–184.

Loftus, G. R., Oberg, M. A., and Dillon, A. M. (2004). Linear theory, dimensional theory, and the face-inversion effect. *Psychological Review, 111,* 835–865.

Macmillan, N. A., and Creelman, C. D. (2005). *Detection Theory: A User's Guide* (2nd edn). Mahwah, NJ: Lawrence Erlbaum.

Maltz, M., and Shinar, D. (2003). New alternative methods in analyzing human behavior in cued target acquisition. *Human Factors, 45*, 281–295.

Meyer, J. (2001). Effects of warning validity and proximity on responses to warnings, *Human Factors, 43*, 563–572.

Meyer, J. (2004). Conceptual issues in the study of dynamic hazard warnings. *Human Factors, 46*, 196–204.

Parasuraman, R., and Riley, V. (1997). Humans and automation: Use, misuse, disuse, and abuse. *Human Factors, 39*, 230–253.

Parasuraman, R., Sheridan, T. B., and Wickens, C. D. (2000). A model for types and levels of human interaction with automation. *IEEE Transactions on Systems, Man, and Cybernetics, 30*, 286–297.

Rice, S., and McCarley, J. (2008). The Effects of Automation Bias and Saliency on Operator Trust: *Proceedings of the International Congress of Psychology.* Berlin, Germany.

Sorkin, R. D., and Woods, D. D. (1985). Systems with human monitors, a signal detection analysis. *Human-Computer Interaction, 1*, 49–75.

Wickens, C. D., and Dixon, S. (2007). The benefits of imperfect diagnostic automation: A synthesis of the literature. *Theoretical Issues in Ergonomic Science, 8*, 201–212.

Wickens, C. D., Dixon, S. R., Goh, J., and Hammer, B. (2005). Pilot dependence on imperfect diagnostic automation in simulated UAV flights: An attentional visual scanning analysis. *Proceedings of the 13th Annual International Symposium of Aviation Psychology.* Dayton, Ohio.

Wickens, C. D., and Hollands, J. G. (2000). *Engineering Psychology and Human Performance* (3rd edn). Upper Saddle River, NJ: Prentice Hall.

Yeh, M., and Wickens, C. D. (2001). Display signaling in augmented reality: Effects of cue reliability and image realism on attention allocation and trust calibration. *Human Factors, 43*, 355–365.

Young, L. R. (1969). On adaptive manual control. *Ergonomics, 12*, 635–675.

Chapter 18
An Examination of the Social, Cognitive, and Motivational Factors that Affect Automation Reliance

Mary T. Dzindolet
Cameron University

Linda G. Pierce
Army Research Institute

Hall P. Beck
Appalachian State University

Fratricide is estimated to have caused nearly one-fifth of the deaths in the Persian Gulf War and may be responsible for even more deaths in Operation Iraqi Freedom (Greitzer and Andrews, Chapter 11, this volume). One U.S. Army's analysis of fratricide concluded that nearly one-third of the incidents were caused by target identification errors (OTA, 1993 as cited in Wilson, Salas, Priest, and Andrews, 2007). According to Wilson et al., the most common approach to reducing fratricide is the development of combat identification systems. These automated-decision aids provide soldiers with the ability to 'interrogate' a potential target by sending a signal that, if returned, identifies the target as a 'friend.' Unanswered signals produce an 'unknown' response. The underlying assumption in providing these automated aids to military personnel is that the human-computer 'team' will commit fewer fratricide errors than the human working alone. However, this assumes that the human operator will appropriately rely on the automated combat identification aid. Research has found human operators often under-utilize (disuse) or overly rely on (misuse) automated systems (cf. Parasuraman and Riley, 1997).

Ideally, human operators will rely on the combat identification system when doing so will maximize gains and/or minimize losses. Failure to rely on the aid in this situation constitutes disuse (Parasuraman and Riley, 1997). Disuse is defined as 'underutilization of automation' (Parasuraman and Riley, 1997, p. 233). Anecdotal evidence supports disuse; Parasuraman and Riley (1997) described many real-world incidences in which disastrous results occurred due to people ignoring automated warning signals. In addition, several experiments have shown that human operators disuse automated decision aids in a laboratory context (e.g., Dzindolet, Pierce, Beck, and Dawe, 2002, Study 2; Moes, Knox, Pierce, and Beck, 1999; Riley, 1996).

Alternatively, when ignoring the combat identification system will maximize gains and/or minimize losses, the human operator should ignore the aid and rely on his or her decisions. Relying on the automated aid in this circumstance would constitute misuse (Parasuraman and Riley, 1997). Misuse is defined 'as overreliance on automation' (p. 233). Parasuraman, Molloy, and Singh (1993) and Singh, Molloy, and Parasuraman (1997) found misuse among operators performing monitoring functions. They labeled this behavior complacency and defined it as 'a psychological state characterized by a low index of suspicion' (Wiener, 1981, p. 117). Misuse has also been found with automated decision aids (Dzindolet, Pierce, Beck, Dawe, and Anderson, 2001; Layton, Smith, and McCoy, 1994; Mosier and Skitka, 1996).

In order to design a combat identification system that encourages the human operator to appropriately rely on the aid (i.e., avoid disuse and misuse), it is important to view the human operator and the combat identification system as a human-system team. This paper will review some of the laboratory studies designed to understand human operator's automation use decisions in terms of the Framework of Automation Use Model (Dzindolet, Beck, and Pierce, 2006; Dzindolet, Beck, Pierce, and Dawe, 2001).

Appropriate automation reliance decisions may be particularly difficult with combat identification due to the nature of the decisions. When the hardware on the combat identification system is working properly, soldiers should rely on the automated aid's 'friendly' decisions. It is unlikely the combat identification system will make a decision of 'friendly' to an enemy. However, human operators must be careful not to interpret the automated aid's 'unknown' decision as 'enemy.' Friendlies who are not provided with the combat identification system or whose system has been damaged in battle will lead the combat identification system to make a decision of 'unknown.' An 'unknown' decision should be interpreted by the soldier as an indicator to search for more information.

The probability that a 'friendly' response identifies a friendly is likely to be very high. However, the probability that an 'unknown' response identifies an enemy is only high when all the friendlies in the area are equipped with properly functioning systems. If the hardware does not function properly or if it is lost or damaged in battle, or if a large number of allied forces not supplied with combat identification systems enter the battle area, then the probability that an 'unknown' response identifies an enemy is much lower.

Therefore, the combat identification system's reliability will, to some extent, vary based on the situation. Most likely, when the aid reaches a 'friendly' decision, the vehicle transmitting the response will be a friendly. Thus, soldiers should rely on the combat identification system. However, when the aid reaches an 'unknown' decision, the likelihood that the responding vehicle will be friendly will vary based on a number of factors. Under these conditions, soldiers should ignore the combat identification system. Unfortunately, human operators may have difficulty adjusting to the varying reliability of an automated decision aid (Dzindolet, Pierce, Pomranky, Peterson, and Beck, 2001; Parasuraman et al., 1993; Singh et al., 1997; see Wiegmann, Rich, and Zhang, 2001). Several laboratory studies indicate that sub-optimal use of the combat identification system has the potential of increasing, rather than decreasing, fratricide (Dzindolet, Pierce, Beck, et al., 2001).

Experimental Paradigm

In order to better understand this sub-optimal usage, Dzindolet, Pierce, Pomranky, et al. (2001) conducted a laboratory study to determine whether human operators would adjust their automation reliance to varying levels of automation reliability. Regardless of the condition to which the participant was assigned, they viewed 300 slides displaying pictures of Fort Sill terrain on a computer screen (see Figure 18.1).

Half of the slides contained one soldier in various levels of camouflage; the remaining slides were of terrain only. Participants were told that sometimes the soldier would be rather easy to spot; other times he would be more difficult to find. Each slide was presented on the computer screen for about ¾ of a second. After viewing each slide, the participants had the option of viewing the slide a second time or making their decision as to whether or not they believed the soldier was in the slide. Participants were given three seconds to make their decision.

Figure 18.1 Sample slide containing a soldier

Provision of an Automated Aid

Participants were randomly assigned to work with or without an automated decision aid. Some participants performed the task without an automated aid. However, other participants were told that a computer routine had been written to assist them in performing their task. They were told that the routine performed a rapid scan of the photograph looking for contrasts that suggested the presence of a human being. The decision of the contrast detector appeared on the computer screen. For the first 200 trials, the contrast detector's decision was provided to participants only after they had made their decision; participants were not able to rely on the aid's decision. These 200 trials provided the aided participants with experience with their automated aid.

However, for the last 100 trials, aided participants were given the opportunity to rely on their automated aid. For these trials, the decision reached by the aid was provided after the viewing of the slide but *before* they were given the opportunity to view the slide a second time or make their absent-present decision. We examined participants' automation reliance decisions on these 100 trials.

Provision of Training

Participants who were provided with an automated decision aid were randomly assigned to receive or not to receive training that paralleled that of combat identification systems. Specifically, participants who received training were informed that whenever the target was present, the aid was always correct. In other words, the contrast detector would always make a 'present' decision when the target was present; it would never make an 'absent' decision when the target was present. However, when the target was absent, the aid would reach the correct answer only half the time. In other words, the contrast detector would make a 'present' decision for half of the slides in which the target was absent; it would make an 'absent' decision for half of the slides in which the target was absent. The automated aid would operate no better than chance on the target-absent trials (see Figure 18.2).

During training, participants were also told that if they focused on the column in Figure 18.2 labeled 'Aid Says,' it would become clear to them that the contrast detector would never give an 'absent' decision if the target was present. Therefore, on the slides for which the automated aid reached an 'absent' decision, the probability of the aid being incorrect was zero. On such slides, the participant should rely on the decision reached by the automated aid. However, it was possible for the contrast detector to reach a 'present' decision when the target was absent. In fact, this would happen for one-third of the slides in which the target was absent (see Figure 18.3). Therefore, on the slides for which the aid reached a 'present' decision, the probability of the aid being incorrect was .33.

Participants who were randomly assigned to the condition in which they were not provided with training were not explicitly given this information. In addition, they were not given Figures 18.2 or 18.3 to examine. Were they able to figure out the reliability levels of the contrast detector when it reached a 'present' or 'absent' decision during the first 200 trials?

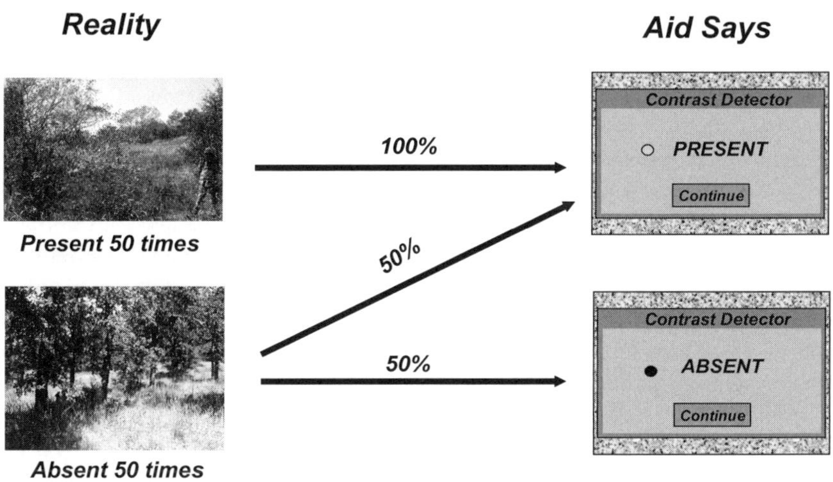

Figure 18.2 Reliability rates focused on slide

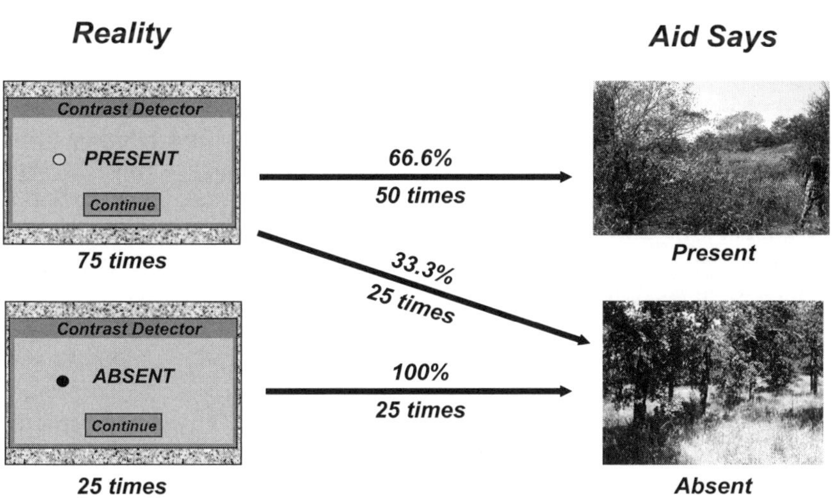

Figure 18.3 Reliability rates focused on contrast detector's decision

Results indicated that participants provided with the contrast detector were less likely than those not provided with the contrast detector to ask to re-examine a slide after the contrast detector reached a decision known to be perfectly reliable (i.e., 'absent'). However, both the trained and untrained participants were unable to adjust their reliance strategy to the reliability of the automated aid. In fact, participants provided with the automated decision aid made *more* errors than those not provided with the automated aid.

The participants who were explicitly trained that the aid was always correct when it reached an 'absent' decision but only correct $\frac{2}{3}$ of the time when it reached a 'present' decision were *not* more likely to rely on the aid when it reached an 'absent' decision than when it reached a 'present' decision. Thus, experience with the automated aid, alone, was sufficient to produce the superior re-examination strategy. Explicit training of the aid's reliability did not improve this strategy.

Clearly, an understanding of human-automation decision-making is necessary. The purpose of this paper is to apply the Framework of Automation Use (Dzindolet, Beck, et al., 2006; Dzindolet, Beck, et al., 2001) to understand the conditions which encourage soldiers to appropriately rely on combat identification systems. We believe the Framework will be useful in guiding combat identification system design and human operator training. In addition to describing the elements of the Framework, we will discuss some experimental support for the Framework from studies that utilized a paradigm similar to the Dzindolet, Pierce, Pomranky, et al. (2001) study described above.

Framework of Automation Use

What are the processes leading to appropriate use of an automated aid? Drawing from the work of Mosier and Skitka (1996) as well as the group dynamics literature, Dzindolet, Beck, et al. (2001) created the Framework of Automation Use (see Figure 18.4). The Framework hypothesizes that cognitive, social, and motivational processes combine to predict automation use. Thus, a discussion of the three processes as it relates to automation use decisions will be presented in the next three sections of this chapter.

Cognitive Processes

Mosier and Skitka (1996) hypothesized that faulty cognitive processing may lead people to overly rely on automated systems. In addition to the large body of literature examining errors due to flawed cognitive processing in individual decision-making (Tversky and Kahneman, 1973), the social cognition literature is replete with examples of less-than-ideal cognitive processing in teams or groups. Additionally, a variety of errors and biases have been identified in various domains of social psychology, such as attributing causality (Doosje and Branscombe, 2003; Jones and Nisbett, 1972; Karasawa, 1995; Ross, 1977) and impression formation (Frey and Schulz-Hardt, 2001; Karasawa, 2003). Overall, the literature suggests that people often adopt effort-saving strategies called heuristics rather than logically processing relevant pieces of information. Tversky and Kahneman (1973) suggested that people use heuristics, cognitive 'rules of thumb,' due to the fact that they quickly lead to solutions that are often reasonably close to optimal with little cognitive effort. Mosier and Skitka (1996) defined 'automation bias' as 'the tendency to use automated cues as a heuristic replacement for vigilant information seeking and processing' (p. 205). This suggests that the use of the automation bias heuristic is a cognitive process through which misuse may occur.

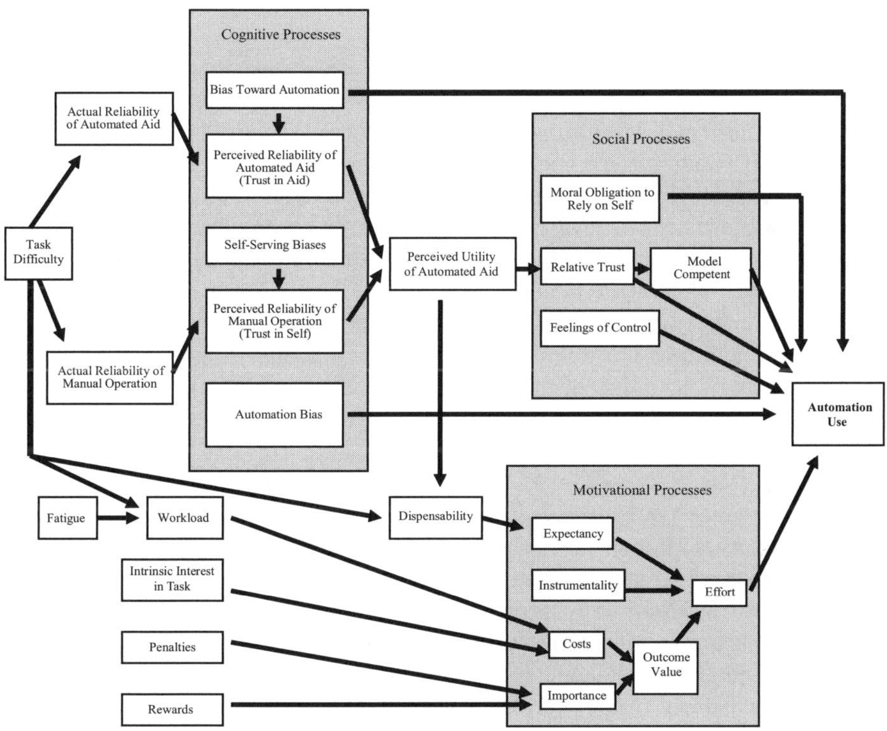

Figure 18.4 Framework of Automation Use

The fact that the automated system provides a decision may lead the decision-maker to rely on this information in a heuristic manner (see Figure 18.5). Rather than going through the cognitive effort of gathering and processing information, the information supplied by the automated systems is misused (i.e., overreliance; Mosier and Skitka, 1996; Mosier, Skitka, Dunbar, and McDonnell, 2001). Conceivably, this may occur in various degrees. In its most extreme form, the decision reached by the automated aid is immediately adopted. In a less extreme

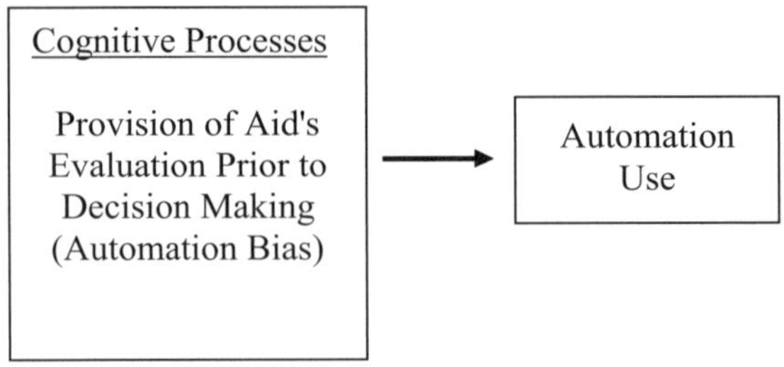

Figure 18.5 Cognitive processes

form, the decision reached by the aid may be given an inappropriately large role in the human's decision-making process. For example, Layton et al. (1994) found many pilots provided with an automated aid's poor en-route flight plan did not explore other solutions (e.g., they did not generate actual flight plans on screen) as much as pilots who were not provided with the automated aid's decision. Thus, the automation bias will lead people to rely on the automated aid. Often, this strategy will be appropriate. However, under certain conditions, this reliance may be inappropriate, leading to misuse.

Indirect evidence to support the existence of automation bias can be found by examining automation reliance across studies with paradigms similar to that described at the beginning of this chapter (e.g., Dzindolet et al., 2002, Study 2; Dzindolet, Pierce, Beck, et al., 2001).

Automation Bias is possible: View aid's decision before making decision In one of the studies (Dzindolet et al., 2001), participants were provided with the absent-present decision of the contrast detector known to be accurate either 60 percent, 75 percent, or 90 percent of the time (based on the condition to which the participant was randomly assigned) *before* making their own absent-present decision. To examine misuse and disuse, the probability of error was determined for two subsets of the data. Misuse, or overreliance on the contrast detector, was operationally defined as the p(error | aid error). Disuse, or underutilization of the contrast detector, was operationally defined as the p(error | aid correct). Analyses indicated that regardless of the reliability of the automated aid, participants were more likely to err by relying on their decision aids than by ignoring them, p(error | aid error) = .27; p(error | aid correct) = .13. Therefore, when provided with the aid's decisions first, thereby allowing the automation bias to occur, participants were more likely to misuse than disuse their automated aids.

Automation bias eliminated: View aid's decision after making decision However, in other studies (e.g., Dzindolet, Pierce, Beck, et al., 2002, Study 2), the automation bias was eliminated. Participants were *not* provided with the decisions reached by the contrast detector until *after* they had indicated their decision and their level of confidence in their decision. After participants had completed 200 trials in which they could view the automated aid's decision only after making their own decision, we told them that they could earn rewards for every correct decision made on ten trials randomly chosen from the 200 trials. Participants had to choose whether the performance would be based on their decisions or on the decisions of their automated aid. After making their choice, participants were asked to justify their choice in writing. Without the automation bias, would participants still be more likely to misuse than disuse their automated aids?

Rather than misusing the contrast detector, participants in these studies disused the automated aid. Even among participants provided with feedback that their aid's performance was far superior to their own, the majority (81 percent) chose to rely on their own decisions rather than on the decisions of the automated aid!

Therefore, when the automation bias could play a role in the decision to rely on automation, misuse occurred more than disuse. However, when the automation bias was eliminated by providing the automated aid's decisions only *after* participants recorded their decision, disuse, not misuse, was found on a subsequent task allocation decision.

Although we hypothesize that the automation bias plays a role in automation use and misuse, we do not believe it is the only important variable. By itself, it cannot account for disuse. Analyses of the justifications of the task allocation decisions provided by participants in one of the Dzindolet et al. (2002, Study 2) experiments indicated that trust might play a role. For example, nearly one-quarter (23 percent) of the participants justified their disuse by stating they did not trust the automated aid as much as they trusted themselves.

Social Processes

According to the Framework of Automation, several social factors are also important in automation use decisions. One important social factor is trust. According to Mosier and Skitka's (1996) authority hypothesis, people rely on the automated system's decision because they believe it to be more reliable, and thus place greater trust in it. Trust has been found to affect reliance in many domains, such as car's navigation system (Kantowitz, Hanowski, and Kantowitz, 1997), flight cockpits (Tenney, Rogers, and Pew, 1988), and with a teleoperated robot (Dassonville, Jolly, and Desodt, 1996). Moreover, trust has been recognized by many researchers to be important in automation reliance decisions (Bailey and Scerbo, 2007; Cohen, Parasuraman, and Freeman, 1998; Jian, Bisantz, and Drury, 2000; Lee, 2001; Lee and Moray, 1992; 1994; Lee and See, 2004; Lewandowsky, Mundy, and Tan, 2000; Liu and Hwang, 2000; Merritt and Ilgen, 2008; Moray, 2001; Moray, Inagaki, and Itoh, 2000; Muir, 1987; 1994; Rovira, McGarry, and Parasuraman, 2007; Tan and Lewandowsky, 1996; Wiegmann et al., 2001). Overly trusting an automated aid will lead human operators to misuse; lack of trust in a superior aid will lead to disuse.

According to the Framework of Automation, trust is determined from the outcome of a comparison process between the perceived reliability of the automated aid (trust in aid) and the perceived reliability of manual control (trust in self). We call the outcome of the decision process the perceived utility of the automated aid (see Figure 18.6). If one perceives the ability of the aid to be greater than one's own ability, then perceived utility of the aid will be high. If one perceives the ability of the aid to be inferior to one's own ability, then perceived utility of the aid will be low. People may misuse an automated aid because the perceived utility of the aid is overestimated, or they may disuse an aid when the perceived utility of the aid is underestimated.

Since accurately perceiving the utility of the aid will lead to appropriate automation use, it is very important that we understand how this perception is formed. The perceived utility of the aid will be most accurate when the *actual* ability of the aid and *actual* ability of the manual operator are compared. However, the actual ability is often unknown. Perceived ability is determined through a function of actual ability and error. The larger the error, the more likely misuse and disuse is to occur. We suspect that at least two types of errors occur.

One type of error occurs when human operators estimate their own performance. Human operators tend to overestimate their own ability. Social psychological literature is fraught with examples of biases in which people overestimate their ability (self-serving biases). For example, humans exaggerated their contribution to a group product (appropriation of ideas, Wicklund, 1989), overestimated the number of tasks they can complete in a given period of time (planning fallacy, Buehler, Griffin, and Ross, 1994), were overconfident in negotiations (Neale and Bazerman, 1985), and inflated their role in positive outcomes (Whitley and Frieze, 1985). Thus, according to the Framework of Automation Use, human operators will be likely to overestimate their manual ability.

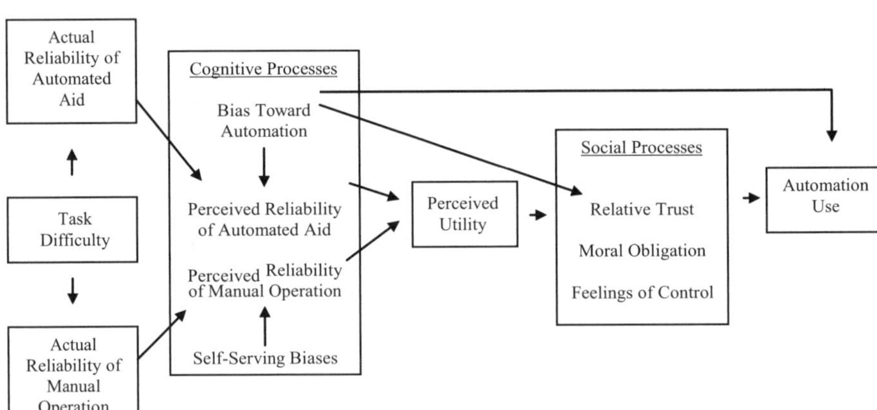

Figure 18.6 Social processes

The other type of error occurs when human operators estimate the performance of their aid. Prior to working with the aid, the human must rely on stereotypes formed concerning the performance of the aid. Although individual differences exist, a bias toward automation leads many people to predict near-perfect performances from automated aids. In the Dzindolet et al. (2002; Study 1) study, half of the participants were told that they would be provided with the decisions reached by the contrast detector before they made their soldier absent/present decision for each of 200 slides. Other participants were told they would be provided with the decisions reached by the prior participant before making their soldier absent/present decision for each of the 200 trials. The instructions informed the participants that their aid (human or automated) was *not* perfect. When asked to estimate the number of errors the automated aid would make in the upcoming 200 trials, participants predicted, on average, the aid would make only 24.79 errors (i.e., correct nearly 88 percent of the time). When asked to estimate the number of errors the human aid would make, participants predicted, on average, 51.26 errors (i.e., correct only 74 percent of the time—only 24 percent better than chance!). Although prior researchers have reported that people tend to have negative initial expectations of automation (e.g., Halpin, Johnson, and Thornberry, 1973), our studies reveal a strong bias toward automation.

For this reason, Madhavan and Wiegmann (2007a) suggested that one important variable in determining trust in automation use is whether or not the human operator knows if the aid is an automated system or another human. There is evidence that there are individual differences in bias toward automation. Singh, Molloy, and Parasuraman (1993) created a scale to determine individual differences in the propensity to misuse automated aids that we suspect is due to inflated estimates of the automation's reliability. The more inflated one's estimate of the automated aid's reliability, the more likely one is to trust the automated aid and rely on it.

The bias toward automation may exist because people expect automated aids to be experts (Dzindolet, Pierce, and Beck, 2006; Lerch, Prietula, and Kulik, 1997; Mosier and Skitka, 1996). To examine this hypothesis, Madhavan and Wiegmann (2007b) explicitly provided human operators with information about the pedigree of the automated and human aid. Whether described as a 'novice' or 'expert,' automated aids were perceived to be more reliable than human aids *prior* to working with the aid.

Madhavan and Wiegmann (2007a) hypothesize that the visible 'behavior' of the aid affects the perceived reliability of the aid. For example, conspicuity, easiness, the type of errors generated by the aid (FA vs. misses), and the extent to which the reliability of the aid changes during a task will affect perceived reliability of the aid. In one study, Madhavan, Wiegmann, and Lacson (2006) found trust was lost in automated aids that made errors on easy slides more than on difficult slides. In addition, Dzindolet, Peterson, Pomranky, Pierce, and Beck (2003) found that trust was better calibrated for participants who could not view errors made by their automated aids. In summary, the perceived reliability of the automated aid is determined by the actual reliability of the automated aid and by the bias toward automation and the visibility of the aid's behavior. The perceived reliability of manual operation is determined by the actual reliability of the human operator and by self-serving biases.

According to the Framework of Automation Use, what is important in determining automation use, though, is not the perceived reliability of the aid or the perceived reliability of manual control, but the result of a comparison process between the two, perceived utility. Increasing the reliability of the aid will not increase automation use unless the aid's perceived reliability surpasses that of manual operations.

In summary, we hypothesize that soldiers may misuse their combat identification system when the perceived utility of the combat identification aid is overestimated and not use it when its perceived utility is underestimated. The perceived utility of the combat identification system results from a comparison between the combat identification system's perceived ability and one's own perceived ability. Perceived ability is hypothesized to be affected by actual ability and various biases (self-serving and bias toward automation).

Reducing the biases should decrease inappropriate automation use. Beck, Dzindolet, and Pierce (2007) found that disuse could be reduced by providing participants multiple forms of feedback of the aid's performance. Behaving in expected ways is important and has been researched within the context of 'etiquette' by Miller (2002, 2004) and others.

Miller (2002) defined etiquette in this way:

> By 'etiquette', we mean the defined roles and acceptable behaviors and interaction moves of each participant in a common 'social' setting—that is, one that involves more than one intelligent agent. Etiquette rules create an informal contract between participants in a social interaction, allowing expectations to be formed and used about the behavior of other parties, and defining what counts as good behavior. (Miller, 2002, p. 2)

Understanding the 'informal contract' the human operators have with their automated decision aids should help researchers to predict when operators will trust their automated aids (Parasuraman and Miller, 2004). Understanding why an automated aid might err has also been found to affect trust and reliance (Dzindolet et al., 2003).

In addition to trust (through perceived utility) affecting automation use, we suggest that two other social processes may affect automation use: feelings of control and moral obligation to rely on self. Analyses of the justifications of the task allocation decisions provided by participants in one of the Dzindolet et al. (2002; Study 2) experiments revealed that 71 percent of the students, who were provided cumulative feedback that indicated that the aid made about an equal number of errors as the participant, justified self-reliance with statements indicating they would not earn more rewards if they relied on the aid. Since the task allocation decision would not affect the size of their rewards, why did participants opt for self-reliance? We hypothesize that self-reliance provides participants with an illusion of control. Langer (1983) has found that people often behave illogically in order to have an illusion of control.

In addition, many participants (though more working with human aids (n = 24, 43.64 percent) than automated aids (n = 9, 16.67 percent), $\chi^2 = 8.58$, $p < .01$) justified self-reliance with statements concerning a moral obligation to rely on oneself. One student wrote, 'I would rather the amount of coupons I receive be based on my performance—it seems more 'fair' to myself.' Another wrote, 'I feel anything earned should be based on how well I did or didn't do.'

Beck, Dzindolet, and Pierce (2002) explained that these action errors (misusing an aid when one is aware that the aid is inferior or disusing an aid when one is aware that the aid is superior) may be due to a John Henry Effect; the aid is thought of as a competitor rather than a teammate (Beck et al., 2007). Beck, Dzindolet, Pierce, and McKinney (2003) found that students taking a multiple choice test who *could* request help from an automated aid known to be correct about 70 percent of the time performed better than those who could not request to view the automated aid's responses. What is surprising about this result was that the performance difference existed *even for the trials in which the students did not view the aid's help*. Just knowing that one could be aided by automation led to better performance! Whether this was due to social facilitation, motivation, or other factors has yet to be explored. Only with a more clear understanding of these processes will we be able to suggest ways that misuse and disuse can be reduced. The variables which affect perceived utility are of special interest to us because perceived utility is not only predicted to affect trust, but also to affect the last of the processes, motivational processes.

Motivational Processes

A third explanation of the over-reliance on automation discussed by Mosier and Skitka (1996) involves the idea that when working in a group, the responsibility for the group's product is diffused among the group members. Several researchers have thought of the human-computer system as a dyad or team in which one member is not human (Bowers, Oser, Salas, and Cannon-Bowers, 1996; Scerbo, 1996; Woods, 1996). Thus, the human may feel less responsible for the outcome when working with an automated system than when working without one. The person may not be as motivated to extend as much effort when paired with an automated system as when working alone. In the social psychological literature, this phenomenon has been labeled social loafing (cf. Latane, Williams and Harkins, 1979) or free riding (Kerr and Bruun, 1983).

One theory which has been successful in accounting for much of the findings in the social loafing literature is Shepperd's Expectancy-Value Theory (1993; 1998). The Framework of Automation Use applies Shepperd's Expectancy-Value Theory in order to explain the motivational processes used in automation reliance decisions. According to this theory, motivation is predicted from a function of three factors: expectancy, instrumentality, and outcome value.

Expectancy The first factor, expectancy, is the extent to which members feel that their efforts are necessary for the group to succeed (see Figure 18.7). When members feel their contributions are dispensable, or when one's individual contribution is unidentifiable or not evaluated, one is likely to free ride or work less hard (Kerr and Bruun, 1983; Williams and Karau, 1991). With a human-computer system, individual contributions tend to be identifiable and evaluated, thus these variables are not thought to affect motivational processes. However, when the perceived utility of a system is high, one is likely to feel his or her efforts are more dispensable when working with a system low in perceived utility. Thus, we would expect human operators to be likely to misuse a combat identification system deemed more reliable than themselves in the same way people free ride on group members deemed more reliable than themselves. In summary, the more dispensable the human operator feels, the lower expectancy will be; effort will likely be low and the likelihood that the automated aid will be relied upon will be high. In some instances, this will lead to automation misuse.

Counter-intuitively, aids with especially high perceived utility may actually be disused. Steel and Konig (2006) suggested that when people estimate the

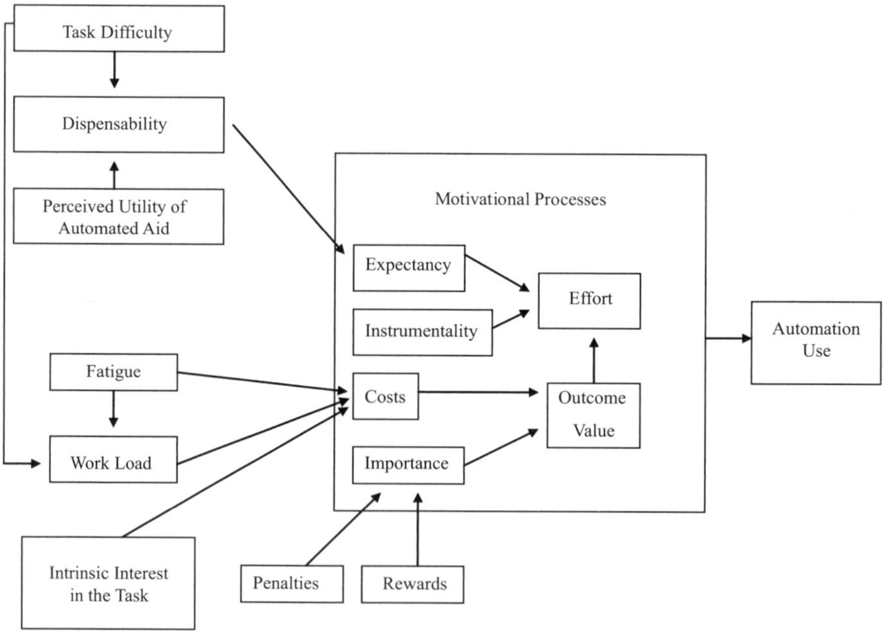

Figure 18.7 Motivational processes

probability that an outcome will occur, they tend to overestimate low probability events and underestimate high probability events. Thus, the human operator may underestimate the probability of a near-perfect combat identification system making a correct decision. This may actually lead to disuse.

Instrumentality The extent to which members feel that the group's successful performance will lead to a positive overall outcome (instrumentality) is also predicted to affect effort. Members who feel the outcome is not contingent on the group's performance are less likely to work hard. Thus, inappropriate use should be high among members who feel their group's performance is irrelevant. On the battlefield, there will be many soldier-computer teams. If the human determines that the overall outcome is not contingent on his or her human-computer team (either because he estimates other teams are more able to do the task or that his human-computer team is dispensable), then the human will put little effort into the task.

Outcome value Finally, the value of the outcome is predicted to affect motivation. Outcome value is the difference between the importance of the outcome and the costs associated with working hard. Increasing the costs or minimizing the importance of the reward will lead members to put forth less effort. More effort will be extended toward tasks that lead to valuable outcomes without requiring much cost. Costs vary with the number of other tasks one must perform, fatigue, intrinsic interest of the task, and cognitive overhead. Importance is predicted to be affected by the rewards of successful task completion and the penalties of task failure. Steel and Konig (2006) suggested that there are differences in motivation depending on whether the outcome is perceived negatively or positively. People's motivations are more affected by negative outcomes than positive ones. Given the lethal effects of misidentification on the battlefield, outcome value is assumed to be extremely high and negative; therefore, the amount of effort expended on the task is predicted to be high. Thus, the likelihood for automation use is low, and the potential for disuse will be great. Among such highly motivated people, misuse may not exist at all! This is consistent with findings from some interviews with Gulf War Soldiers, who indicated they turned off their automated systems.

For this reason, it is imperative that some of the research testing the model be performed in more combat-like environments. At the very least, researchers should examine automation reliance while varying the consequences for successful task performance.

Conclusions

Understanding automation usage decisions for even a very simple target detection task is complex. Such decisions are often counter-intuitive. For example, providing people with a highly reliable automated decision aid does not always improve the performance of the human-automated team. Increasing the reliability of the automated aid does not always improve the performance of the human-automated team. Training humans to know the conditions under which the automated aid's performance will be perfect does not always improve the performance of the human-automated team.

The Framework of Automation Use predicts that cognitive, social, and motivational processes work together to cause misuse, disuse, and appropriate automation use. Many factors affect each of the processes, and may therefore affect automation use. The reliability of the automated aid, the reliability of manual operation and several cognitive biases (including self-serving and the bias toward automation) combine to affect the perceived utility of the aid. When the perceived utility of the aid is high, the operator is likely to trust the aid and feel dispensable; his or her efforts are not necessary for the task to be completed. Automation use is predicted to be high through both social and motivational processes. Fatigue, number of tasks, intrinsic interest in the task, cognitive overhead, penalties for task failure, and rewards for task completion combine to affect the outcome value. The outcome value will affect the effort the human will expend on the task and the likelihood he or she will rely on the automated aid.

Future research needs to be conducted in order to determine the effect of each of the cognitive, social, and motivational processes on automation use. In addition, the interaction of the three processes must be examined. Specifically, the relative importance of each process needs to be determined and understood within a military context. The effectiveness of various training procedures designed to encourage human operators to appropriately rely on combat identification aids should be determined. For example, an examination of the usefulness of training automation users to overcome common biases and errors committed by those using automation should prove fruitful. Clearly, much research is needed to examine this framework. We believe the framework will prove useful to researchers interested in reducing automation misuse and disuse.

Acknowledgements

Portions of this paper were presented in a previous technical report: Dzindolet, M. T., Beck, H. P., Pierce, L. G., and Dawe, L. A. (2001). *A Framework of Automation Use* (Rep. No. ARL-TR-2412). Aberdeen Proving Ground, MD: Army Research Laboratory and at a conference: Dzindolet, M. T., Pierce, L. G., and Beck, H. (2003, May). *Understanding the Human-Computer Team.* Paper presented at the North Atlantic Treaty Organization's (NATO) Critical Design Issues for the Human-Machine Interface, Prague, Czech Republic.

References

Bailey, N. R., and Scerbo, M. W. (2007). Automation-induced complacency for monitoring highly reliable systems: The role of task complexity, system experience, and operator trust. *Theoretical Issues in Ergonomics Science, 8,* 321–348.

Beck, H. P., Dzindolet, M. T., and Pierce, L. G. (2002). Applying a decision-making model to understand misuse, disuse, and appropriate automation use. In E. Salas, C. A. Bower, N. Cooke, J. Driskell, and D. Stone (eds), *Advances in Human Factors and Cognitive Engineering* (Vol. 2, pp. 37–78). Boston: JAI Press.

Beck, H. P., Dzindolet, M. T., and Pierce, L. G. (2007). Automation usage decisions: Controlling intent and appraisal errors in a target detection task. *Human Factors, 49,* 429–437.

Beck, H. P., Dzindolet, M. T., Pierce, L. G., and McKinney, J. B. (2003). *Looking to the Future: A Simulation of Decision Aids in Tomorrow's Classroom.* Paper presented at the Human Factors and Ergonomics Society Meeting, Denver, CO.

Bowers, C. A., Oser, R. L., Salas, E., and Cannon-Bowers, J. A. (1996). Team performance in automated systems. In R. Parasuraman, and M. Mouloua (eds), *Automation and Human Performance: Theory and Applications. Human Factors in Transportation* (pp. 243–263). Mahwah, NJ: Lawrence Erlbaum.

Buehler, R., Griffin, D., and Ross, M. (1994). Exploring the 'planning fallacy': Why people underestimate their task completion times. *Journal of Personality and Social Psychology, 67,* 366–381.

Cohen, M. S., Parasuraman, R., and Freeman, J. T. (1998). Trust in decision aids: A model and its training implications. *Proceedings of the 1998 Command and Control Research and Technology Symposium.* Washington, DC: CCRP.

Dassonville, I., Jolly, D., and Desodt, A. M. (1996). Trust between man and machine in a teleoperation system. *Reliability Engineering and System Safety, 53,* 319–325.

Doosje, B., and Branscombe, N. R. (2003). Attributions for the negative historical actions of a group. *European Journal of Social Psychology, 33,* 235–248.

Dzindolet, M. T., Beck, H. P., and Pierce, L. G. (2006). Adaptive automation: Building flexibility into human-machine systems. In C. S. Burke, L. G. Pierce and E. Salas, (eds), *Advances in Human Factors and Cognitive Engineering: Understanding Adaptability: A Prerequisite for Effective Performance within Complex Environments* (Vol. 6, pp. 213–248). Boston: Elsevier Press.

Dzindolet, M. T., Beck, H. P., Pierce, L. G., and Dawe, L. A. (2001). *A Framework of Automation Use* (Rep. No. ARL-TR-2412). Aberdeen Proving Ground, MD: Army Research Laboratory.

Dzindolet, M. T., Peterson, S. A., Pomranky, R. A., Pierce, L. G., and Beck, H. P. (2003). The role of trust in automation reliance. *International Journal of Human Computer Studies: Special Issue on Trust and Technology, 58*, 697–718.

Dzindolet, M. T., Pierce, L. G., and Beck, H. P. (2006). Misuse of human and automated decision aids in a soldier detection task. In *Proceedings of the Human Factors and Ergonomics Society 50th Annual Meeting*, Santa Monica, CA: Human Factors and Ergonomics Society.

Dzindolet, M. T., Pierce, L. G., Beck, H. P., and Dawe, L. A. (2002). The perceived utility of human and automated aids in a visual detection task. *Human Factors, 44,* 79–94.

Dzindolet, M. T., Pierce, L. G., Beck, H. P., Dawe, L. A., and Anderson, B. W. (2001). Predicting misuse and disuse of combat identification systems. *Military Psychology, 13*, 147–164.

Dzindolet, M. T., Pierce, L. G., Pomranky, R. A., Peterson, S. A., and Beck, H. P. (2001). Automation reliance on a combat identification system. *Proceedings of the Human Factors and Ergonomics Society 45th Annual Meeting*, Santa Monica, CA: Human Factors and Ergonomics Society.

Frey, D., and Schulz-Hardt, S. (2001). Confirmation bias in group information seeking and its implications for decision making in administration, business, and politics. In F. Butera, and G. Mugny (eds), *Social Influence in Social Reality: Promoting Individual and Social Change* (pp. 53–73). Ashland, OH: Hogrefe and Huber.

Halpin, S. M., Johnson, E. M., and Thornberry, J.A. (1973). Cognitive reliability in manned systems, *IEEE Transactions on Reliability, R-22,* 165–170.

Jian, J., Bisantz, A. M., and Drury, C. G. (2000). Foundations for an empirically determined scale of trust in automated systems. *International Journal of Cognitive Ergonomics 4,* 53–71.

Jones, E. E., and Nisbett, R. E. (1972). The actor and the observer: Divergent perceptions of the causes of behavior. In E. E. Jones, D. Kanouse, H. H. Kelley, R. E. Nisbett, S. Valins, and B. Weiner (eds), *Attribution: Perceiving the Causes of Behavior* (pp.79–94). Morristown, NJ: General Learning Press.

Kantowitz, B. H., Hanowski, R. J., and Kantowitz, S. C. (1997). Driver acceptance of unreliable traffic information in familiar and unfamiliar settings. *Human Factors, 39,* 164–176.

Karasawa, K. (1995). An attributional analysis of reactions to negative emotions. *Personality and Social Psychology Bulletin, 21,* 456–467.

Karasawa, M. (2003). Projecting group liking and ethnocentrism on in-group members: False consensus effect of attitude strength. *Asian Journal of Social Psychology, 6,* 103–116.

Kerr, N. L., and Bruun, S. E. (1983). Dispensability of member effort and group motivation losses: Free-rider effects. *Journal of Personality and Social Psychology, 44,* 78–94.

Langer, E. J. (1983). *The Psychology of Control.* Beverly Hills, CA: Sage.

Latane, B., Williams, K. D., and Harkins, S. (1979). Many hands make light the work: The causes and consequences of social loafing. *Journal of Personality and Social Psychology, 37,* 822–832.

Layton, C., Smith, P. J., and McCoy, C. E. (1994). Design of a cooperative problem-solving system for en-route flight planning: An empirical evaluation. *Human Factors, 36,* 94–119.

Lee, J. D. (2001). Emerging challenges in cognitive ergonomics: Managing swarms of self-organizing agent-based automation. *Theoretical Issues in Ergonomics Science, 2,* 238–250.

Lee, J. D., and Moray, N. (1992). Trust, control strategies and allocation of function in human-machine systems. *Ergonomics, 35,* 1243–1270.

Lee, J. D., and Moray, N. (1994). Trust, self-confidence, and operators' adaptation to automation. *International Journal of Human-computer Studies, 40,* 153–184.

Lee, J. D., and See, K. A. (2004). Trust in automation: Designing for appropriate reliance. *Human Factors, 46*, 50–80.

Lerch, F. J., Prietula, M. J., and Kulik, C. T. (1997). The turing effect: the nature of trust in expert system advice. In: P. J. Feltovich, K. M. Ford, and R. R. Hoffman (eds), *Expertise in Context: Human and Machine.* (pp. 417–448). Menlo Park, CA: AAAI Press.

Lewandowsky, S., Mundy, M., and Tan, G.P.A. (2000). The dynamics of trust: Comparing humans to automation. *Journal of Experimental Psychology: Applied, 6,* 104–123.

Liu, C., and Hwang, S.L. (2000). Evaluating the effects of situation awareness and trust with robust design in automation. *International Journal of Cognitive Ergonomics, 4,*125–144.

Madhavan, P., and Wiegmann, D. A. (2007a). Similarities and differences between human-human and human-automation trust: An integrative review. *Theoretical Issues in Ergonomics Science, 8,* 277–301.

Madhavan, P., and Wiegmann, D. A. (2007b). Effects of information source, pedigree, and reliability on operator interaction with decision support systems. *Human Factors, 49*, 773–785.

Madhavan, P., Wiegmann, D. A., and Lacson, F. C. (2006). Automation failures on tasks easily performed by operators undermine trust in automated aids. *Human Factors, 48,* 241–256.

Merritt, S. M., and Ilgen, D. R. (2008). Not all trust is created equal: Dispositional and history-based trust in human-automation interactions. *Human Factors, 50,* 194–210.

Miller, C.A. (2002). Definitions and dimensions of etiquette. Paper presented at the *AAAI Fall Symposium on Etiquette and Human-Computer Work*, North Falmouth, MA.

Miller, C. A. (2004). Human-computer etiquette: Managing expectations with intentional agents. *Communications of the ACM, 47*, 31–34.

Moes, M., Knox, K., Pierce, L. G., Beck, H. P. (1999). *Should I Decide or Let the Machine Decide for Me?* Poster presented at the meeting of the Southeastern Psychological Association, Savannah, GA.

Moray, N. (2001). Humans and machines: Allocation of function. In J. Noyes, and M. Bransby (eds.), *People in control: Human Factors in Control Room Design*. IEE control engineering series, 60. Edison, NJ: Institution of Electrical Engineers.

Moray, N., Inagaki, T., and Itoh, M. (2000). Adaptive automation, trust, and self-confidence in fault management of time-critical tasks. *Journal of Experimental Psychology: Applied, 6,* 44–58.

Mosier, K. L., and Skitka, L. J. (1996). Human decision-makers and automated decision aids: Made for each other? In R. Parasuraman, and M. Mouloua (eds), *Automation and Human Performance: Theory and Applications. Human Factors in Transportation* (pp. 201–220). Mahwah, NJ: Lawrence Erlbaum.

Mosier, K. L., Skitka, L. J., Dunbar, M., and McDonnell, L. (2001). Aircrews and automation Bias: The advantages of teamwork? *The International Journal of Aviation Psychology, 11,* 1–14.

Muir, B. M. (1987). Trust between humans and machines, and the design of decision aids. *International Journal of Man-Machine Studies, 27,* 527–539.

Muir, B. M. (1994). Trust in automation: Part I. Theoretical issues in the study of trust and human intervention in automated systems. *Ergonomics, 37,* 1905–1922.

Neale, M. A., and Bazerman, M. H. (1985). The effects of framing and negotiator overconfidence on bargaining behaviors and outcomes. *Academy of Management Journal, 28,* 34–49.

Parasuraman, R., and Miller, C.A. (2004). Trust and etiquette in high-criticality automated systems. *Communications of the ACM, 47*, 51–55.

Parasuraman, R., Molloy, R., and Singh, I. L. (1993). Performance consequences of automation-induced 'complacency.' *International Journal of Aviation Psychology, 3,* 1–23.

Parasuraman, R., and Riley, V. (1997). Humans and automation: Use, misuse, disuse, abuse. *Human Factors, 39,* 230–253.

Riley, V. (1996). Operator reliance on automation: Theory and data. In R. Parasuraman and M. Mouloua (eds.), *Automation and Human Performance: Theory and Applications* (pp. 19–35). Hillsdale, NJ: Lawrence Erlbaum.

Ross, L. (1977). The intuitive psychologist and his shortcomings: Distortions in the attribution process. In L. Berkowitz (ed.), *Advances in Experimental Social Psychology* (Vol. 10, pp. 174–221). New York: Academic Press.

Rovira, E., McGarry, K., and Parasuraman, R. (2007). Effects of imperfect automation on decision making in a simulated command and control task. *Human Factors, 49,* 76–87.

Scerbo, M. W. (1996). Theoretical perspectives on adaptive automation. In R. Parasuraman, and M. Mouloua (eds), *Automation and Human Performance: Theory and Applications. Human Factors in Transportation* (pp. 37–63). Mahwah, NJ: Lawrence Erlbaum.

Shepperd, J. A. (1993). Productivity loss in performance groups: A motivation analysis. *Psychological Bulletin, 113,* 67–81.

Shepperd, J. A. (1998). *Expectancy Value Theory.* Paper presented at the Midwestern Psychological Association, Chicago, IL.

Singh, I. L., Molloy, R., and Parasuraman, R. (1993). Automation-induced 'complacency': Development of the complacency-potential rating scale. *International Journal of Aviation Psychology, 3,* 111–122.

Singh, I. L., Molloy, R., and Parasuraman, R. (1997). Automation-induced monitoring inefficiency: Role of display location. *International Journal of Human-Computer Studies, 46,* 17–30.

Steel, P., and Konig, C. J. (2006). Integrating theories of motivation. *Academy of Management Review, 31,* 889–913.

Tan, G., and Lewandowsky, S. (1996). *A Comparison of Operator Trust in Humans Versus Machines.* Presentation of the First International Cyberspace Conference on Ergonomics: http://www.curtin.edu.au/conference/cyberg/centre/paper/tan/paper.html.

Tenney, Y. J., Rogers, W. H., and Pew, R. W. (1998). Pilot opinions on cockpit automation issues. *International Journal of Aviation Psychology, 8,* 103–120.

Tversky, A., and Kahneman, D. (1973). Availability: A heuristic for judging frequency and probability. *Cognitive Psychology, 5,* 207–232.

Whitley, B. E., Jr., and Frieze, I. H. (1985). Children's causal attributions for success and failure in achievement settings: A meta-analysis. *Journal of Educational Psychology, 77,* 608–616.

Wicklund, R. A. (1989). The appropriation of ideas. In P. B. Paulus (ed.), *Psychology of Group Influence* (2nd edn, pp. 393–423). Hillsdale, NJ: Lawrence Erlbaum.

Wiegmann, D., Rich, A., and Zhang, H. (2001). Automated diagnostic aids: The effects of aid reliability on users' trust and reliance. *Theoretical Issues in Ergonomics Science, 2,* 352–367.

Wiener, J. L. (1981). A theory of human information processing for economists. *Dissertation Abstracts International, 42,* 809.

Williams K. D., and Karau, S. J. (1991). Social loafing and social compensation: The effects of expectations of coworker performance. *Journal of Personality and Social Psychology, 61,* 570–581.

Wilson, K.A., Salas, E., Priest, H. A. and Andrews, D. H. (2007). Errors in the heat of battle: Taking a closer look at shared cognition breakdowns through teamwork. *Human Factors, 49,* 243–256.

Woods, D. D. (1996). Decomposing automation: Apparent simplicity, real complexity. In R. Parasuraman and M. Mouloua (eds), *Automation and Human Performance: Theory and Applications* (pp. 3–17). Mahwah, NJ: Lawrence Erlbaum.

Chapter 19

On Fratricide and the Operational Reliability of Target Identification Decision Aids in Combat Identification

John K. Hawley
Anna L. Mares
U.S. Army Research Laboratory

Jessica L. Marcon
University of Texas at El Paso

Introduction

Military systems increasingly rely on various classes of technology-based aids in engagement decision-making. Such decision aids are intended to assist weapon system operators in answering the question, 'Should I or should I not engage this target?' In general, there are two broad classes of combat identification decision aids. These are (1) target identification (TI) decision aids, and (2) technical methods to support enhanced situation awareness (SA). Endsley (1996) defines SA as the *perception* of elements in the environment, the *comprehension* of their meaning, and the *projection* of their status in the near future. Identification Friend or Foe (IFF) query systems and so-called battlefield combat identification systems (BCIS) are representative of the former class, while the Army's current Blue Force Tracker and related battle command tracking enhancements are examples of the latter.

IFF and BCIS systems are based on a query-response framework in which a query signal (Who are you?) is sent by one of the systems and a response that permits identification as friend, foe, or neutral is sent in reply. SA enhancement systems such as Blue Force Tracker rely on the Global Positioning System to record their location on friendly battle command displays. Both Blue and Red systems (to the extent they are known) are displayed to battle command operators, and this information is used to enhance operator SA.

Research and operational experience have indicated that a major factor in the successful use of TI decision aids is their operational reliabilities. Operational reliability, in present usage, refers to the percentage of total engagement opportunities in which use of the aid results in correct target classification and identification.

TI decision aids and systems supporting enhanced SA are often discussed and approached from both system development and later test and evaluation perspectives as if these are two separate, unrelated processes. Recent experience suggests, however, that TI decision aids and SA 'enhancers' are best used together in a complementary fashion (see Hawley and Mares, 2006). Both classes of aids provide information that when fused by a competent operator can improve the overall reliability of the combat identification process. The focus of the discussion to follow is, however, human performance issues in TI decision aiding. TI decision aiding might be referred to as the 'low-hanging fruit' of the combat identification problem in the sense that it is the most straightforward aspect of the targeting problem and the topic for which there is the most operational experience and relevant data.

The next section discusses the historical background and importance of the combat identification problem. Two questions are addressed to focus this material: 'Is the combat identification problem important?' and 'What is the historical locus of the misidentification problem?'

Background, Importance, and Locus of the Combat Identification Problem

To put the issues of fratricide and the importance of combat identification in proper context, it is instructive first to consider historical fratricide rates across various types of military systems. The conventional wisdom in military circles has been that fratricide is rare—a nominal two percent rate of friendly casualties (Steinweg, 1995). System performance and operational expectations tacitly are set at this low level. Recent research suggests, however, that fratricide may not be as rare as historically assumed. For example, a study performed by the Congressional Office of Technology Assessment (OTA) in the aftermath of Operation Desert Storm (ODS) in the early 1990s, where 24 percent of coalition casualties were attributable to fratricide, suggested that 15–20 percent of total losses may be the historical norm (OTA, 1993). A more recent report on the same subject prepared for the Joint Staff (Sparta, 2002) put the figure at 11–16 percent of total casualties. The OTA report further concluded that the rate observed during ODS may also be representative of future conflicts. This increased rate of fratricide beyond nominal expectations was hypothesized to be a function of (1) the increased lethality of precision munitions, and (2) increasing reliance on imperfect sensor data and imperfect classification algorithms in engagement decision-making. However, the Sparta report advises caution on the 24 percent rate observed during ODS, remarking that this figure may be biased by a very low overall casualty count. The OTA report concluded that reducing fratricide is desirable and feasible, but eliminating it is not; the Sparta report concurred with this conclusion.

Sheridan (2002) remarked that nearly any automated or partially automated system can be made more reliable by restricting its range of operating circumstances. In the case of combat identification, restrictive rules of engagement

(ROE) might lessen the likelihood of adverse events, but these limitations could take some systems 'out of the fight,' so to speak. It is thus appropriate to ask whether restrictive ROE are the best course of action across all situations. The OTA's (1993) report cautions, for example, that overly restrictive ROE may reduce combat effectiveness to such an extent that that casualties inflicted by the enemy could increase more than friendly fire losses are reduced. Command actions in the aftermath of the fratricides committed by the Army's Patriot air defense missile system during the major combat operations phase of Operation Iraqi Freedom (OIF) clearly illustrate this possibility. Following Patriot's second fratricidal engagement (a Navy F/A-18), restrictive ROE were imposed on all Army Patriot units in theater. Patriot units were not permitted to engage any targets—even in self defense—without the direct permission of the Air Force controlling authority. During combat operations, Patriot and most other Army air defense systems are under the operational control of the Air Force. And it is not known whether such permission could have been obtained in time to successfully engage incoming tactical ballistic missiles. It turned out that the second portion of OTA's caution—increased casualties inflicted by the enemy because of restrictive ROE—did not materialize during OIF, but the possibility existed. With Patriot effectively out of the fight, the U.S. led coalition had a significantly degraded defense against tactical ballistic missiles.

The answer to the first issue framing the present discussion 'Is the combat identification problem important?' obviously is 'yes.' Whether one accepts OTA's estimate of 15–20 percent of total losses or Sparta's more conservative estimate of 11–16 percent of casualties, neither is an insignificant percentage; they are both well beyond a casual reference to fratricide as unfortunate but part of the larger cost of war. That being the case, the obvious follow-on question concerns the dynamics of fratricide: 'What is the locus of the problem?'

The most recent comprehensive sampling of data and experiences on the topics of combat identification and fratricide was prepared by Sparta (2002) for the Joint Staff J-8. These data are taken from operational experiences during Operations Desert Shield and Desert Storm, the Army's instrumented combat training centers (CTCs—such as the National Training Center at Fort Irwin, California), and the various service safety centers. The emphasis on results since ODS is justified by the OTA's observation concerning elevated fratricide rates resulting from an increased use of precision munitions (i.e., you're going to hit what you shoot at) and increased reliance on machine-aided target classification and identification. Results from Sparta's (2002) study are summarized in the following points:

- Ground-to-ground (G-G) and air-to-ground (A-G) mission areas encompassed over 97 percent of fratricide incidents.
- The principal causes of G-G fratricide, in decreasing order, were direct fire (59.3 percent), indirect fire (24.4 percent), and mines (16.3 percent).

- The principal cause of G-G direct fire fratricide and A-G close air support fratricide was TI error—misidentification of a friendly or neutral/noncombatant system as enemy.
- The principal cause of G-G indirect fire and mine loss fratricides was SA error.
- G-G direct fire fratricides tended to cluster among a small number of shooters: Three systems—M1 Tank/M2 Bradley/LAV (Light Armored Vehicle)—accounted for 59 percent of fratricide incidents.
- G-G direct fire fratricide victims tended to cluster among a small number of systems. Five victim categories accounted for 83 percent of fratricide incidents: (1) dismounted individuals, (2) HMMWVs (High Mobility Multipurpose Wheeled Vehicles, or HumVees), (3) M113 armored personnel carriers, (4) M2 Bradleys/M3 Cavalry Fighting Vehicles/BSFVs (Bradley Stinger Fighting Vehicles), and (5) M1 Tanks.
- Ground to Air (G-A) and Air-to-Air (A-A) fratricides did not represent significant problems within the operational or training contexts studied. It should be noted, however, that G-A engagements are not exercised to any significant extent at the Army's instrumented training centers. For example, Patriot units typically do not accompany their supported maneuver forces to National Training Center exercises. Patriot's OIF experience discussed later in the chapter illustrates that while there might not be a large number of G-A fratricides, their operational impact is significant well beyond their actual numbers.

The Sparta report concluded that there is no 'silver bullet' solution to all fratricides. Reliable TI systems have the potential to reduce G-G direct fire and A-G fratricide. Furthermore, improved SA on the part of shooters might lessen the potential for indirect fire fratricides. Improved SA on the part of maneuver support forces such as combat engineers is necessary to reduce mine-related incidents. It should be emphasized that the term SA involves more than a simple display of icons on a situation display. The full meaning of SA also includes comprehension and projection of future status, and these are decidedly human-centered activities.

Target Identification Decision Aid Reliability and Human Performance

The reader should recall that, according to recent data, 97 percent of fratricide incidents occur in the G-G and A-G mission areas. Moreover, the largest cause of G-G direct fire fratricide and A-G fratricides is TI error—misidentification of a friendly or neutral/noncombatant system as enemy. These data suggest that TI decision aid reliability is a major consideration in fratricide reduction.

One of the more relevant studies looking at the relationship between direct-fire fratricide rates and TI decision aid reliability is discussed by Kogler (2003). Kogler's study was performed on the live-fire ranges at the Army's Aberdeen

Proving Ground, Maryland, using soldiers from Military Occupational Specialty 95B (Military Police). Test subjects fired a standard M16A2 assault rifle equipped with a TI system at simulated friendly and threat targets. Two levels of TI reliability were used in the research: 100 percent and 60 percent. Kogler's results indicated that, as expected, the highly reliable TI system virtually eliminated friendly fire engagements (fratricides). Not surprisingly, the TI system operating at 60 percent reliability did not produce the same level of fratricide reduction. In his discussion section, Kogler speculated that overconfidence in an unreliable TI system might actually increase fratricide rates above unaided levels. Results reported by Dzindolet, Pierce, Beck, Dawe, and Anderson (2001) reinforced Kogler's contention. In a study of the misuse and use of combat identification decision aids, Dzindolet and her colleagues concluded that over reliance on automated combat identification decision aids (termed misuse) exceeded disuse even when the reliability of the system was known to be very low. Taken together, these results support the conclusion that users will act upon the results provided by a TI decision aid even if it is known that the aid is unreliable. Kogler (2003) concluded that designers of future TI systems must consider their reliability in not only a benign laboratory environment but also in all operational settings. Without such consideration, operational reliability (reliability when applied in an operational environment) may drop to levels that adversely impact the overall goal of fratricide reduction. Unreliable TI decision aids might make things worse than in an unaided situation.

Kogler's study also explored the impact of TI use on engagement latencies—defined as the time between target acquisition and engagement added through a requirement to use the TI system. The TI device used in Kogler's study was relatively simple, involving only a small number of steps to employ. Consequently, he did not find a significant latency effect attributable to use of the TI system. In his discussion, Kogler speculated that if fielded TI systems are difficult or time-consuming to operate the reduction in fratricide could be suboptimal and lead to an increase in missed targets. This speculation was based on after-action interviews with test subjects.

The Army's experience with the Stinger man-portable air defense missile system speaks directly to the problem hypothesized by Kogler. The Stinger missile system itself is rather simple to operate and highly effective against close-in air targets. Engagement problems began, however, when an IFF interrogation component was added to the baseline missile system. The additional steps necessary to conduct an IFF interrogation on an incoming air threat took so much time that many threats were no longer engageable by the time the process was completed (Barber, Drewfs, and Johnson, 1987). This resulted in a significant increase in the number of missed threats, as Kogler speculated would result from deployment of a complex or time-consuming TI system.

Automation reliability and its impact on operator performance has been the subject of a wide range of recent studies. Results from a number of these studies are relevant to the issue of TI reliability and its impact on human users. Rovira,

McGarry, and Parasuraman (2002) remarked that one of the true 'ironies' of automation is that the more reliable the automation, the greater its detrimental effects when it fails. Repeated successful use lulls users into a false sense of security or complacency regarding the automation's performance. Consequently, they tend not to use other information at their disposal when making an engagement decision.

Dixon and Wickens (2006) present results indicating that imperfect automation is *manageable*, but users must be pre-warned of the nature and source of the automation's imperfections. These authors cautioned, however, that reliabilities less than 75 percent are worse than no automation at all, and, in fact, can provide users with what they termed a 'concrete life preserver.' Their results are consistent with findings reported by Kogler (2003) and Dzindolet et al. (2001).

Cohen, Parasuraman, and Freedman (1997) and Masalonis and Parasuraman (1999) argued that trust in automation should not be all-or-nothing, but graded and differentiated according to the operational context. These authors refer to this as *situation-specific* trust. The automation may work very reliably in certain contexts, in which the user should use it and trust it. But in other contexts, the automation's recommendations may be suspect and users should be trained to recognize and react to those situations. Users should be instructed to assess the situation and take the action that best suits the context, *in their judgment*. If users can be trained to recognize the appropriate context, then they can learn when to trust the automation and when its recommendations should be discounted.

Similarly, Lee and See (2004) asserted that automation should be designed for *appropriate* as opposed to *greater* trust. These authors stated that in situations involving imperfect automation, user training must emphasize:

1. Expected system unreliability.
2. The mechanisms underlying potential reliability problems.
3. How usage situations interact with the automation's technical characteristics to affect reliability.

In conclusion, developers, testers, trainers, and users must be honest regarding TI aid reliability. Extensive tests must be performed to determine those field situations in which the aid does not meet design criteria for reliability. Boundaries of successful system performance must be pushed. Moreover, the mechanisms underlying unreliable performance must also be explored. Users must then be apprised of system unreliability patterns and trained in situations that will expose them to system imperfections.

These results provide a useful theoretical background to the TI reliability problem. But can such expectations about user performance be realized? Is it reasonable to expect that in the heat of combat when a soldier's life is on the line and engagement decisions (to shoot or not to shoot) must be made quickly, that he or she will be able to apply the criteria underlying *appropriate* or *system-specific* trust? In this respect, we think that Kogler is spot on: TI decision aids must be operationally reliable. Anything less than near-perfect operational reliability is likely to result in problems. Research and experience suggest that there is little margin for user override of an unreliable TI decision aid in a 'hot' combat environment.

OIF Patriot Fratricides: The Operational Impact of Unreliable Target Identification

During the major combat operations phase of the Second Gulf War (Operation Iraqi Freedom)—March-April 2003, U.S. Army Patriot air defense missile units were involved in two fratricides. In the first incident, a British Tornado was misclassified as an anti-radiation missile (ARM) and subsequently engaged and destroyed. The second fratricide involved a Navy F/A-18 that was misclassified as a tactical ballistic missile (TBM) and also engaged and destroyed. Three flight crew members lost their lives in these incidents. OIF involved a total of 11 Patriot engagements by U.S. units. Of these, nine resulted in successful missile engagements; the other two (18 percent) were fratricides.

A team from the Army Research Laboratory (ARL) began looking into Patriot system performance at the invitation of the then Fort Bliss, Texas, commander, Major General (MG) Michael A. Vane. Fort Bliss is the site of the Army's air defense artillery center and school. General Vane was interested in operator vigilance and SA as they relate to the performance of automated air defense battle command systems. He was particularly concerned by what he termed a 'lack of vigilance' on the part of Patriot operators along with an apparent 'lack of cognizance' of what was being presented to them on situation displays with an ensuing 'unwarranted trust in automation.' The project team spent most of the summer and fall of 2004 reviewing the OIF fratricide incidents—reading documents, interviewing knowledgeable personnel in the Fort Bliss area, and observing Patriot training and operations. An initial report was delivered to MG Vane in October 2004. ARL's assessment was not intended as another exercise in 'Monday morning quarterbacking.' Instead, the focus was to look into the deeper story behind the events leading to the OIF fratricides from a human performance perspective and to identify actionable solutions.

The ARL assessment team organized its presentation to MG Vane around two central themes, denoted (1) *undisciplined automation* during Patriot development and (2) *automation misuse* on the part of Patriot crews (see Hawley and Mares, 2006). The first contributing factor was termed undisciplined automation, defined as the automation of functions by designers and subsequent implementation by users without due regard for the consequences for human performance (Parasuraman and Riley, 1997). Undisciplined automation tends to define the operators' roles as by-products of the automation. Operators are expected to 'take care of' whatever the system cannot handle. In the case of Patriot, little explicit attention was paid during system design and subsequent testing to determining (1) what these residual operator functions were, (2) whether operators actually could perform them, (3) how operators should be trained to perform properly, or (4) the impact of operator performance deficiencies on the system's overall decision-making reliability.

The downstream impact of undisciplined automation was exacerbated by two additional factors: (1) unacknowledged system fallibilities, and (2) a fascination with and 'blind faith'[1] in technology. A series of Patriot operational tests going back to the 1980s indicated that the system's engagement logic was subject to track misclassification problems—system fallibilities. However, these sources of automation unreliability were not satisfactorily addressed during system software upgrades, nor did information about them find its way into operator training, battle command doctrine, operating procedures, or unit standard operating procedures. System developers continued to pursue technology-centric solutions to automation reliability problems (increased use of artificial intelligence, non-cooperative target recognition, etc.), but the basic problem remained: The system was not completely reliable in critical functional areas, most notably track classification and identification. To make matters worse, users were not informed regarding these reliability problems, or if they were informed, little effective responsive action was identified for them to take.

In the aftermath of the First Gulf War (ODS), the air defense user community acquiesced to the developmental community's apparent lack of concern for problems with Patriot's track classification accuracy. Emboldened by Patriot's seeming success in engaging the Iraqi TBM threat during ODS, Patriot's organizational culture emphasized 'Reacting quickly, engaging early, and *trusting the system without question*.' This cultural norm was exacerbated by the air defense branch's traditional training practices, which were criticized as emphasizing 'rote drills' versus the 'exercise of high-level judgment.' The Patriot user community continued to approach training for air battle operations in much the same manner as less cognitively oriented tasks such as system movement and set-up. The emphasis during training was on mastering routines rather than critical thinking and adaptive problem solving.

A second detrimental factor was the branch's traditional personnel assignment practices which tended to place inexperienced personnel in key battle command crew positions. Before the first round was fired during OIF, the stage was thus set for the second major contributor, termed automation misuse (Parasuraman and Riley, 1997). Automation misuse took the form of extensive automation bias on the part of Patriot crews. Automation bias is defined as unwarranted over-reliance on automation. It has been demonstrated to result in failures of monitoring (vigilance problems) and accompanying decision biases (an unwarranted and unthinking trust in automation—let's do what the system recommends). Recall that these are

1 Note: Several terms used here and elsewhere in quotes without a reference citation are taken directly from the Army Board of Inquiry investigation report on the OIF Patriot fratricides.

the very concerns expressed by MG Vane in his kick-off discussion with ARL's project staff.

One must be careful not to lay too much blame for these shortcomings at the feet of the offending Patriot crews. The roots of these apparent human shortcomings could be traced back to systemic problems resulting from decisions made years earlier by concept developers, software engineers, procedures developers, testers, trainers, and commanders. In one sense, the OIF Patriot crews did what they had been trained to do, what Patriot's culture emphasized and reinforced, and what the automation literature suggested they likely would do in such circumstances.

Patriot is a very lethal system. It can be argued, however, that the system was not properly managed during OIF. Driven by technology and mission expansion, the Patriot crew's role changed from traditional operators to supervisory controllers whose primary role was the supervision of subordinate automatic control systems. But this change in role was not reflected in the air defense culture. Nor was it reflected in design and evaluation practices, battle management concepts, operational procedures, training practices, or personnel usage patterns. System management issues (doctrine, battle command concepts, and user procedures) and crewmembers' ability to execute procedures supporting system management were not addressed with the same rigor during system development and testing as were hardware and software capabilities. As the lessons of OIF suggest, these aspects of human-in-the-loop system performance are as important to later operational effectiveness as hardware and software capabilities.

An obvious follow-on question to the previous discussion concerns the impact of TI decision aids—IFF query systems—on the OIF fratricide incidents. IFF query systems have been in routine use in air defense operations for more than 50 years. In the case of the British Tornado, the engaging Patriot unit is thought to have queried the incoming track, but did not receive a response. The Tornado was equipped with an IFF transponder. Either it did not function properly, the Patriot IFF query system did not operate properly, or the response to the query was invalid. The Navy F/A-18 was classified by the Patriot system as a tactical ballistic missile. Under rules of engagement in effect at that time, no IFF challenge was issued. The operating assumption was that the Patriot system would not confuse air breathing threats and TBMs—in spite of the fact that operational test results indicated that such misclassifications were possible (an 'unacknowledged' system fallibility). ROE have now changed, however, and all engageable tracks are queried, regardless of their classification.

On the subject of IFF and other query-type TI systems, Steinweg (1995, p. 16) remarked that 'Technology will have to provide an answer to this problem [TI] for surface warfare, much as it has solved the problem for air warfare.' As the situation with Patriot during OIF illustrates, IFF query systems have *solved* the target identification problem, but only from a theoretical perspective. Such miscues should not occur, but they do. The problem is one of *operational* reliability. Anecdotal reports and conversations with people knowledgeable about IFF systems and their field use reveal a number of problems with such systems that lessen their operational reliability. Some of the more frequently mentioned issues are: (1) pilots do not always turn them on, (2) invalid responses are not uncommon, (3) aircraft IFF transponders are unknowingly inoperative (the likely situation with the British Tornado), (4) ground system IFF interrogators unknowingly do not function properly, and (5) other technical issues impact IFF operational reliability. The aspect angle between the ground interrogator and the aircraft's IFF antenna is an example of the latter type of problem. It should be noted that during OIF Patriot crews reported that roughly 70 percent of the tracks on their situation displays were classified as unknown.

The combat identification lessons that can be derived from the Patriot fratricides during OIF are complex. Even a mature TI technology like the IFF systems employed in G-A query is not perfect. Given the relatively short engagement timelines involved in G-A missile engagements, it might be unreasonable to expect users to actualize situation-specific trust and compensate for TI decision aid unreliability.

The Path Forward

Steinweg (1995, p. 4) cautioned that 'the expected conditions of the future battlefield can only aggravate present fratricide rates.' Combat identification truly is a challenging problem both from technical and human performance perspectives. On balance, however, there are three potential ways to improve operational combat identification. All three paths forward involve the interaction of users with technical systems:

1. Increased emphasis on visual combat identification.
2. More reliable target identification decision aids.
3. A greater emphasis on improved situation awareness in combination with the use of TI decision aids.

Visual target classification (What is it?) and identification (Whose is it?) have been the historical foundations of effective combat identification. For example, the Army's air defense branch has historically emphasized visual aircraft recognition skills during initial and unit training (see Frederickson, Dawdy, and Carter, 1982). However, the advancing technology of weapons systems and the larger geographic scope of contemporary combat operations are making visual combat identification a less viable option. Steinweg (1995) noted, for example, that target acquisition ranges during ODS in the early 1990s were consistent at 3000 meters, with first round hits at 2500 meters. At these ranges, tank crews cannot reliably identify targets. Future technologies will permit target acquisition at 4000 meters, with a potential for acquisition at 5000–7000 meters. These ranges are well beyond human visual identification capabilities, even with visual aids. Visual combat identification is training intensive, operationally unreliable, and increasingly unrealistic at current engagement ranges (Barber, Drewfs, and Lockhart, 1987). This is not to imply that visual target classification and identification should be ignored, but visual methods cannot be counted upon to contribute much to solving the contemporary combat identification problem.

The second potential solution to the combat identification problem is more reliable TI decision aids for G-G, G-A, and A-G operations. In principle, systems such as IFF, BCIS, and those proposed for individual soldiers have the potential to greatly reduce TI errors. As emphasized, however, the key issues in effective TI decision aiding are operational reliability and ease of use. Even mature TI technologies such as IFF are subject to operational reliability problems that can lead to unacceptable rates of friendly fire engagements.

Developers of TI decision aiding systems must not leave users in the Catch-22 of unreliable decision aiding (see Reason, 1990; Sheridan 2002). The Catch-22 bind in which operators find themselves can be described succinctly in terms of five assertions:

1. TI decision aiding has been introduced because it is thought to be able to do the job better than a human user alone.
2. Humans are kept in the shoot-or-not-to-shoot decision loop to verify that the TI decision aid's conclusions are correct and override the aid when it is wrong.
3. The tacit assumption is that users can properly decide when the aid's decisions should be overridden (i.e., users will have achieved situation-specific trust).
4. Human users are expected to compensate for the aid's unreliability.
5. Humans suffer from a variety of physical and cognitive limitations that make it virtually impossible to meet this expectation.

The research and operational experiences discussed in previous sections all highlight the extreme hazard of expecting users in the heat of combat to provide a satisfactory resolution to the Catch-22 of unreliable TI decision aiding. In most cases, users will go with the aid's recommendations—right or wrong. In a comment on automation use in future systems, Norman (2007, p. 113) argued that the worst of all possible situations is to leave users somewhere between manual operation and full automation—the 'dangerous middle ground.' Any such in-between state leaves users at the mercy of the Catch-22 of unreliable automation. And as research and operational experience indicate, that situation is not likely to be satisfactorily resolved.

If there is a partial solution to the Catch-22 of unreliable TI decision aiding, it likely involves the interplay of TI decision aids and improved SA. Users must develop some level of situation-specific trust in the aid, and the key to developing that trust is improved SA—better knowledge of context, the current state of things, and what might happen next. The sensemaking process that leads to SA in situations of uncertainty might provide users of combat identification systems with the *opportunity* to better understand what is going on around them now and anticipate future situations (see Weick, Sutcliffe, and Obstfeld, 2005). An example from Patriot during OIF illustrates this point. Recall that during the ARL project team's initial conversation with MG Vane, he alluded to 'lack of vigilance' on the part of the offending Patriot crew as a major factor in the Tornado fratricide. General Vane also expressed a general concern about 'lack of comprehension of the tactical situation' and a resulting 'unwarranted trust in automation.' ARL's assessment suggested that he was correct on all of these points. Post-mortem analyses suggested that the Tornado should have been on the Patriot unit's situation display before its classification was changed from unknown air breathing threat to anti-radiation missile. The crew apparently did not notice the change in classification from air-breathing threat to ARM—lack of vigilance—and did not challenge or override that system decision. It should be noted, however, that Patriot crew shifts during OIF were quite long (12 hours or more) and frequent unit moves were the norm. The unit's personnel reportedly were exhausted by near 24-hour operations, and it is not reasonable to expect high levels of vigilance under such circumstances.

The crew also was criticized after the fact for not being more aware of the tactical situation—not being more context sensitive and situationally aware: 'If that track really was an ARM, where was the ARM carrier? Is there any reason to expect that a hostile ARM carrier would be here at this stage of the combat operation? The Iraqi air force hasn't flown since the operation started.' These comments were made by a senior air defense officer far removed from the stress of the moment and thus reflect considerable hindsight bias—of course they should have known better and acted accordingly. However, is it reasonable to expect junior personnel with limited training and minimal experience to have properly overridden the system's recommendations under such circumstances? The key to developing better sensemaking abilities and enhanced SA is a higher level of crew expertise: skills, knowledge, aptitude, and relevant experiences (Endsley, 2006). And even then, one must ask whether that expertise can reliably be applied in a stressful, time-sensitive combat setting.

The OTA's (1993) fratricide assessment in the aftermath of the First Gulf War (ODS) concluded that reducing fratricide is feasible, but eliminating it is *operationally unrealistic*. The previous discussion strongly supports this conclusion. That same report also cautioned that focusing on fratricide reduction alone could be counterproductive in the sense that overly restrictive rules of engagement might increase losses due to missed threats. All of the aforementioned paths forward—better visual combat identification, more reliable TI systems, and improved SA—have the potential to contribute to a reduction in fratricide. But none of these solutions alone or in combination holds the potential for a silver bullet solution to the combat identification problem. Fratricide is and will continue to be an unfortunate fact of life in contemporary combat operations. The only practical course of action going forward is to try to reduce the rate through a combination of more reliable technical means coupled with higher levels of operator expertise.

There is no technical silver bullet on the immediate horizon. As has been found in many other situations, infusions of technology often are accompanied by increased aptitude and training requirements for users of that technology. Woods and Hollnagel (2006, p. 167) asserted, for example, that new technologies often do not 'reduce workload or simplify tasks for operators.' Instead, new capabilities are exploited by commanders, and users are asked 'to do more, to do it faster, and to do it in more complex ways' (Cordesman and Wagner, 1996, p. 25).

References

Barber, A. V., Drewfs, P. R., and Johnson, D. M. (1987). *Performance of Stinger Teams using the RADES (Realistic Air Defense Engagement Simulation) Multiple Weapon Simulation* (WP-FB-87–09). Alexandria, VA: U.S. Army Research Institute for the Behavioral and Social Sciences.

Barber, A. V., Drewfs, P. R., and Lockhart, J. M. (1987). *Effective Stinger Training in RADES (Realistic Air Defense Engagement Simulation)* (ARI-FB-87–02). Alexandria, VA: U.S. Army Research Institute for the Behavioral and Social Sciences.

Cohen, M. S., Parasuraman, R., and Freeman, J. T. (1997). *Trust in Decision Aids: A Model and its Training Implications* (Technical Report). Arlington, VA: Cognitive Technologies, Inc.

Cordesman, A. H., and Wagner, A. R. (1996). *The Lessons of Modern War, Vol. 4: The Gulf War.* Boulder, CO: Westview Press.

Dixon, S. R., and Wickens, C. D. (2006). Workload and automation reliability in unmanned air vehicle control: A reliance-compliance model of automation dependence in high workload. *Human Factors*, *48*, 474–486.

Dzindolet, M. T, Pierce, L. G, Beck, H. P., Dawe, L. A., and Anderson, B. W. (2001). Predicting misuse and disuse of combat identification systems. *Military Psychology, 13*, 147–164.

Endsley, M. R. (1996). Automation and situation awareness. In R. Parasuraman and M. Mouloua (eds), *Automation and Human Performance: Theory and Applications*. Hillsdale, NJ: Erlbaum.

Endsley, M. R. (2006). Expertise and situation awareness. In K. Ericsson, N. Charness, P. Feltovich, and R. Hoffman (eds), *Expertise and Human Performance*. New York: Cambridge University Press.

Frederickson, E. W., Dawdy, E. D., and Carter, R. L. (1982). *Feasibility of a Realistic Air Defense Experimentation System for Evaluating Short-range and Man-portable Weapon System Operators* (ARI-RR-1340). Alexandria, VA: U.S. Army Research Institute for the Behavioral and Social Sciences.

Hawley, J. K., and Mares, A. L. (2006). *Developing Effective Human Supervisory Control for Air and Missile Defense Systems* (ARL-TR-3742). Adelphi, MD: U.S. Army Research Laboratory.

Kogler, T. M. (2003). *The Effects of Degraded Vision and Automatic Combat Identification Reliability on Infantry Friendly Fire Engagements* (ARL-RP-0065). Adelphi, MD: U.S. Army Research Laboratory.

Lee, J. D., and See K. A (2004). Trust in automation: Designing for appropriate reliance. *Human Factors, 46*, 50–80.

Masalonis, A. J., and Parasuraman, R. (1999). *Applying the Trust Concept to the Study of Human-automation Interaction*. Washington, DC: Cognitive Sciences Laboratory, The Catholic University of America.

Norman, D. A. (2007). *The Design of Future Things*. New York: Basic Books.

Office of Technology Assessment (1993). *Who Goes There: Friend or Foe?* (OTA-ISC-537). Washington, DC: US Government Printing Office.

Parasuraman, R., and Riley, V. (1997). Humans and automation: Use, misuse, disuse, abuse. *Human Factors, 39*, 230–252.

Reason, J. T. (1990). *Human Error*. New York: Cambridge University Press.

Rovira, E., McGarry, K., and Parasuraman, R. (2002). Effects of imperfect automation on decision making in command and control. In *Proceedings of the Human Factors and Ergonomics Society 46th Annual Meeting* (pp. 428–432). Santa Monica, CA: Human Factors and Ergonomics Society.

Sheridan, T. B. (2002). *Humans and Automation: System Design and Research Issues*. New York: Wiley.

Sparta, Inc. (2002). *Combat Identification Fratricide Research Study* (Final Report). Arlington, VA: Author.

Steinweg, K. K. (1995). Dealing realistically with fratricide. *Parameters*, Spring, 4–29.

Weick, K. E., Sutcliffe, K. M., and Obstfeld, D. (2005). Organizing and the process of sensemaking. *Organizational Science*, *16*, 409–421.

Woods, D. D., and Hollnagel, E. (2006). *Joint Cognitive Systems: Patterns in Cognitive Systems Engineering*. New York: Taylor and Francis.

Chapter 20

The Case for Active Fratricide Avoidance in Net-Centric C2 Systems

John Barnett
U.S. Army Research Institute for the Behavioral and Social Sciences

Combat identification, the discrimination between friendly and enemy units in combat, is becoming more and more important as modern combat becomes more complex. As warfare has evolved, distances between combatants have increased. Engagements have moved from face-to-face close combat, to lines of musket men, to direct fire at the limits of visual range, to beyond visual range (BVR), where friendly and enemy units are represented iconically on display screens. This distance has complicated combat identification and made it more difficult for soldiers to identify friend from foe.

Net-centric command and control (C2) systems bring additional 'distance' because soldiers see only icons generated by the computer rather than units, vehicles, or individual soldiers. Instead of seeing things with their own eyes, they must rely on data that has been filtered and processed by an electronic system. Soldiers must trust the system to correctly classify the target and display the correct icon to distinguish between targets, friendly forces, and civilians.

In addition, the availability of electronic systems makes automation more readily available, including automated systems which can alert users to potential fratricide situations. Automated alerting systems can be a valuable aid for users to recognize and avoid potential fratricide. However, other areas which use automation, such as aviation, have found that automation sometimes has unintended consequences which make situations worse rather than better. Therefore, the designers of future automated alerting systems should carefully consider the abilities of the user to prevent user-automation interface problems that could, at worst, increase the potential for fratricide.

This chapter discusses factors to consider in the application of automation to aid in reducing fratricide. The purpose of the chapter is to consider how automation could be used to aid in fratricide avoidance, and what factors should be taken into account in the design of such systems.

Combat Identification and Fratricide Avoidance

The benefit of combat identification is primarily to prevent fratricide; the accidental attack of friendly forces by other friendly forces. Fratricide often causes more destruction than a similar enemy attack because it is essentially the same as an ambush–soldiers are not prepared to receive attacks from friendly units, and thus are caught unawares. For a number of reasons, fratricide can have a more devastating effect on combat effectiveness than a similar attack by the enemy. Fratricide affects soldiers' morale and their confidence in leadership and other friendly units. It also causes confusion that often disrupts operations.

Besides the casualties inflicted and equipment damaged, soldier morale takes heavy damage from fratricide. One incident during World War II during the Italian campaign illustrates how friendly fire can affect morale. A platoon of infantry received a single mortar round in their position. Believing it to be friendly fire, the soldiers abandoned their positions and scattered down the hillside. Such was the effect on their morale that the officers had an extremely difficult time reorganizing the platoon and getting the soldiers to return to their positions (Shrader, 1982).

After a fratricide incident, soldier confidence in both leadership and other friendly units drops considerably. A fratricide incident that occurred in Europe during World War II between American infantry and armor units caused ill will between the units. Fights between infantry and tankers occurred in hospitals where casualties were sent and in rest areas behind the lines (Shrader, 1982). Such feuds can have a devastating affect on unit cohesion.

Fratricide causes considerable confusion because of the efforts required to identify attacking units and stop the attacks. A fratricide incident can reduce combat power by making soldiers reluctant to fire on an enemy until they have double- and triple-checked their identity (Shrader, 1982).

Fratricide incidents have frequently disrupted planned operations. A major operation in the European Theater in World War II, operation COBRA, was delayed because a scheduled bombing of enemy lines actually landed inside friendly lines among the units scheduled to attack. The operation was delayed until the units could be reorganized (Shrader, 1982).

There is currently more emphasis on fratricide prevention because beliefs are changing about the nature of fratricide. It is no longer seen as an unavoidable artifact of warfare, but instead, as something that is well worth the effort to reduce as much as possible. Fratricide can also have detrimental political consequences, especially with a coalition force of different nations. Distrust between the military forces of different nations can weaken the coalition.

Combat identification is also important to prevent the unintentional or incidental injury of civilians or other neutrals, known as collateral damage. For obvious humanitarian reasons, it is important to minimize injuring civilians and damage to civilian property. However, there are also political and operational consequences of collateral damage. In situations where civil affairs personnel and military forces are trying to build trust with civilians, collateral damage can undermine their efforts.

As with any safety issue which involves human error, it is unreasonable to believe that fratricide can be prevented entirely. Rather, the objective of fratricide avoidance is to reduce the number of fratricide incidents to the lowest level possible while still maintaining effective combat power. At some point there is a trade-off between safety and effectiveness. As an extreme example, a commander could greatly reduce the probability of fratricide by prohibiting the employment of any weapon for any reason. While this would increase safety, it would destroy the unit's combat effectiveness. Therefore, the ideal fratricide avoidance measures are those which reduce the probability of fratricide while having a minimal effect on combat effectiveness. For this reason, the term 'fratricide avoidance' is used as opposed to 'fratricide prevention.'

Net-centric C2 systems have the potential both to help and hinder combat identification and subsequently fratricide avoidance. On one hand, net-centric C2 operations occur in what is essentially virtual space and are therefore once-removed from the battlespace. This has the potential to make it that much harder to discriminate between friend and foe. On the other hand, the automation inherent in net-centric C2 could be used to support friend-or-foe decisions.

Background

Net-Centric Command and Control Systems

C2 is necessary to synchronize the efforts of military forces. This is done by communicating orders, reports, graphics, and other data throughout the military command structure. Modern C2 often moves data over a network, much like information moves over the Internet. A typical net-centric C2 system might consist of computer workstations that are connected by a web of servers and communications nodes. Although there are a number of different net-centric C2 systems with unique features, all systems tend to have certain similarities. They all tend to be computer systems linked together over a network, and share C2 information between the various nodes. Most transmit information wirelessly, using line-of-site radio or satellite transmissions. Combat related information is typically shared across units of the same echelon, as well as being distributed to higher and lower echelons. User controls and displays often mirror personal computer (PC) workstations, and may in fact be PC or laptop computers connected to a network and running specialized software.

On land, workstations and displays may be in fixed or mobile command centers, or may be in individual vehicles such as tanks, infantry fighting vehicles, or smaller vehicles such as jeeps and HMMWVs (High-Mobility Multi-Wheeled Vehicles). Naval net-centric C2 systems may connect a ship's command centers with other ships and aircraft to coordinate anti-air, anti-surface, and anti-submarine warfare.

In the air, net-centric C2 systems typically coordinate the defensive counter-air mission, but may support other missions as well.

Individual workstations get their information from a network that is normally providing data continuously. The information from the network is processed and displayed to the operator. Data on friendly forces can come from inputs to the network from friendly units, such as Identification Friend or Foe (IFF) transmissions of aircraft or 'heartbeat' signals from ground vehicles. Information on enemy forces is less certain, and is normally either entered manually or detected electronically, such as by radar or sonar. When detected electronically, the classification of units as enemy forces is normally verified by an operator before being released to the network. This chain of information can be interrupted for a number of reasons. Communications failures, problems with the interoperability of different systems, or the inclusion of non-networked forces can make it difficult to get the information to the correct display.

Automation Aiding in Fratricide Avoidance

As net-centric C2 systems operate in computer networks, automation is often used to provide alerts to operators. These alerts are designed to direct the operator's attention to important information, such as safety and fratricide concerns. Alerting software can monitor the data stream transmitted across the network and provide alerts when certain trigger conditions are met.

The types of alerts available might include, for example, when a ground vehicle is approaching a hazardous area such as a mine field or area contaminated with chemical weapons. Another alert might be when a vehicle strays from its own unit's area of responsibility (AOR) and into the area of another unit. Straying across unit boundaries has the potential to increase the likelihood of fratricide as a vehicle in an unexpected location is often considered an enemy. In the stress and confusion of combat operations correct identification can be difficult. Therefore, ensuring unit members stay within their own boundaries facilitates fratricide avoidance.

Ideally, the types of automated alerts most needed are those where the human operator has difficulty identifying when dangerous conditions occur. In this case, the alert would direct the operator's attention to those conditions. For example, a person might have difficulty remembering the times when a restricted zone, such as a restricted fire area or no-fly area, is active and when it is not. Automation has no such problem, and can assist the user by highlighting the area on the display when it is active and de-emphasizing it when it is not.

The Need for Adaptable Automation

One of the complexities of designing automated aids, such as automated alerts, is that people have differing requirements of the automation at different times and in different situations. For example, the level of automation needed may vary depending on the levels of workload and performance stress. During periods of low workload and stress, people tend to become bored and attention wanders. At these times, automation can help direct the user's attention to important information or events they might otherwise miss. Conversely, during periods of high workload and stress, people's attention tends to focus on a central task or event and disregard other information, even important information. Also, there may be so much information presented to the user that it cannot all be mentally processed. In this case, the task of the automated aiding system is to present important information to the user in such a way that it does not interfere with the user's task, nor does it increase their already overburdened mental workload.

A further consideration is that the relative importance of information presented to the user may change with the situation. Information that the user might consider important during tedious parts of a mission might be trivial during active combat.

Change Blindness

One of the reasons that alerts are necessary is that sometimes people working with a computer display will miss changes that occur in the display, a phenomenon called change blindness (Durlach and Chen, 2003; Durlach, 2004). Change blindness can occur for a number of reasons, the most common of which is that the change occurred when the operator was distracted, or during eye blinks or saccades (eye movements). Minor changes in on-screen icons during such distractions can be very difficult to notice. A number of factors influence whether the operator notices the change. Icons which move or simulate movement (e.g., blink on-and-off) are more noticeable than static icons (Wickens, 1992), and are therefore less likely to be missed due to change blindness. Icons which change color are also fairly noticeable. However, icons which change shape or are near the periphery of the display are less salient.

Also, recent research suggests that people are better at correctly detecting item appearance than disappearance. Durlach, Kring, and Bowens (2008) found that, once the false alarm rate was corrected for, people correctly detected the appearance of icons on a display more often than they detected the icon's disappearance. This means that, even in a moderately cluttered display screen, unit icons that disappear from the screen due to loss of signal have less of a chance of being noticed by the operator than units that join the net and appear on the screen.

Considerations for the Use of Automation in Fratricide Avoidance

As a form of automation, an alerting system used to reduce fratricide would have many of the benefits and drawbacks of other forms of automation. However, there are a number of issues to consider when employing automation.

Automation's Effect on Situation Awareness

Situation awareness (SA) has been defined by Endsley as 'the perception of elements in the environment within a volume of time and space, the comprehension of their meaning, and the projection of their status in the near future' (Endsley, 1995, p. 96). Lack of SA is frequently cited as the cause of accidents, particularly in aviation (Jones and Endsley, 1996), and good SA is critical for safe driving (Gugerty, 1997). Logically, poor SA would increase the chances of fratricide, while good SA would enable a reduction. For this reason, any method that increases an operator's SA would logically reduce the probability of fratricide incidents.

Alerts can be used to direct the user's attention to important events and hopefully increase SA. If the user happens to miss an important change in the display, through change blindness for example, an alert can direct the user's attention to the change to ensure the user attends to the information. In this way, automated alerts can improve the user's SA by guiding the user's attention to important information about the situation. However, if an alert occurs at the wrong time, it can distract the user from more important information and actually contribute to change blindness.

Automation's Effect on Mental Workload and Work Flow

Alerts can sometimes interfere with the task the operator is performing and can actually degrade performance. An example of such intrusive automation which should be familiar to anyone who uses common office software suites is the animated figure which offers to provide 'help' to the user when the software determines the user is performing tasks incorrectly or inefficiently. When the animation appears, it prevents the user from continuing to work on the task and forces the user to attend to the animation. Many users find this frustrating and deactivate the automated help function. While such automated help is valuable in theory, in practice it can become intrusive. For example, during one experiment with a highly automated future combat system, observers noted that system operators who were allowed to set a number of automated alerts frequently set the majority of the alerts at first, but during operations found them to be intrusive and either ignored or disabled them (P. Durlach, personal communication, March 26, 2008).

How Automation Affects Decision-making and User Confidence

Decision-making How automation affects people's decision-making is not entirely clear. Under certain conditions, automation aids decision-making, but under different conditions people make worse decisions. In one case, the research showed decision-making by air crews was aided when automation presented accurate status or recommendation information (Crocoll and Coury, 1990). On the other hand, other research showed air crews using automation displayed worse decision-making than non-automated crews (Bowers, Oser, Salas, and Cannon-

Bowers, 1996). Also, when people are presented with decisions by automation, they tend to accept it at face value and curtail any further data collection (Mosier, Heers, Skitka, and Burdick, 1997). For example, Chappell (1997) found that when displays offer redundant data, users will tend to use the easiest source, and may not cross-check the other source to determine if there is a mismatch.

Reliability and user confidence Automation is often seen as a method for reducing human error. However, like humans, automation is never 100 percent reliable. Automation may fail due to a straightforward malfunction, or it may fail because of an unforeseen situation for which it was not programmed. This has implications for automated systems where the consequences of failure can be dire, such as automated target detection systems which are currently under investigation, or systems which operate autonomously, such as certain air defense systems. However, there are also implications for how humans interact with automation.

People's attitudes towards automation reliability, and consequently their confidence in the automation are somewhat complex. People's confidence in automation is affected by its reliability. Bowers, Oser, Salas, and Cannon-Bowers (1996) pointed out that people's trust in automation follows a continuum from too little trust, which discourages people from using the automation, to too much trust, which encourages complacency. When people avoid using the automation, they are failing to exploit a possibly advantageous tool. On the other hand, sometimes when people have automated systems to assist them, they can become complacent and allow the automation to perform the task without sufficient supervision (Parasuraman, Molloy, and Singh, 1993). When operators are complacent, they tend not to monitor the operation of the automation sufficiently, so that they are surprised when the automation fails. While this may not be a problem as long as the automation is correctly performing the task, it can be a serious challenge if the automation were to malfunction. Finding the right balance between caution and trust for a particular system can sometimes be quite challenging.

Muir (1994) and Muir and Moray (1996) found that people tend to trust automation the same way they would trust people, in that they place more trust in entities that are predictable and dependable. Lee and Moray (1994) and Eidelkind and Papantonopoulos (1997) found that people's confidence in automation dropped dramatically once they experienced a failure of the system. Following this drop their confidence returned slowly, provided the automation performed correctly, but was never as high as before the failure.

Automation failure and failure recovery An important issue to consider is how automation failure, or the possibility of failure, affects human-automation interaction. Although systems are most often designed based on perfect performance, in fact, all systems perform incorrectly some percentage of time. The potential for malfunction affects how people interact with automated systems in a number of ways. The most obvious is that when a system fails, the operator must scramble to switch to manual operation in order to continue to perform necessary tasks. If the operator is not aware of how much of the task has been performed by the automation (known as the 'out of the loop' problem [McClumpha, James, Green, and Belyavin, 1991]), there may be a significant disruption in task performance while the operator attempts to complete the task manually. While this may be tricky during normal operations, it becomes particularly difficult when things are not going well (Sarter, 1996).

For example, some Tactical Operations Centers (TOC), use an electronic map display which shows the movements of ground forces in near real time. If the TOC happens to lose the display due to a malfunction, there would have to be considerable effort expended to create the same information picture on a common map board. There would also be a disruption in operations while the map board is updated and unit positions verified. If the disruption occurred at the wrong time, it could severely impact the ongoing operation.

Another problem with automation failure is the system may fail without the operator being aware of the malfunction. People typically consider automation to be highly reliable, and are often surprised when it fails (Sarter and Woods, 1997). This may cause them to fail to monitor the automation sufficiently. In addition, when there is a discrepancy between themselves and the automation, people often assume the automation is correct, a phenomenon known as 'automation bias' (Mosier, Skitka, Heers, and Burdick, 1998).

Automation bias can be a potential problem for an automated system designed to aid the operator in distinguishing between friendly and enemy targets. Normally, the automation would present a classification which would be verified by the operator. However, an operator who experiences automation bias may not sufficiently monitor the automation and may not catch mis-categorizations, allowing friendly forces to be classified as enemy and enemy as friendly. Such mistaken classification can get friendly forces fired upon by other friendlies, while enemy forces can operate without hindrance.

Automation control tradeoffs Although systems can be set to run autonomously, when safety is a concern the final decision should be made by a human. Billings (1997) listed an automation control hierarchy which describes how people and automation can share control of a system:

- autonomous operation;
- management by exception;
- management by consent;
- management by delegation;
- shared control;
- assisted manual control;
- direct manual control.

In this list, the highest level is fully autonomous operation, where there is no input by the human, and the lowest level is direct manual control, where there is no input by the automation. In between these extremes are various levels of shared control. Each level of control has tradeoffs between workload and control. The lowest level has the most human control, but normally the highest workload, while the highest level reduces both human workload and control to zero. Since the human has the ultimate responsibility for the actions of the system, the logical way to choose the appropriate level of control is to choose the option which gives the automation the highest level of control which is acceptable to the user.

Automated Aiding in Combat Identification and Fratricide Avoidance: Recommendations

The use of automated systems on the battlefield provides an opportunity to use this automation to aid combat identification and thus reduce fratricide incidents. Automation is used in commercial aviation to aid human performance and reduce aircraft accidents (Wiener and Nagel, 1988). In the same way, automation can be used to reduce fratricide incidents. However, given some of the unintended consequences of automation experienced in other domains, it is important to carefully design automated aids so that they help reduce fratricide incidents and do not add to the user's task complexity.

Active versus Passive Automation

One of the considerations for automated aids for fratricide avoidance is whether the automation should be passive or active. Used in this context, passive automation would present timely information to the user that is pertinent to avoiding fratricide, but would require no action from the user. On the other hand, active fratricide avoidance automation would present information to the user and require the user to take action to acknowledge the information. There are advantages and disadvantages to each type of automation, however, because of the criticality of fratricide avoidance, active measures may have more advantages than disadvantages in most, though not all, situations.

The primary reason for suggesting active alerts over passive alerting is that the same situations where alerts would be beneficial, such as high stress environments, are those where the user could miss important information through attentional narrowing and change blindness. Attentional narrowing, focusing on a central task at the expense of peripheral information, tends to be more prevalent in stressful situations (Wickens, 1992). Coupled with what is known about change blindness, we could theorize that in many military situations where fratricide is a possibility, a user could easily miss important information displayed by the automated system if it was presented passively. There is a greater chance of directing the user's attention to fratricide related information if the user is obliged to take some action in response to the information, such as pressing a button to acknowledge it.

For example, an alert notifying the operator that a new friendly unit has joined the network is less critical than an alert saying a friendly unit has been selected for targeting. For the new friendly unit alert, a simple notice on the display that requires no action by the operator would suffice. However, targeting a friendly unit has dire enough consequences that the automation should doubly ensure the operator is aware of their actions. One method of ensuring the operator's active participation would be to constrain the operator from targeting a unit classified as friendly. Because there is the possibility of mis-classifying the unit by the automation, the operator should be allowed to override the constraint if the operator determined the unit is indeed an enemy. The operator's actions to override the constraint imposed by the automation would ensure the operator was actively involved in the process and reduce the chance that the operator would act unwittingly. Besides active versus passive automation, there are also a number of other considerations to developing a usable automated fratricide avoidance aid.

Other Considerations

Reduce intrusiveness There are some cautions to using active alerts. In some cases, forcing the user to acknowledge alerts can actually interfere with correcting the problem the alert is warning about. In a high stress and workload environment, interrupting the user's work flow by forcing them to attend to an alert can lengthen the time it takes to correct the problem. In extreme cases, where multiple alerts are competing for the user's attention, they can further hinder the user from correcting the problem (Wickens, Gordon, and Liu, 1998).

Systems can be designed such that the alerts do not become overly intrusive. In one experiment, Ross, Barnett, and Meliza (2007) had participants monitor a simulation of a network-enabled C2 system which had an automated alerting system to notify the operator of various events that might be fratricide related. They compared the participant's SA and workload with the alerting system enabled and disabled. They found whether the alerting system was enabled or not had no significant effect on SA, but self-reported workload was significantly lower with the system enabled.

Automation should be adaptable Even though an active alerting system might be preferable; it should also be carefully designed so that it does not become overly intrusive. One means to do this is to allow users to modify the level of interaction required by the automated aid, which could allow for adaption to changes in the environment and user priorities. Although requiring the user to acknowledge important information presented by the aid ensures that the user is aware of it, the user should be able to quickly and simply modify when the interaction is required and when it is not. For example, the user could push a 'don't bother me' switch to set the alerts from active to passive for a set period of time so that he or she could finish a task without interruption. After a set period of time, the aid would automatically revert to the active state. The time limit is to prevent the user from forgetting to change back to the active state and consequently missing important information.

Automation's actions should be visible The automation should keep the user informed about what it's doing and its 'health' status in a non-intrusive way. Failures should not be silent or hidden. This is important not only as a means for the user to estimate the automation's reliability, but also to help the user revert to manual operation should the automation malfunction. For example, a simple display that shows the current state of the automation, possibly in symbolic form, would be most helpful to the user and would keep the user in the loop. For example, a 'stoplight' display that shows either green (proper operation), yellow (partial malfunction), or red (inoperative) would provide status information in a way that doesn't overburden the user's mental workload.

The human operator should be in charge Since the human operator will ultimately be held accountable for safety and fratricide incidents, the human must be given ultimate authority over the automation. Thus, the highest acceptable level of automation in the automation control hierarchy shown above should be 'management by exception,' where the system works autonomously unless the operator overrides it. The operator must also be able to easily monitor the automation to be able to override it if required, which provides another reason why that actions of automation should be visible. This means the operator must have access to sufficient information, such as the state of the automation, to be able to make an informed decision quickly, and the automation should provide as much aid to the operator as necessary for the operator to make a correct decision. When necessary, the operator must be able to override the automation easily and quickly.

As an example of why this is necessary, suppose an armored vehicle had a fratricide avoidance system which prevented the crew from firing on another vehicle assessed as friendly. A situation could arise where the automation erroneously determined a vehicle as friendly, when actually it was an enemy preparing to fire on a friendly vehicle. Under fully autonomous operation, if the system prevented the crew from targeting the enemy vehicle (erroneously determined to be friendly), the crew would be unable to fire on the enemy, who would be fully capable of firing on them. Thus the automation, rather than increasing safety, would put the crew in danger with no way of recovering.

Alternately, a system which automatically targeted and fired on vehicles determined to be enemies, or one with an active protection system like those currently under development, could commit fratricide if it erroneously classified a friendly unit as enemy. Again, under full autonomy the human operators would have no way of preventing fratricide.

Automation's usability should be verified Systems are always designed to work flawlessly, but there are often extraneous variables that creep in when the system is fielded causing unintended consequences. Some users may not employ the systems the way the designers had intended. The environment the system was designed to operate in may be different, it may change periodically, or it may be more complex than originally envisioned. All of these factors may cause the system to operate in ways that were not intended.

The way to verify that the system will operate correctly in the field is to test it, either in the field or under realistic field conditions. In addition, operators that accurately represent the population expected to operate the system in the field should be used. Such user testing is one of the cornerstones of good human-automation interaction, but often time or budgetary constraints minimize or eliminate this vital step in systems design. Given that the consequences of failure of the human-automation system to avoid fratricide are dire, thorough user testing is critical in this area.

Conclusion

Automation has the potential to be a valuable aid to fratricide avoidance. Automated alerts could be designed to actively engage operators in fratricide avoidance. However, research has shown that there are a number of issues dealing with automation that must be addressed to ensure automation is a benefit and not a hindrance. There must be an understanding of how automation affects SA, team interaction, and decision-making. There must also be an understanding of how humans work with automation, in particular, how this is different from how people work with other people, as well as considerations about how people deal with automation failure. While some of these concerns can be complex, with a good understanding of the issues and careful design and testing, automated systems can be developed which will work with the operator to reduce the probability of fratricide to its lowest level.

References

Billings, C. E. (1997). *Aviation Automation: The Search for a Human-centered Approach*. Mahwah, NJ: Lawrence Erlbaum.

Bowers, C. A., Oser, R. L., Salas, E., and Cannon-Bowers, J. A. (1996). Team performance in automated systems. In R. Parasuraman and M. Mouloua (eds), *Automation and Human Performance: Theory and Applications* (pp. 243–261). Mahwah, NJ: Lawrence Erlbaum.

Chappell, S. L. (1997, September). Cross checked but not seen: The effect of automation and reliable systems. *Poster Presented at the 41st Annual Meeting of the Human Factors and Ergonomics Society*, Albuquerque, NM.

Crocoll, W. M., and Coury, B. G. (1990). Status or recommendation: Selecting the type of information for decision aiding. *Proceedings of the Human Factors Society 34th Annual Meeting* (pp. 1524–1528). Santa Monica, CA: The Human Factors Society.

Durlach, P. J. (2004). Change blindness and its implications for complex monitoring and control systems design and operator training. *Human-Computer Interaction*, 19, 423–451.

Durlach, P. J., and Chen, J. Y. C. (2003). Visual change detection in digital military displays. *2003 Proceedings of the Interservice/Industry Training, Simulation, and Education Conference (I/ITSEC)*. Arlington, VA: National Training Systems Association.

Durlach, P. J., Kring, J. P., and Bowens, L. D. (2008). Detection of icon appearance and disappearance on a digital situation awareness display. *Military Psychology*, 20, 81–94.

Eidelkind, M. A., and Papantonopoulos, S. A. (1997). Operator trust and task delegation: Strategies in semi-autonomous agent system. In M. Mouloua and J. M. Koonce (eds), *Human Automation Interaction: Research and Practice*. Mahwah, NJ: Lawrence Erlbaum.

Endsley, M. R. (1995). Toward a theory of situation awareness in dynamic systems. *Human Factors*, 37, 32–64.

Gugerty, L. J. (1997). Situation awareness during driving: Explicit and implicit knowledge in dynamic spatial memory. *Journal of Experimental Psychology: Applied*, 3, 42–66.

Jones, D. G., and Endsley, M. R. (1996). Sources of situation awareness errors in aviation. *Aviation, Space, and Environmental Medicine*, 67, 507–512.

Lee, J. D., and Moray, N. (1994). Trust, self-confidence and operators' adaptation to automation. *International Journal of Human-Computer Studies*, 40, 153–184.

McClumpha, A. J., James, M., Green, R. G., and Belyavin, A. J. (1991). Pilot's attitudes to cockpit automation. *Proceedings of the Human Factors Society 35th Annual Meeting*. Santa Monica, CA: The Human Factors Society.

Mosier, K. L., Heers, S., Skitka, L. J., and Burdick, M. (1997). Patterns in the use of cockpit automation. In M. Mouloua and J. Koonce (eds), *Human-automation Interaction: Theory and Applications*. Mahwah, NJ: Lawrence Erlbaum.

Mosier, K. L., Skitka, L. J., Heers, S., and Burdick, M. (1998). Automation bias: Decision making and performance in high-tech cockpits. *International Journal of Aviation Psychology*, 8, 47–63.

Muir, B. M. (1994). Trust in automation: Part I. Theoretical issues in the study of trust and human intervention in automated systems. *Ergonomics*, 37, 1905–1922.

Muir, B. M., and Moray, N. (1996). Trust in automation. Part II. Experimental studies of trust and human intervention in a process control simulation. *Ergonomics*, 39, 429–460.

Parasuraman, R., Molloy, R., and Singh, I. L. (1993). Performance consequences of automation-induced 'complacency.' *International Journal of Aviation Psychology*, 3, 1–23.

Ross, J. M., Barnett, J. S., and Meliza, L. L. (2007, October). Effect of audio-visual alerts on situation awareness and workload in a net-centric warfare scenario. *Poster Presented at the 51st Annual Meeting of the Human Factors and Ergonomics Society*, Baltimore, Maryland.

Sarter, N. B. (1996). Cockpit automation: From quantity to quality, from individual pilots to multiple agents. In R. Parasuraman and M. Mouloua (eds), *Automation and Human Performance: Theory and Applications*. Mahwah, NJ: Lawrence Erlbaum.

Sarter, N. B., and Woods, D. D. (1997). Team play with a powerful and independent agent: Operational experiences and automation surprises on the Airbus A-320. *Human Factors*, 39, 553–569.

Shrader, C. R. (1982). *Amicicide: The Problem of Friendly Fire in Modern War*. Fort Leavenworth, KS: Combat Studies Institute.

Wickens, C. D. (1992). *Engineering Psychology and Human Performance* (2nd edn). New York: HarperCollins.

Wickens, C. D., Gordon, S. E., and Liu, Y. (1998). *An Introduction to Human Factors Engineering*. New York: Longman.

Wiener, E. L., and Nagel D. C. (1988). *Human Factors in Aviation*. San Diego, CA: Academic Press.

Chapter 21
Mitigating Friendly Fire Casualties through Enhanced Battle Command Capabilities

Jean W. Pharaon
U.S. Army Research Laboratory

Introduction

Background

The reality of any armed conflict is the possible yet evitable incidence of friendly fire casualties, commonly referred to as fratricide. Fratricide is as old as warfare itself. Fratricide can be broadly defined as the employment of friendly weapons and munitions, with the intent to kill the enemy or destroy enemy equipment or facilities, which results in the unforeseen and unintentional death or injury to friendly personnel. This definition of fratricide is found in several Army Field Manuals (FM) and Pamphlets (PAM) including FM 3.0 (Operations), FM 17–97 (Cavalry Troop), and TRADOC Pam 525–58 (Military Operations: U.S. Army Operations Concept for Combat Identification).

The modern battlefield is just as complex as in the past, but more dynamic and lethal. Although modern warfare has thrived significantly with extraordinary technological and tactical advances, fratricide prevention is still considered one of the colossal challenges facing all combat leaders. Although progress has been made in the development of friendly forces visibility, tracking systems such as blue force (BLUFOR) trackers, and other Combat Identification devices, fratricide remains a persistent problem that generations of warriors will have to address.

According to FM 17–97, situational awareness (SA) is the key to combat identification, and must be maintained throughout an operation to avoid fratricide. However, neither visual identification nor situational awareness has been enough to completely eliminate fratricide. Table 21.1 is a breakdown of the number of friendly fire casualties of some of the past conflicts involving the United States.

Casualty data for Operation Iraqi Freedom (OIF) during the initial fighting phase of the conflict (March 21 to April 30, 2003) are presented in Table 21.2. Note that for a little more than one month of fighting, about 14 percent of the casualties occurred as a result of friendly fire.

Table 21.1 Historical American Fratricide rates in past United States conflicts (adapted from Steinweg, 1995)

Conflict	Source of Data	Fratricide Rate
World War I	Besecker Diary (Europe)	10% Wounded in action (WIA)
World War II	Hopkins New Georgia Burma Bougainville Study	14% Total casualties 14% Total casualties 12% WIA 16% Killed in action (KIA)
Korea	25th Infantry Division	7% Casualties
Vietnam	WEDMET (autopsy) WEDMET (autopsy) WEDMET Hawkins	14% KIA (rifle) 11% KIA (fragments) 11% Casualties 14% Casualties
Just Cause	U.S. Department of Defense	5–12% WIA 13% KIA
Desert Storm	U.S. Department of Defense	15% WIA 24% KIA

Table 21.2 Attributed causes of U.S. deaths during the initial phase of OIF (adapted from Caseley et al., 2006)

Cause	Number	Percentage
Accident	30	22
Killed in Action	90	65
Friendly Fire	19	14

Friendly fire accidents may be grouped into three main categories: human, environmental, and technological (Kulsrud, 2003). Human causes include a lack of training, situational awareness failures, lack of discipline, and combat stress (Kulsrud, 2003). 'Given the clear preponderance of direct human error as the source of most [fratricide] incidents, it is manifest that preventative measures must be directed toward the correction or improvement of human frailties, and these, as always, are the factors least amenable to correction' (Stein and Fjellstedt, 2006). 'Weather, darkness, terrain, and visual obscurants affect the ability to locate and identify friendly or enemy combat vehicles' (Kulsrud, 2003, p. 6). 'During Desert Storm, 11 of the 13 fratricide incidents reported by the Army were attributed to environmental factors' (p. 6).

The U.S. Army has made great strides in the development of situational awareness tools to give commanders a digital picture of the battlefield. Aircraft have Identify

Friend or Foe (IFF) systems on board and some ground vehicles are equipped with Enhanced Position Location and Reporting System (EPLRS) radios; both are designed to identify aircraft and vehicles as friendly. Nevertheless, technology must be correctly maintained, operated, and used to be effective (Kulsrud, 2003), keeping in mind that technology has its constraints and limitations.

With fratricide being an undesirable outcome in armed conflicts, numerous initiatives have been taken to eliminate or at least reduce it. From a tactical and doctrinal standpoint, lessons learned from past conflicts have influenced the development of new tactics, techniques, and procedures (TTPs). For instance, the U.S. Army Training and Doctrine Command (TRADOC) has developed a Fratricide Action Plan with the Combined Arms Command Training (CAC-T) at Fort Leavenworth as the overall proponent. This action plan coordinates and directs service schools' efforts to resolve recognized shortfalls in TTP products relating to the prevention of fratricide on the battlefield. The proponents review and update this action plan every six months until changes in doctrine, TTP, and course Programs of Instruction (POI) are accomplished (Department of the Army, 1992).

From a technological standpoint, the U.S. Department of Defense (DoD) embraced the Network Centric Warfare (NCW) concept in the mid-1990s. One application of NCW is the U.S. Army Battle Command System (ABCS), which is the backbone of the Army's Force XXI transformation initiative. One of the systems comprising ABCS is the Force XXI Battle Command, Brigade and Below (FBCB2).

The FBCB2 forms the principal Digital Command and Control System for the Army at brigade and below. All FBCB2/Embedded Battle Command (EBC) systems are interconnected through a communications infrastructure called the Tactical Internet to exchange Situation Awareness data and conduct Command and Control. Blue Force Tracking employs L Band satellite communication links that have proven to be reliable over long distances and throughout mountainous terrain (PM FBCB2, 2008). The FBCB2 system includes many desirable features:

- exchanges 40 different command and control messages (PM FBCB2, 2008);
- communicates with other ABCS systems (PM FBCB2, 2008);
- operates at temperatures from -40° to 120° Fahrenheit (PM FBCB2, 2008);
- integrates into 114 platforms ranging from M1A2SEP tank to the A2C2S command helicopter (PM FBCB2, 2008);
- implements all MIL STD 2525B symbology (PM FBCB2, 2008;

The prime contractor for the FBCB2 is Northrup Grumman Mission Systems with Raytheon Ruggedized Computer Division action as a sub-contractors. DRS (hardware supplier) and Comteck (satellite connection) are also prime contractors (PM FBCB2, 2008). Figure 21.1 provides a picture of the FBCB2 system.

Figure 21.1 The FBCB2 system hardware

Advances in the development of Combat Identification (CID) and BLUFOR Tracking systems such as the FBCB2 have yielded some rather promising results. The incidence of fratricide has been greatly reduced in many recent engagements and even completely eliminated in others. For instance, a 2006 Mitre Corporation presentation (Network Centric Warfare [NCW] in Western Iraq) credited the BLUFOR Tracking capability of NCW for the absence of fratricide in the Western Iraq Theater in the beginning phase of Operation Iraqi Freedom (Stein and Fjellstedt, 2006).

Purpose

Ideally, a battlefield completely free of fratricide is desirable. In essence, this vision is not impossible to achieve. The main question is how can it be achieved? Can a viable fratricide avoidance system be developed to mitigate human, environmental, and technological fratricide induced factors? The purpose of this study is to attempt to objectively answer these questions by first examining the FBCB2's capabilities and limitations through the analysis of Operation Iraqi Freedom (OIF) and Operation Enduring Freedom (OEF) user data. Furthermore, this chapter seeks to make some recommendations on the possible effective employment of the FBCB2 in conjunction with other available CID systems. An important point of this chapter is to delineate the limitations of the FBCB2 being restricted only to vehicle platforms. The argument points to the feasibility of the FBCB2 incorporating currently available portable systems that could be deployed with dismounted troops. This chapter presents the results from a comprehensive literature review and a survey of veteran OIF/OEF FBCB2 users.

Study 1: Literature Review

Method

Procedures As the main focus of this chapter is fratricide prevention or reduction from a technological perspective, namely BLUFOR trackers, the literature review effort was Network Centric Warfare (NCW) intensive as the NCW concept is the adopted approach used by the United States military to optimize command and control and to help reduce fratricide. To obtain the pertinent information, numerous online materials were explored. The following list of online resources (see below) enumerated the most useful websites used in the literature review. The websites listed in order of the quantity and quality of the information on reducing incidents of fratricide:

- The Defense Technical Information Center (http://www.dtic.mil)
- Google (http://www.google.com)
- The Program Executive Office Command, Control, and Communications Tactical (PEO C3T at http://peoc3t.monmouth.army.mil)
- PM FBCB2 (http://peoc3t.monmouth.army.mil/FBCB2/)
- RAND Corporation (http://rand.org/)
- Defense Update – The International Online Defense Magazine (http://www.defense-update.com)
- The U.S. Army Training and Doctrine Command (TRADOC at http://www.tradoc.army.mil/)
- The Program Executive Office Soldier (PEO Soldier at http://www.peosoldier.army.mil)
- The Program Executive Office for Simulation, Training, and Instrumentation (PEO STRI at http://www.peostri.army.mil/)
- Raytheon (http://www.raytheon.com/)
- Northrup Grumman Mission Systems (http://www.ms.northropgrumman.com)
- The MITRE Corporation (http://www.mitre.org/)
- The Program Executive Office Enterprise Information Systems (PEO EIS at http://www.eis.army.mil/)
- Military Information Technology (http://www.military-information-technology.com/)
- U.S. Army Natick Soldier Systems Center (http://www.natick.army.mil)
- The Federation of American Scientists (http://www.fas.org)
- The U.S. Army Research Laboratory (http://www.arl.army.mil)
- Jane's (http://www.janes.com)
- General Dynamics C4 Systems (http://www.gdc4s.com/)
- Defense News (http://www.defensenews.com/)
- ACM Technews (http://technews.acm.org/)
- Global Security (http://www.globalsecurity.org)

- The Defense Information Systems Agency (DISA at http://www.disa.mil/)
- The Armed Forces Communications and Electronics Association (AFCEA International at http://www.afcea.org)
- The U.S. Air Force Research Laboratory (http://www.wpafb.af.mil/AFRL/)
- The Defense Advanced Research Projects Agency (DARPA at http://www.darpa.mil)

Results and Discussion

Literature review The FBCB2 has proven to be an effective tool from the user's perspective as suggested by the results of the Fort Hood, TX, survey presented later. These results have only reinforced data from previous reports, studies, and surveys (Gonzales et al., 2007) that have likewise pointed to the overall satisfaction of FBCB2 users and the efficacy of this system on the battlefield. The FBCB2 system is not perfect, but it has performed according to its design purposes and expectations. However, such a system would be more potent in the prevention of fratricide was it not limited only to vehicle platforms.

As indicated earlier, a purpose of this chapter is to advocate for the enhancement of the FBCB2 BLUFOR Tracker capabilities with the addition or implementation of tracking capability for individual dismounted Soldiers. With current technological advances in computer memory capacity and size of portable or hand held computers, it will be possible for dismounted troops to accomplish a variety of tasks with relative ease while on the go. A sub-system of the FBCB2 such as a Personal Digital Assistant (PDA) is an excellent and feasible addition to the system. Efforts have been made in this regard. The U.S. Army has been particularly active in working to field such systems. The Land Warrior project envisions having dismounted Soldiers equipped with portable computers capable of performing most if not all of the functions of the current vehicle platform FBCB2 system. Some commanders in Iraq are already using what amounts to Force XXI Battle Command Brigade and Below (FBCB2) commander's digital assistants (CDAs; Ackerman, 2005). The Rand study referred to earlier, made the following recommendation, 'Provide battle command devices or at least BLUFOR Tracker (BFT) devices to dismounted units. Stryker Brigade Combat Team (SBCT) Soldiers requested the auto-population of dismount locations on FBCB2, at least down to the team level' (Gonzales et al., 2007, p. xxxiv).

The U.S. Army' s commander's digital assistant (CDA) Leader Planning Tool is an application built for militarized personal digital assistant (PDA), which provides situational awareness and mission planning capabilities for field commanders. It is intended for the battalion commander and staff, company commanders, and platoon leaders. CDA provides dismounted troops the same functionality of the U.S. Army Battle Command, Brigade-and-Below (FBCB2) command and control system. FBCB2 is the principal element in the 'Blue Force Tracking' situational awareness system, and integrates dismounted elements into the complete situational picture making it a key element in the system (Defense Update, 2007). With the software loaded and the system connected to a satellite, the CDA is a dismounted FBCB2 Blue Force Tracking system (Boland, 2007).

The CDA is designed for automatic communication in a peer-to-peer formation with other CDAs and supported communications devices. It can also operate in a network when required. Other communications features support 'Blue Force Tracking' facility to support situational awareness of all friendly troops (Defense Update, 2007). Developers designed the system primarily for the dismounted Soldier. The CDA fills a gap at the company level and below (Boland, 2007). The CDA is manufactured by Tallahassee Technologies, Inc. (Talla Tech), a U.S. division of Tadiran Communications, an Israeli-based defense corporation. Figure 21.2 presents Tadiran's RPDA–39 CDA model.

In 2006, a new version of Commander's Digital Assistant (CDA) was introduced by General Dynamics for the Land Warrior program. The system, also manufactured by Talla Tech, is currently available as Version 5, offering a larger color touch screen, hard disk, integral GPS and built-in satellite voice communications, and the capability to exchange voice messaging with other CDAs. The system uses U.S. Army Standard Battle Command software to provide dismounted leaders with a situational awareness display derived by FBCB2. The system also maintains constant position reporting for non-line-of-sight blue force tracking (Defense Update, 2007). Figure 21.3 presents General Dynamics' Version 5 CDA, the RPDA-57.

The systems can use various batteries installed in the front, in an expansion sleeve, or even externally to the units, and 'pool' the power to distribute it where it is needed (see Figure 21.4, Defense Update, 2007).

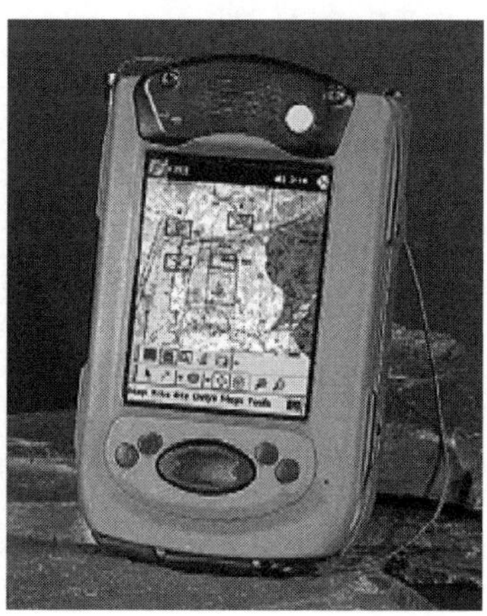

Figure 21.2 Tadiran's RPDA-39 CDA version available to both U.S. and Israeli forces

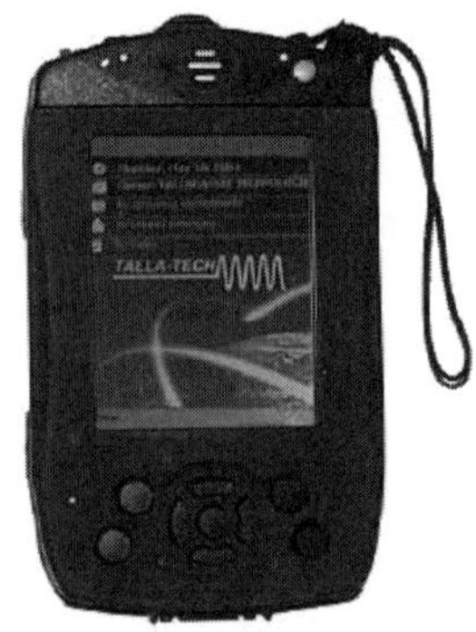

Figure 21.3 General Dynamics' RPDA-57 CDA for the Land Warrior program

Figure 21.4 A CDA with external battery adapter

At the Association of the United States Army (AUSA) 2007, Tadiran announced a new order worth $14.4 million for the supply of 5th generation RPDA (RPDA-57) systems (shown previously in Figure 21.3) for multiple programs run by the U.S. Army (Defense Update, 2007). Also at A.U.S.A 2007, Raytheon unveiled some details about its Commander's Digital Assistant (CDA). Raytheon's new device establishes the smallest, lightest package currently available for dismounted 'blue force tracking' applications. The new device weighs 4.5 to 5.6 pounds (depending on configuration) using an internal, rechargeable 10.8 VDC Lithium-ion battery pack sustaining five hours of operation (Defense Update, 2007).

The CDA also offers improved commonality with the Army's Air Warrior Electronic Data Manager, further improving its application for aviators and ground troops. The CDA communicates with existing networks such as the FBCB2 or Interactive Situational Awareness System (ISAS), using an integral satellite communications L band transceiver and GPS receiver set with anti-spoofing capability (SAASM). Both antennae are combined into a single, external device. It also interfaces smoothly with most tactical radios (Defense Update, 2007). The Raytheon's CDA version is shown in Figure 21.5.

Figure 21.5 Raytheon's CDA version

CDAs have certain limitations, however. In urban terrains, signals from CDAs to CDAs could be blocked by surrounding buildings. Like the FBCB2, position data of CDAs are not updated in real time. This drawback can hamper the very fratricide prevention efforts the system is designed to perform. Therefore, the Army is implementing improvements for the next fielding. These are taking the form of handheld personal digital assistants (PDAs), with the first device drawn from the INTER-4 Tacticomp 1.5 (Ackerman, 2005).

The Tacticomp 1.5 PDA is manufactured by Sierra Nevada Corporation (SNC). At two pounds, the system features a 3.5-inch color liquid crystal display. Unlike the CDA, the Tacticomp 1.5 does not have an L-band antenna. To keep the unit as small and as lightweight as possible, L-band connectivity is provided through vehicle-mounted systems to which the handheld unit is networked. The Tacticomp 1.5 connects through local links to the vehicle-mounted FBCB2 unit (Ackerman, 2005). Figure 21.6 displays the Tacticomp 1.5.

The next step up is the Tacticomp 6, which is larger than the Tacticomp 1.5 (8.4-inch color liquid crystal display screen versus 3.5-inch color liquid crystal display). It is designed to be mounted in a vehicle, but a commander can take it out on foot (Ackerman, 2005). Figure 21.7 is a picture of the Tacticomp 6.

The key difference between these units and the CDA is their low-probability-of-intercept spread-spectrum mesh-network transceivers. The mesh network opens up a host of connectivity capabilities hitherto unavailable to foot Soldiers. Not only can they link with individuals carrying identical handsets, but Soldiers on patrol in urban environments are also no longer limited to line-of-sight connectivity or plagued by heavy communications traffic (Ackerman, 2005).

Unlike the CDA wireless networking, the mesh network will not fill up in a dense user environment. Troops will not have to worry about stepping on the signal of other nearby users. And, the mesh network offers another key advantage lacking in the CDA and in other FBCB2 systems. Soldiers at last will be able to link around the corner with another like user. Previously, users could not connect with a nearby fellow Soldier without tying into a larger network via satellite. This mesh network allows a user to link directly with a fellow Soldier out of sight but in the immediate vicinity—and in real time (Ackerman, 2005).

Figure 21.6 SNC's Tacticomp 1.5 ruggedized PDA

Figure 21.7 SNC's Tacticomp 6 ruggedized PDA

Study 2: User Survey

Method

Participants Soldiers who had direct working experience with the FBCB2 system during Operation Iraq Freedom (OIF) and/or Operation Enduring Freedom (OEF) in Afghanistan were recruited to complete the questionnaire. The intent was to obtain the participation of 20 Soldiers of various ranks and military occupational specialties (MOS) or branch for officers. However, with the current operation tempo (OPTEMPO) at Fort Hood, it proved very difficult to get the number of Soldiers requested. The participation of nine male Soldiers was obtained.

Before starting the survey, participants were briefed on the purpose of the questionnaire and the safeguard of their privacy. Additionally, the voluntary nature of the survey was fully explained to all participants; it was made clear to them that they could refrain from taking the survey or answering any questions. Participants also understood that they were under no obligation to fully complete the survey even after it was already initiated.

Questionnaire The questionnaire was divided into five main sections (the Appendix contains the actual questionnaire used in the survey). The demographics section was formulated to collect basic participants' demographic data and information on their skill levels handling computer systems with emphasis on the FBCB2 system. The system hardware and software overall acceptability section sought to gauge the users' overall satisfaction with the system-user interface. In the FBCB2 training effectiveness section, the type and quality of the training Soldiers received on the FBCB2 was addressed. The emphasis was on the extent to which the training prepared Soldiers for mission execution and accomplishment. The FBCB2 operational effectiveness section sought to determine the impact of the FBCB2 system on mission execution enhancement and overall mission accomplishment. In the improving the FBCB2's capabilities section, participants were provided with enhancement ideas of the FBCB2 system. This section also encouraged participants to provide their own suggestions on how to improve the FBCB2 system.

Procedures The FBCB2 user survey was prepared and administered at Fort Hood, Texas. The survey examined the effectiveness of the FBCB2 in terms of the system's hardware and software usability, training, and operation. Soldiers with direct experience on the FBCB2 system during combat operations in either Iraq or Afghanistan were requested through a formal request process to Headquarters, 13th Sustainment Command (Expeditionary) (13th SC (E)) at Fort Hood, Texas. Coordination was made with the 13th SC (E) for a time and place to conduct the survey.

Results and Discussion

Analysis of the survey data collected Of the nine Soldiers surveyed, four were non-commissioned officers (NCOs) and five were junior Soldiers. The mean time in service was eight years. Overall, the participants reported having average computer and FBCB2 user skills and using the FBCB2 often. The mean length of time since their last FBCB2 use was five months. As for the Army handheld GPS receiver, the Plugger, participants, on average, also reported average skill level operating the system. Four of the nine participants reported possessing a Personal Digital Assistant (PDA) and being proficient at using it. All of the participants had been deployed to either OIF or OEF. One was deployed to OEF only, one was deployed to both OIF and OEF, and three had been deployed twice to OIF. The mean length of time since the participants returned from their last deployment was eight months.

As indicated earlier, the questionnaire was divided in five main sections. The questions pertaining to hardware and software usability, training, and operational effectiveness were rated on a scale of 1 to 5. The participants were to answer these sets of survey questions from their experience with the FBCB2 system. The participants were also given the opportunity to comment at the end of each section. Tables 21.3, 21.4, and 21.5 outline the questions used in the respective sections of the questionnaire and the percentage of participants with respect to the ratings selected for each question.

As can be observed in Table 21.3, participants mostly gave questions a 4 (effective) rating for the FBCB2 system hardware and software overall acceptability.

As with the data of Table 21.3, it can be observed in Table 21.4 and Table 21.5 that the questions were also mostly rated 4 (effective) respectively for training and operational effectiveness of the FBCB2 system. Once again, the questions were mostly rated 4 (effective). This trend suggests that, for the most part, the participants considered the FBCB2 a good and useful tool to help in their job performance and mission accomplishment. These results are consistent with results of a RAND study that examined the general contributions of FBCB2 to the overall job performance of three units which operated in northern Iraq during the 2003–2005 timeframe. According to the study, 'FBCB2 was praised for making significant contributions to the quality of information available, especially for blue force and engagement-related information' (Gonzales et al., 2007, p. 56). Figures 21.8, 21.9, and 21.10 show the mean rating number associated with each question.

As illustrated in Figure 21.8, the FBCB2 system hardware and software overall acceptability was rated above 3 (adequate) in almost all of the categories. The qualitative responses were also revealing. One participant commented, 'System did hold up to the environment. It would still operate even in 120-degree temperature. Timely start up phase needed. On certain missions where we would [be waiting] so system would boot up. We had to get to the location, so we ended up using a map while FBCB2 was booting up.' Another participant commented, 'The FBCB2 system is the best navigational system I have used. It is a good system. The only problem I faced was the bouncing inside the cab trying to touch the screen to send messages to battle desk and having a 3 to 5 min time delay to present position.' Yet another participant pointed out that, 'The FBCB2 is well rounded; just seemed a little lag in the updating process.'

Table 21.3 FBCB2 system hardware and software overall acceptability questions

FBCB2 system hardware and software overall acceptability (data as %)							
Questions		**Questions' ratings**					
		1	**2**	**3**	**4**	**5**	**N/A**
1a	Ability to operate	0	11	22	**56**	11	0
1b	Ease to perform basic functions	0	0	33	**45**	22	0
1c	Interoperability with other digital systems	0	22	33	**45**	0	0
1d	Access to controls	0	0	44	**56**	0	0
1e	Ease of controls operation	0	22	11	**56**	11	0
1f	Access to hookups	0	11	**45**	33	11	0
1g	System performance in hot weather	0	0	22	**67**	11	0
1h	System performance in cold weather	0	0	22	**56**	11	11
1i	Overall comfort with system hardware	0	12	25	**38**	25	0
1j	Reliability of system hardware	11	0	33	**45**	11	0
1k	Overall comfort with system software	0	11	22	**45**	22	0
1l	Reliability of system software	11	11	33	**34**	11	0
1m	Perform PMCS and repair	0	0	33	**56**	11	0
1n	Time spent in the shop for repair	11	22	22	**45**	0	0
1o	Time to receive ordered parts	11	11	22	**45**	11	0

Table 21.4 FBCB2 training effectiveness questions

FBCB2 training effectiveness (data as %)							
Questions		**Questions' Choice Selections**					
		1	**2**	**3**	**4**	**5**	**N/A**
2a	Usefulness of the training in theater	0	0	**63**	25	12	0
2b	Relevance of the training to actual theater operations	0	0	25	**63**	12	0
2c	Ability to operate the FBCB2 from the training received.	0	12	25	**63**	0	0
2d	Ability to operate the FBCB2 from your own intuition or computer skills.	0	12	25	**50**	13	0

Table 21.5 FBCB2 operational effectiveness questions

FBCB2 operational effectiveness (data as %)							
Questions		**Questions' Choice Selections**					
		1	**2**	**3**	**4**	**5**	**N/A**
3a	Obtain knowledge of own location.	0	0	11	**56**	33	0
3b	Obtain knowledge of other friendly locations	0	0	12	**75**	13	0
3c	Obtain knowledge of enemy locations	0	25	25	**50**	0	0
3d	Provide physical protection from enemy fire	0	29	**42**	29	0	0
3e	Provide sufficient and manageable information	0	0	33	**67**	0	0
3f	Provide accurate information	0	12	38	**50**	0	0
3g	Provide timely information	0	22	**45**	33	0	0
3h	Establish situational understanding	0	25	12	**63**	0	0
3i	Maintain situational understanding	0	0	37	**63**	0	0
3j	Ease of communication with friendly at all echelons	0	0	33	**45**	22	0
3k	Overall ease of operation of the FBCB2	0	0	11	**67**	22	0

These three comments indicate that the time taken for the FBCB2 system software upload is a serious concern. These precious few minutes of not knowing a friendly element's location can be disastrous. For fratricide prevention, this crucial system flaw should be addressed.

As in Figure 21.8, the ratings in Figure 21.9 are all above 3 (moderately effective) for the FBCB2 system training effectiveness. The following comments were recorded from some of the participants, 'I was never really trained, but the good thing, it was not really hard to figure out.', 'Need more hands on training.', and 'FBCB2 is a great system, but nothing prepares you like hands on. As for training, it builds a good base on how and where to start so you make minimal errors.' These comments indicate that the FBCB2 system software was user friendly and that hands-on training was the preferred training method.

In Figure 21.10, participants indicate overall satisfaction with the FBCB2 operational effectiveness. The ratings were almost all above 3 (moderately effective) with questions 3a (obtain knowledge of own location) and 3k (overall ease of operation of the FBCB2) being at or above 4 (effective). Comments from some of the participants pointed to concerns of system information overload. One participant noted, 'The screen can get a little busy with a considerable amount of traffic. Mostly not being able to see other systems.' Other participants wrote similar comments on their questionnaires. From a human factors perspective, these issues should be addressed and some mitigation measures should be taken. However, these issues do not negate the overall consensus among the participants that the FBCB2 was an adequate system to help with navigation and situational awareness, two important ingredients in fratricide prevention.

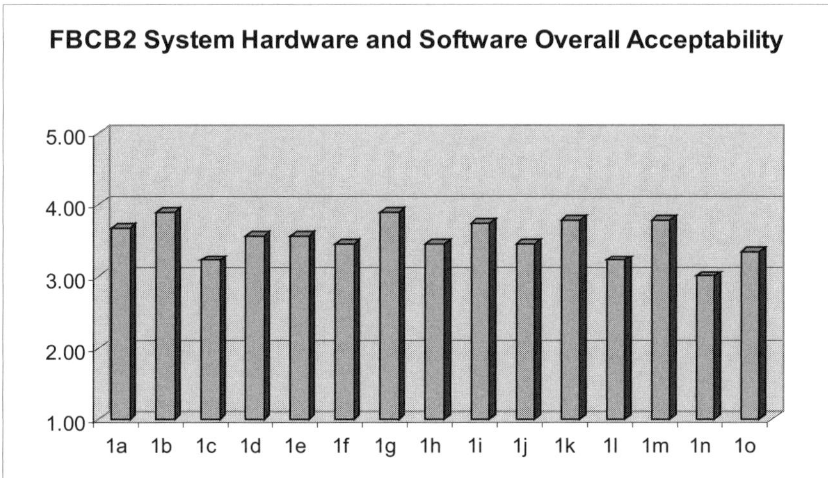

Figure 21.8 Mean rating of each question for the FBCB2 system hardware and software overall acceptability

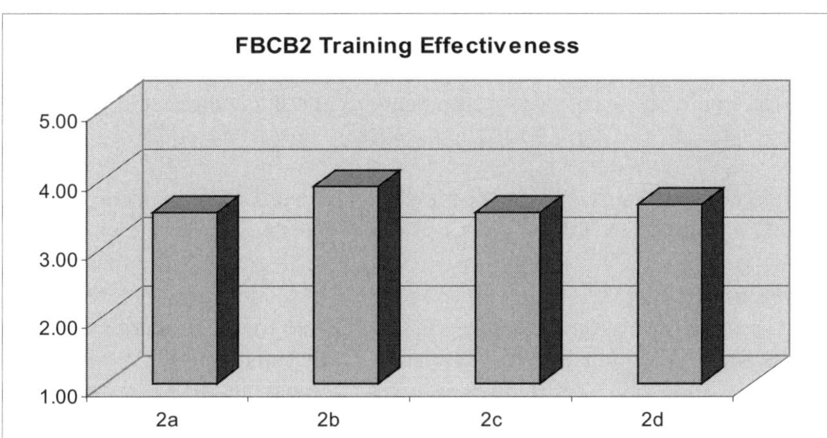

Figure 21.9 Mean rating of each question for the FBCB2 system training effectiveness

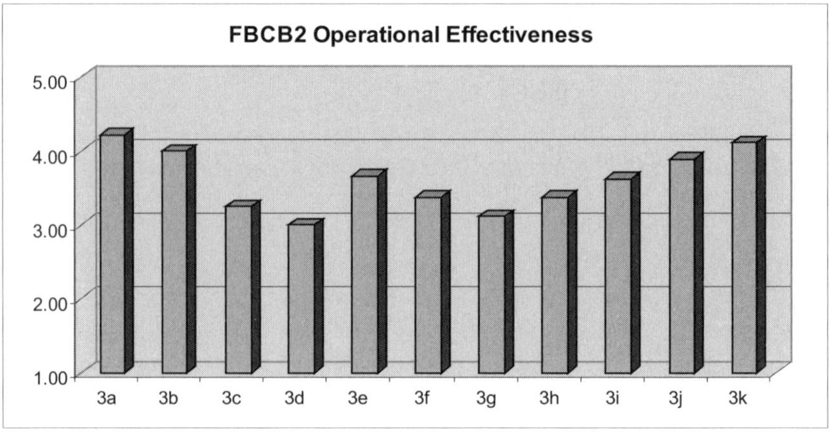

Figure 21.10 Mean rating of each question for the FBCB2 operational effectiveness

Conclusions

Major Findings and their Implications

Some of the known capabilities and flaws of the FBCB2 and the portable systems are compiled in this section. With this knowledge placed in perspective, the integration of these two systems to complement each other is further discussed in terms of the overall effort to prevent fratricide by enhancing blue force tracking, situational awareness, and battle command for both mounted and dismounted troops.

The CDA can show the user's location relative to other forces' locations, and it quickly can communicate data such as situational awareness (Ackerman, 2005).

Implication The CDA/PDA presents a very promising tool in the effort to prevent, reduce, or eliminate fratricide. The amazing results obtained from the FBCB2 system for mounted troops in previous and current operations suggest that the portable systems will provide similar results for dismounted troops at lower unit levels as an extension of the FBCB2.

General Dynamics has demonstrated integrating the CDA with the Stryker combat vehicle, to provide a dismounted capability for FBCB2, and exchange C2 and situational awareness information between mounted and dismounted units (Defense Update, 2007).

Implication The FBCB2 and the CDA being made compatible with each other would simplify the blue force tracking capabilities of these two systems and reinforce the argument of the CDA being an extension of the FBCB2.

The CDA automatically networks via standard Army radios, embedded GPS, wireless LAN, and tactical modem, and operates over more than one communications means simultaneously. The CDA is designed to self-form in peer-to-peer networks, but also to participate as either a client or server as required (Chisholm, 2004).

Implication This capability makes the CDA a very versatile system with the necessary redundancy to keep blue force tracking going at all times for fratricide prevention.

Whereas the Land Warrior system is targeted for the individual Soldier, the CDA is targeted for the dismounted leader, ranging from the brigade commander and staff, to the battalion commander and staff, to the company commander, to the platoon leader, and possibly down to squad leaders. CDA is the dismounted equivalent of Force XXI Battle Command, Brigade-and-Below (FBCB2), but not tied exclusively to combat platforms or the tactical Internet (Defense Update, 2007).

Implication Extending blue force tracking capabilities to the lowest echelon possible should be the aim of the CDA program. In other words, each Soldier should be equipped with a CDA/PDA as projected by the Land Warrior program. Individual Soldiers with CDA/PDAs ensures that commanders have complete visibility on every single Soldier on the battlefield. Although such an ambition would prove cumbersome, the hierarchy of the CDAs and FBCB2s should be built similar to the established command and control echelons of the units. Team leaders would have visibility of their team members on their respective CDAs while squad leaders would have visibility of the teams in their squads. Likewise, the platoon leaders would have visibility of the squads in their platoons, and so on. The command posts would then have general visibility of the main units operating in their respective areas of responsibility (AOR). With this setting, CDAs and FBCB2s screens would not be too crowded with friendly icons. Clicking on a unit's icon would then show another screen displaying the sub-units of that unit at the desired pre-set level or echelon. Unit leaders at each echelon having visibility over every individual Soldier in their respective units would absolutely be beneficial at preventing fratricide on the battlefield.

Drawbacks of the Portable Systems

With all the benefits of the CDA, it also presents many drawbacks. For instance, the following issues are some of those that were identified during testing and actual combat application of the portable systems:

- Information assurance concerns given the small size of the portable systems (Chisholm, 2004) and the potential to lose or misplace them.
- Wireless LAN as a security issue (Chisholm, 2004).
- Battery life (Chisholm, 2004).
- Condensed version of training not feasible (Chisholm, 2004).
- Information overload.
- Screen sizes relative to the amount of information to be displayed.
- Vulnerability to the threat of electromagnetic pulse (EMP) attack.

Recommendations

Ideally, the CDA should be implemented to complement the FBCB2. As indicated earlier, the FBCB2/CDA system architecture can be designed to keep continuous connection between members of a unit from the highest to the lowest echelon. A RAND study of fratricide at the National Training Center (NTC) revealed that the causes of ground to ground fratricides tended to be for the most part command and control problems rather than simple vehicle identification (Goldsmith et al., 1993). As argued before, command and control problems can be alleviated with unit leaders at each echelon having visibility over every individual Soldier in their respective units.

Additional research should be conducted to further assess the feasibility of the CDA as an extension of the FBCB2. Furthermore, additional research is needed to address and mitigate the flaws of the FBCB2 and the drawbacks identified for the CDA or the FBCB2/CDA system. As new technologies emerge, it may be possible to fit Soldiers with implanted GPS Transceiver on their uniforms or equipment for tracking purposes should equipping each one of them with a CDA prove unfeasible or impractical. The ability to keep a tab on each Soldier on the battlefield not only would allow commanders to prevent fratricide, but also to quickly rescue missing or kidnapped Soldiers.

Lessons learned from OIF and OEF have shown the ever increasing complexity of the battlefield. Although it may be inconceivable to completely eliminate fratricide on the battlefield, the FBCB2 system has proven effective at tracking mounted units, promoting situational awareness, and facilitating command and control. Once again, should dismounted units be afforded the same capabilities, fratricide will be prevented and greatly diminished on the battlefield.

Acknowledgments

I would like to thank Dr. Sam Middlebrooks (ARL-HRED) for his guidance and assistance with some of the literature used as references and to Mrs. Regina Pomranky (ARL-HRED) for collecting and analyzing the user survey data at Fort Hood, Texas. I would like to also extend my gratitude to Mr. Charles Augustus (ARL-HRED) for making it possible to travel to Fort Hood to collect the user survey data and to Mesa, Arizona, to attend the conference. Finally, I would like to thank ARL–HRED in general for its support and for giving me the opportunity to answer the CID call for paper.

To my wife Felicita for being so supportive and so wonderful to me. You are the apple of my eyes.

References

Ackerman, R. K. (2005, July 1). Army Intelligence Digitizes Situational Awareness. *SIGNAL Magazine.*

Boland, R. (2007, August 1). Little computer produces big results. *SIGNAL Magazine, 61*, 29–32.

Caseley, P., Dean, D., Gadsden, J., and Houghton, P. (2006). *Concepts of network enabled capability-safety issues and potential solutions*. Defence Science and Technology Laboratory.

Chisholm, P (2004, August 17). Handheld net-centricity. *Military Information Technology Online Edition, 8.*

Chisholm, P (2003, December 31). Info warfare in the palm of the hand. *Special Operations Technology Online Edition, 1.*

Defense Update International, Online Defense Magazine (2007). *Commander's Digital Assistant (CDA).* Retrieved March 16, 2009, from http://www.defense-update.com/products/c/cda.htm

Goldsmith, M., Grossman, J., and Sollinger, J. (1993). *Quantifying the Battlefield, RAND Research at the National Training Center* (Tech. Rep. No. MR-105-A). Santa Monica, CA: RAND Corporation.

Gonzales, D., Hollywood, J., Sollinger, J., McFadden, J., DeJarnette, J., Harting, S., and Temple, D. (2007). *Networked Forces in Stability Operations: 101st Airborne Division, 3/2 and 1/25 Stryker Brigades in Northern Iraq.* Santa Monica, CA: RAND Corporation.

Kulsrud, L. (January). *Fratricide: Reducing Self-inflicted Losses* (Tech. Rep. No. 92–4). Fort Leavenworth, KS: U.S. Army Combined Arms Command.

Project Manager, Force XXI Battle Command Brigade and Below (PM, FBCB2). FBCB2 system description, retrieved March 16, 2009, from http://peoc3t.monmouth.army.mil/fbcb2/about.html

Shrader, C. R. (1982). *Amicicide: The Problem of Friendly Fire in Modern War* (Research Survey No. 1) Fort Leavenworth, KS: U.S. Army Command and General Staff College.

Stein, F., and Fjellstedt, A. (2006, June). *Network Centric Warfare in Western Iraq.* Paper presented at CRP, San Diego, CA.

Steinweg, K. (1995). Dealing realistically with fratricide. *Parameters: Army War College Quarterly, Spring*, 4–29.

Steinweg, K., and Bowman, S. (1994). *Piercing the Fog of War Surrounding Fratricide: The Synergy of History, Technology, and Behavioral Research.* Carlisle Barracks, PA: U.S. Army War College.

U.S. Department of the Army, Military Operations (1993). *U.S. Army Operations Concept for Combat Identification* (TRADOC Pam 525–58). Fort Monroe, VA: Training and Doctrine Command.

U.S. Department of the Army (1992). *Fratricide: Reducing Self-inflicted Losses* (No. 92–4). Fort Leavenworth, KS: U.S. Army Combined Arms Command.

Appendix: Survey Questionnaire

Instructions

This questionnaire should take no more than 15 minutes to complete. The purpose of this questionnaire is to collect background information on the functionality, usefulness, and reliability of the Force XXI Battle Command Brigade-and-Below (FBCB2). Your answers will be kept confidential and will not be given to or shown to anyone except those who are evaluating the FBCB2 for the Army. The information you provide will not be given in whole or in part to your chain of command or put in your personnel file. Please fill out the questionnaire carefully. Please provide comments for responses in columns 1 or 5. If comments are provided for responses in columns 1, 2, or 3 please provide suggested correction. If you need additional space to answer a question, indicate by an arrow (→) and continue on the back of the page. Please be sure to number the item on the back of the page. If you absolutely don't know the answer to a question or if the question does not apply, please check the N/A box and indicate in the comments section that you don't know the answer or that the question does not apply to you. Please do not skip any question. Your participation in this survey is absolutely voluntary. You do have the right to return a partially completed or a blank questionnaire if you do not want to be in the study. If you have any questions concerning this questionnaire, please contact a test team representative for help. Thank you for your assistance in this endeavor.

Demographics

1. Date: __ __ / __ __ __ __ __ (DD/MMMYY Example: 19 Oct 07)
2. Position: ☐ Officer ☐ NCO ☐ Soldier ☐ DA Civilian/Contractor
3. Prior Service ☐ Yes ☐ No
4. Dates of Prior Service From_______ to _______
5. Time in service _____ year (s) _____ month (s)
6. Time in grade _____ year (s) _____ month (s)
7. Primary Branch/MOS_______
8. Time in primary Branch/MOS _____ year (s) _____ month (s)
9. Secondary Branch/MOS_______
10. Time in secondary Branch/MOS _____ year (s) _____ month (s)
11. Height in inches (optional) _____ 12. Weight in pounds (optional) _____
12. Gender _____
13. Computer literacy level (Please circle): Beginner Average user Proficient Expert
14. FBCB2 comfort level (Please circle): Beginner Average user Proficient Expert
15. Frequency of your FBCB2 usage (Please circle): Never Rarely Often Very Often
16. Length of time since you last used the FBCB2: _____ Months _____ Weeks _____ Days
17. 18. Plugger comfort level (Please circle): Beginner Average user Proficient Expert
18. Do you own a PDA? (Please circle): Yes No
19. If you own a PDA, circle comfort level: Beginner Average user Proficient Expert
20. Do you own a personal portable GPS device? (Please circle): Yes No
21. If you do, circle comfort level: Beginner Average user Proficient Expert
22. Have you ever deployed to OIF? (Please circle): Never Once Twice More than twice
23. 2ave you ever deployed to OEF? (Please circle): Never Once Twice
24. More than twice
25. Length of time since your last deployment: _____ Months _____ Weeks _____ Days

FBCB2 (BluFor Tracker) Short Survey

1. FBCB2 system hardware and software overall acceptability. Use the scale below and mark an X in the appropriate box

	Not Acceptable 1	Bad 2	Adequate 3	Good 4	Excellent 5	N/A
a. Ability to operate the FBCB2.						
b. Ease of performing basic functions on the FBCB2.						
c. FBCB2 Interoperability with other digital systems.						
d. Access to FBCB2 controls						
e. Ease of operation of FBCB2 controls						
f. Access to FBCB2 hookups						
g. System performance in hot weather						
h. System performance in cold weather						
i. Overall comfort with the FBCB2 system hardware						
j. Reliability of system hardware (does it break down or overheat frequently during operation?)						
k. Overall comfort with the FBCB2 system software						

	Not Acceptable 1	Bad 2	Adequate 3	Good 4	Excellent 5	N/A
l. Reliability of system software (does it crash or freeze frequently during operation?)						
m. Perform PMCS and repair on the FBCB2.						
n. Length of time the FBCB2 spends in the shop for repair						
o. Length of time to receive FBCB2 parts ordered.						

1. Comments on the FBCB2 system hardware and software overall acceptability.

____

____

____

____

____

____

2. FBCB2 training effectiveness. Use the scale below and mark an X in the appropriate box

	Not Effective 1	Barely Effective 2	Moderately Effective 3	Effective 4	Very Effective 5	N/A
a. Usefulness of the training received on the FBCB2 in theater (Did it really prepare you for application in theater?)						
b. Relevance of the FBCB2 training to actual operations in theater (Practicality of the training)						
c. Ability to operate the FBCB2 from the training received.						
d. Ability to operate the FBCB2 from your own intuition or computer skills.						

1. Comments on the FBCB2 training effectiveness.

____

____

____

____

____

____

3. FBCB2 operational effectiveness. Use the scale below and mark an X in the appropriate box

	Not Effective 1	Barely Effective 2	Moderately Effective 3	Effective 4	Very Effective 5	N/A
a. Obtain knowledge of own location.						
b. Obtain knowledge of other friendly locations.						
c. Obtain knowledge of enemy locations.						
d. Provide physical protection from enemy fire.						
e. Provide sufficient and manageable information (No information overload).						
f. Provide accurate information.						
g. Provide timely information.						
h. Establish situational understanding.						
i. Maintain situational understanding.						
j. Ease of communication with friendly at all echelons						
k. Overall ease of operation of the FBCB2						

l. Comments on FBCB2 operational effectiveness.

____

____

____

____

____

4. Improving the FBCB2's capabilities

In this section, please provide your opinion to questions a and b and provide any additional comments or suggestions you may have on improving the effectiveness of the FBCB2 in question c. This section is very important because your opinion and recommendations will very much count and will be taken seriously. They may even be considered for implementation in future Blufor Tracker systems.

a. The current Army BluFor tracking system (FBCB2) is only vehicle mounted. Should individual Soldiers be equipped with a small PDA-like portable tracking system as well? Please comment on your circled answer below:

   a) YES
   b) NO

Comments:

b. Which you think would make more practical sense? Please circle a letter and then comment:

   a) Small PDA-like portable tracking system
   b) Implanted transmitter on Soldiers' uniform

Comments:

c. Please provide any additional comments you may have on ways to improve the effectiveness of the FBCB2.

Index

2D media 115, 118–19, 123–4
3D models 72, 114–15, 118–19, 123–4, 167–9

A
ability, cognitive 68, 69
action effect 40–2
ADHD 51, 54
affect *see* emotion *see also* negative affect
Afghanistan xxxii, 181, 227, 228, 229, 243, 337
air tasking order 205–6, 209, 210
Airborne Warning and Control Station *see* AWACS
alerting system *see* automation
alerts 267–8, 316, 317, 318, 321, 322, 324
anxiety 178 *see also* neuroticism and worry
Armored Vehicle Recognition cards 120
Army Battle Command System 329
arousal 50, 178
Arousal Theory 49, 50
asymmetric warfare xxxii–xxxiii, 161
ATO *see* air tasking order
attention 47, 95, 116–17, 125
  and stress 178
  sustained 48, 63, 178
attitudes 3, 8, 177
authority hypothesis 285
automaticity 85–6, 87, 88, 95–6, 179, 181, 184
automation xxxiii, 247, 249, 265–6, 277, 299–300, 313, 315–17, 324
  active versus passive 321–2
  control hierarchy 320–3
  disuse, misuse, and reliance 174, 214, 247, 249–52, 255–6, 259, 261–2, 267–8, 272–4, 277–9, 283–92, 303, 305–6, 310
  failure 247, 319–20, 324
  intrusive 318, 322–3
  reliability 249–62, 269–74, 278–9, 281–8, 290–2, 299–311, 319–20, 323, 329
  undisciplined 305–6
  usability 324
automation bias 130, 269–73, 275, 283–8, 292, 306, 320
AWACS 192, 196, 197–8
  synthetic task environment 196–7, 199–200, 207–8, 214, 251
  and weapons directors 192–201

B
BARS 229–36, 238, 241–3
basic features 132–4
Battlefield Combat Identification System 299
battlefield situation-monitoring task 57–9, 61–2
Behaviorally Anchored Rating Scale *see* BARS
beyond visual range 313
BFT *see* Blue Force Tracker
biases 114, 164, 173, 177, 180–6
  automation 130, 269–73, 275, 283–8, 292, 306, 320
  confirmation 180, 183, 219
  irrational escalation 180, 183
  response 180, 185, 250–5, 269–71
  self-serving 283, 286, 288
binocular cues 67
Blue Force Tracker 154, 157, 185, 299, 332
BOLD QUEST 219
Boles Multiple Resource Questionnaire 52
boredom proneness 49, 51, 53–4

C
C2 *see* command and control

CADPAT 100
camouflage 99–101, 106, 279
Canadian Forces Air Warfare Centre 229
CAS *see* close air support
CCO *see* Chief Combat Officer
CDA *see* commander's digital assistants
change blindness 317–18, 321
Chief Combat Officer 205, 208
CID *see* combat identification
clock test 48
close air support 217, 227–8, 302
  low intensity conflict 228–9
coaching 17–18, 21
cognitive biases *see* biases
cognitive failures 49, 51, 53–4, 57
Cognitive Failures Questionnaire 49
cohesion 13, 18, 174, 176–7, 314
collateral damage xxxi–xxxiii, 228, 314–15
combat identification 182
  and automation 249–62, 308, 330
  and basic research 29
  and camouflage 106
  catch-22 309
  and close air support 218–19, 229, 236
  definition xxxi, 129, 173, 205, 228, 247, 313–14
  and human factors xxxii
  and misidentification 113–25
  and preattentive features 133
  and rules of engagement 300
  and situation awareness 147, 150, 154–8, 161–2, 327
  and spatial ability 67–8, 70–1, 78
  and teams 191–202
  and training 182–4, 185
command and control 129, 152, 313, 315, 342
  display 132–40, 151
  net-centric system 313, 315–17, 329–31
commander's digital assistants 332–6, 342–4
communication 1–4, 22, 175, 177, 222–3, 231, 233, 236–40
  closed-loop 2, 5, 18, 175, 177
confirmation bias 180, 183, 219
conjunction targets 133, 135
conjunctive search 85–8, 95
conscientiousness 51
consistency of mapping 86–7, 89, 96 *see also* variable mapping
consistency of responding 87, 92, 96
controlled processing 86
cooperation 1–3, 6, 22, 176, 231, 236–8, 240–1
coordination 1–3, 5, 22, 155–6, 175, 213, 231, 236–8, 240–1
coping 50–2, 54–7, 63, 184
crew resource management 12–13
CRM *see* crew resource management
cross-cultural training 9, 16–17, 18
cross training 9, 11, 13–14, 18, 155

D

data fusion *see* information fusion
decision aids *see* automation
decision-making 88, 118, 149, 180, 182–5, 191–2, 217–20, 225, 282, 284, 299–300, 306, 318, 324
  and stress 114, 130–1, 178–9
  team and group based 169, 179, 182
decrement function *see* vigilance decrement
Defense Science Board xxxi–xxxii
demonstration-based learning 9, 15–16, 18
Desert Shield, Operation 301 *see also* Gulf War
Desert Storm, Operation xxxi–xxxii, 113, 300–1, 306, 308, 310, 328 *see also* Gulf War
design, interface 42–3
discrimination thresholds 103, 107 *see also* just noticeable difference
display-control compatibility 31, 43
Distributed Mission Operations 229
DSB *see* Defense Science Board
DTC *see* dynamic targeting cell
Dual-Process Theory of Automaticity 85–7, 88, 95, 97
Dundee Stress State Questionnaire 52, 56–7
dynamic distributed decision-making task environment 207–8
dynamic targeting cell 205–15

E

emotion 169, 177–8, 183, 185
Enhanced Position Location and Reporting System radios 329
Epworth Sleepiness Scale 53–4
error training *see* guided error training
errors 2, 9, 12, 113–14, 201–2, 205–6, 214, 266, 282, 284, 286–8, 292, 302, 309
  commission, of 198–200, 206, 215
  omission, of 198–200, 205, 215
Exercise Northern Goshawk 229, 234, 237, 242–3
Expectancy-Value Theory 289
experts 22, 206
extraversion 49, 51, 220

F

F2T2EA *see* Find-Fix-Track-Target-Engage-Assess
FAC *see* forward air controllers
FACSE *see* Forward Air Controller Synthetic Environment
fatigue 178
FBCB2 *see* Force XXI Battle Command Brigade-and-Below
Feature Integration Theory 113–17
feedback 12, 17, 19, 250–2, 255–6, 258–60, 262
  vibrotactile 72
field dependence-independence 49
Find-Fix-Track-Target-Engage-Assess 139
FIT *see* Feature Integration Theory
fog of war xxix, 8, 115, 174, 182, 186
Force XXI Battle Command Brigade-and-Below 329–33, 335–44
Forward Air Controller Synthetic Environment 221
forward air controllers 217, 220–5, 227, 229–31, 233–40
fourier spectra 99–100, 102, 107, 109
fractals 99–103, 105–7, 109
Framework of Automation Use 278, 282–92
fratricide (*see also* misidentification) xxix, xxxi–xxxiii, 1–2, 11–13, 21–2, 113, 129–30, 173–5, 180–1, 185–6, 201–2, 228, 241–3, 277, 327–8, 330
  and camouflage 103
  causes of 174–5, 202, 301–2, 310, 344
  definition of xxxi, 173, 327
  incidents xxxii, 1, 130, 180–1, 301, 305–8, 310, 314
  rate xxxi–xxxiii, 129, 173–4, 277, 300–3, 308–9, 327–8
  reduction xxix, xxxii–xxxiii, 113, 174–5, 277, 279, 300–1, 303, 310, 313–24, 330–2
  and teamwork breakdowns 1–8, 22
Fratricide Action Plan 173, 329
friction of war xxix, 213
friendly fire *see* fratricide

G

game-based training 9, 14–15
GDTA *see* goal directed task analysis
geons (geometric icons) 114, 117–18, 125
Gestalt psychology 180
global positioning system *see* GPS
goal directed task analysis 150–2
GPS 163, 191, 299, 333, 335, 338, 343, 344
group cohesion *see* cohesion
guided error training 9, 12, 18
Guilford-Zimmerman Spatial Orientation Test 69–74
Gulf War xxxi, 129, 173, 277, 291, 305–6, 310 *see also* Desert Shield, Operation and Desert Storm, Operation

H

H-space 164
heuristics 178–9, 283
hierarchical task analysis 217, 230
HMI *see* human-machine interface
HTA *see* hierarchical task analysis
human error *see* errors
human-machine interface 129, 256
human soft sensors 161–4, 166
hypoplasia 101, 104

I

i-space 164
Identification Friend or Foe (IFF) 134, 139, 217–18, 249, 299, 303, 307–9, 316, 329
illusory conjunction 116

IMMERSIVE 251
impulsivity 53–5
INCIDER 217–20, 225
Infantry Situation Awareness Training Program 156
information exchange 2–4, 175, 177
information fusion 161–6
information processing 31–2, 49–50, 191–4, 202
instructional strategies 8, 10–11 *see also* training strategies
Instrumented Military Modeling Engine for Research Using Simulation and Virtual Environments *see* IMMERSIVE
Integrative Combat Identification Entity Relationship *see* INCIDER
intelligence 54–8, 62, 68
Intelligence Surveillance and Reconnaissance 205–8, 211, 214
Interactive Situational Awareness System 335
internal-external locus of control 49
interpositional knowledge 13–14
introversion-extraversion *see* extraversion
Iraq xxxii, 227, 243, 332, 337
irrational escalation bias 180, 183
ISAS *see* Interactive Situational Awareness System
ISAT Program *see* Infantry Situation Awareness Training Program
ISR *see* Intelligence Surveillance and Reconnaissance

J
Joint Terminal Attack Controller 227
JTAC *see* Joint Terminal Attack Controller
just noticeable difference 107–8 *see also* discrimination thresholds

K
kill chain 129, 139, 208, 221
knowledge 8, 179, 182
  interpositional 13–14
  knowledge elicitation 10, 164–5, 169

L
Land Warrior project 332–4, 343
learning 8, 22, 35–6, 38–9, 43 *see also* practice and training
  formal 8
  informal 8, 17
learning strategies 2, 8, 22
  formal 2, 8, 22
  informal 2, 17, 22

M
MARPAT 100
memory 119, 124–5, 179, 184
mental models *see* shared cognition
mental practice 18, 21–2
mentoring 19–20
military judgment panel 217, 225
Military Operations in Urban Terrain 74
misidentification *see under* combat identification
mixed mapping 33–4, 39
mixing cost 33, 42
morale 314
motivational processes 283, 289–92
MOUT *see* Military Operations in Urban Terrain

N
NASA TLX 52, 57, 73
NATO 227
negative affect 131
net-centric system *see under* command and control
neuroticism 220 *see also* anxiety and worry
non-traditional warfare *see* asymmetric warfare
North Atlantic Treaty Organization *see* NATO
Northern Goshawk, Exercise *see* Exercise Northern Goshawk

O
object model 114, 117–18, 123
OJT *see* on-the-job training
on-the-job learning *see* on-the-job training
on-the-job training 17–19, 207
Operation COBRA 314
Operation Desert Shield *see* Desert Shield, Operation

Operation Desert Storm *see* Desert Storm, Operation
Operation Enduring Freedom 330, 337–8, 344
Operation Iraqi Freedom 130, 173, 277, 301, 305–8, 310, 327–8, 330, 337–8, 344
Operation Medusa 181

P

partially observable Markov decision process 213
Patriot Air Defense System 130, 301, 302, 305–7, 310
PDA *see* personal digital assistant
perceptual-motor skill 42–3
performance 2, 31, 41–3, 85–9, 92, 95–6, 107, 265, 273–4
  human-automation 265, 273–4, 285, 292, 300–8, 318–21
  and stress 178–9, 182–4
  team 2–3, 152–3, 155, 193–4, 199–201, 206, 243
Persian Gulf War *see* Gulf War
personal digital assistant 332, 335, 338
phraseology 2, 4, 175, 177
Pittsburgh Sleep Quality Index 53–4, 57
POF *see* presence of friendly
pop-out phenomenon 99, 108
post-traumatic stress syndrome 186
practice (*see also* learning and training) 8, 10, 17–19, 36–43, 85, 108, 179, 181, 206, 318
  extended 36–9, 43, 85, 95, 179, 181, 184
  guided 10
  mental 18, 21–2
preattentive processing 99, 106, 131–4, 135–9
presence of friendly 139
primitive features *see* basic features
proximity compatibility principle 256

R

Radar task 87–8
RBC Theory *see* Recognition By Components Theory
Recognition By Components Theory 114, 117–18
reliance *see under* automation
response bias 180, 185, 250–5, 269–71
response mapping 85–7, 92–6
response selection 31, 38–9, 41–3, 97, 180
response time 33, 35, 86, 94, 96, 253–4, 271–3
rules of engagement xxix, 301, 307

S

S-space *see* human soft sensors
safety 63, 315, 320, 323
scenario-based training 9–11, 15, 18, 20
scenarios 10, 181–2, 185, 222
  threat scenarios 181
self-deception 178
self-serving bias 283, 286, 288
Senior Offensive Duty Officer 205, 208
Sensor Aided Vigilance Environments 213
Shape Memory Test 119, 120, 122, 124
shared cognition 2–3, 152–5, 175, 177, 231–3, 235–6, 238–43
  failure of 1
shared knowledge *see* shared cognition
shared mental models *see* shared cognition
short vigilance task 50–2, 54–6, 59–61, 63–4
signal detection theory 195, 250, 266
Simon effect 35–8, 40–3
Simon task 32, 35–6, 38–9, 41–2
simulation (*see also* virtual reality) 14, 21, 43, 206–7, 213, 229, 239, 242, 251
situation awareness 161–2, 165, 239, 327
  and automation 165, 299–300, 318
  and combat identification xxxi–xxxii, 147
  and decision-making 130
  definition 148, 299, 318
  and fratricide 302
  and information processing 194, 202
  levels of 148
  and teams 149, 150–8, 193, 202
  and unmanned systems 67–8
situational awareness *see* situation awareness
skill training 42–3
skills 8, 96
  perceptual-motor 42–3
  and situation awareness 156–7

vigilance 61–2
sleepiness 51, 53–4
Social Decision Scheme Theory 193
social loafing 289
SODO *see* Senior Offensive Duty Officer
soft sensors *see* human soft sensors
sonification 166–7, 169
spatial ability 67–79
measures of 68, 70, 73, 77–9
spatial orientation 69–74, 75–7
visualization 68–70, 72–5, 78, 133–4, 167–9
speed-accuracy tradeoff 33
SRC *see* stimulus-response compatibility
state-trace analysis 268–9, 272–4
stereotypes 33, 287
stimulus mapping 85, 87–8, 92
stimulus-response compatibility (SRC) 31–5, 38, 40, 43
Stinger missile system 302–3
storytelling 17–18, 20–1
stress 11
and automation 316–17
and decision-making 130–1
effect of 177–84, 185
stress exposure training 11–12, 18, 179, 182–4
subjective workload *see* workload
SVT *see* short vigilance task
Synergy 151
synthetic environment 220–3

T

Tactical Decision Making Under Stress 130
tactical internet 329, 343
Tacticomp 335–6
TADMUS *see* Tactical Decision Making Under Stress
target identification *see* combat identification
team situation awareness *see* situation awareness
team training 9, 12–13, 18, 181, 185
teams 2, 8, 14–16, 18, 171, 179, 191–5, 199–202
human-computer 277–8, 291–2
operator 68, 70, 73, 79
and situation awareness 147–56
and training 206–7, 211–15
teamwork 149, 177
breakdown 1, 2–4, 8, 10, 16, 22, 175, 177
time-sensitive targets 205–12, 215, 229
tracks 133, 135–40
track heading 136
track history 135, 138
TRADOC *see* Training and Doctrine Command
training (*see also* practice and learning) xxxi, 8, 17, 42–3, 96–7, 155, 157, 177, 181–6, 206–7, 210–13, 280
2D training aids versus 3D training aids 114–25
coaching 17–18, 21
crew resource management 12–13
cross-cultural training 9, 16–17, 18
cross training 9, 11, 13–14, 18, 155
demonstration-based learning 9, 15–16, 18
game-based training 9, 14–15
guided error training 9, 12, 18
mental practice 18, 21–2
mentoring 19–20
scenario-based training 9–11, 15, 18, 20
skill training 42–3
stress exposure training 11–12, 18, 179, 182–4
team training 9, 12–13, 18, 181, 185
Training and Doctrine Command 173, 329, 331
training strategies 2–3, 9, 17–18, 22, 175, 183–4 *see also* instructional strategies
transfer of learning *see* transfer of training
transfer of training 19, 35–41, 43
specificity of training 36, 38, 41, 43
trust (in automation) 266–8, 271–2, 283, 285–9, 292, 304, 308–9, 319
TSTs *see* time-sensitive targets

U

UAV 47, 71–8, 214, 266
UGV 71–8
unmanned aerial vehicles *see* UAV
unmanned ground vehicles *see* UGV

Unreal Tournament 2004 251
URGENT QUEST 219
U.S.S. *Vincennes* 129–30, 137

V

Vandenberg Mental Rotation test 70–2
variable mapping 85
vigilance 47–52, 56, 58–9, 62–3, 131, 178, 214, 310
  vigilance skill 61–2
vigilance decrement 48–9, 56, 60
Vigilant Warrior 50–1, 56, 59, 62
Vincennes see U.S.S. *Vincennes*
virtual reality 43, 71–2 *see also* simulation
visual perception 68
visual search 85–8, 96, 132, 266
visualization *see under* spatial ability
VR *see* virtual reality

W

work flow 318, 322
working memory 62, 114, 119, 125, 179
workload 11, 49, 52, 73, 86, 88–9, 92, 255–6, 283, 311, 316–17, 321–2
World War I 328
World War II 113, 129, 134, 173, 314, 328
  Italian campaign 314
worry 52, 55, 63, 336 *see also* anxiety and neuroticism